*Fortschritte der Chemie organischer Naturstoffe*

# Progress in the Chemistry of Organic Natural Products

# 58

Founded by L. Zechmeister
Edited by W. Herz, G. W. Kirby, W. Steglich, and Ch. Tamm

Authors:
R. D. H. Murray, J. A. Robinson

**Springer-Verlag**
Wien New York 1991

Prof. W. Herz, Department of Chemistry,
The Florida State University, Tallahassee, Florida, U.S.A.

Prof. G. W. Kirby, Chemistry Department,
The University, Glasgow, Scotland

Prof. Dr. W. Steglich, Institut für Organische Chemie und Biochemie der Universität
Bonn, Bonn, Federal Republic of Germany

Prof. Dr. Ch. Tamm, Institut für Organische Chemie der Universität Basel,
Basel, Switzerland

© 1991 by Springer-Verlag/Wien
Softcover reprint of the hardcover 1st edition 1991
Library of Congress Catalog Card Number AC 39-1015

Typesetting: Macmillan India Ltd., Bangalore-25
Printed in Austria by novographic, Ing. W. Schmid, A-1238 Wien

With 64 Figures

ISSN 0071-7886
ISBN-13: 978-3-7091-9143-9     e-ISBN-13: 978-3-7091-9141-5
DOI: 10.1007/978-3-7091-9141-5

# Contents

List of Contributors . . . . . . . . . . . . . . . . . . . . . .   VII

**Chemical and Biochemical Aspects of Polyether-Ionophore Antibiotic Biosynthesis.**
By J. A. ROBINSON . . . . . . . . . . . . . . . . . . . . . .   1

1. Introduction; What is a Polyether Antibiotic? . . . . . . . . . . . .   1

2. A Polyketide Origin for the Polyether Antibiotics . . . . . . . . . .   6

3. The Mechanisms of Heterocyclic Ring Formation . . . . . . . . . .   15

4. Model Studies Related to the Late Stages of Polyether Biosynthesis . . . .   30

5. Evidence for the Processive Strategy of Polyketide Chain Assembly . . . .   36

6. A Unified Model of Polyether Antibiotic Structure and Biogenesis. . . . .   43

7. Biochemistry of Polyketide Biosynthesis and Fatty Acid Biosynthesis. . . .   50
   7.1. Programming Polyketide Assembly . . . . . . . . . . . . . .   51
   7.2. Stereochemical Aspects of Polyketide Synthase Action . . . . . . .   52
   7.3. The Structure and Function of Fatty Acid Synthases . . . . . . . .   54
   7.4. The Structure and Function of Bacterial Polyketide Synthases . . . .   61

References . . . . . . . . . . . . . . . . . . . . . . . . . .   72

**Naturally Occurring Plant Coumarins.** By R. D. H. MURRAY . . . . . . .   83

I. Scope of the Review . . . . . . . . . . . . . . . . . . . .   84

II. Progress in the Past Decade . . . . . . . . . . . . . . . . . .   84

III. Introduction to Tables . . . . . . . . . . . . . . . . . . .   86
   Table 1. 7-Oxygenated Coumarins . . . . . . . . . . . . . . .   88
        1.1. 6-Substituted-7-Oxygenated Coumarins . . . . . . . . .   101
        1.2. 8-Substituted-7-Oxygenated Coumarins . . . . . . . . .   114
        1.3. 5,6-Disubstituted-7-Oxygenated Coumarins . . . . . . .   131
        1.4. 6,8-Disubstituted-7-Oxygenated Coumarins . . . . . . .   132
   Table 2. 5,7-Dioxygenated Coumarins . . . . . . . . . . . . .   133
   Table 3. 6,7-Dioxygenated Coumarins . . . . . . . . . . . . .   145
   Table 4. 7,8-Dioxygenated Coumarins . . . . . . . . . . . . .   151
   Table 5. 5,6,7-Trioxygenated Coumarins . . . . . . . . . . . .   157
   Table 6. 5,7,8-Trioxygenated Coumarins . . . . . . . . . . . .   159
   Table 7. 6,7,8-Trioxygenated Coumarins . . . . . . . . . . . .   165
   Table 8. 5,6,7,8-Tetraoxygenated Coumarins . . . . . . . . . . .   174
   Table 9. 3-Substituted Coumarins . . . . . . . . . . . . . . .   174
        9.1. 3-Aryl-Substituted Coumarins . . . . . . . . . . . .   178

Table 10. 4-Substituted Coumarins . . . . . . . . . . . . . . . 180
        10.1. 4-Aryl-Substituted Coumarins . . . . . . . . . . 182
Table 11. Miscellaneous Coumarins . . . . . . . . . . . . . . . 195
        11.1. 3-Aryl Oxygenated Coumarins . . . . . . . . . . 221
        11.2. Coumestans . . . . . . . . . . . . . . . . . . . 223
Table 12. Biscoumarins . . . . . . . . . . . . . . . . . . . 230
Table 13. Triscoumarins . . . . . . . . . . . . . . . . . . . 236

Amendments/Additions to Entries in Reference *448*

Table 1. 7-Oxygenated Coumarins . . . . . . . . . . . . . . . 237
        1.1. 6-Substituted-7-Oxygenated Coumarins . . . . . . . . 247
        1.2. 8-Substituted-7-Oxygenated Coumarins . . . . . . . . 250
Table 2. 5,7-Disubstituted Coumarins . . . . . . . . . . . . . 254
Table 3. 6,7-Disubstituted Coumarins . . . . . . . . . . . . . 257
Table 4. 7,8-Disubstituted Coumarins . . . . . . . . . . . . . 259
Table 5. 5,6,7-Trisubstituted Coumarins . . . . . . . . . . . . 260
Table 6. 5,7,8-Trisubstituted Coumarins . . . . . . . . . . . . 261
Table 7. 6,7,8-Trisubstituted Coumarins . . . . . . . . . . . . 262
Table 8. 5,6,7,8-Tetrasubstituted Coumarins . . . . . . . . . . 261
Table 9. 3-Substituted Coumarins . . . . . . . . . . . . . . . 263
Table 10. 4-Substituted Coumarins . . . . . . . . . . . . . . . 264
Table 11. Miscellaneous Coumarins . . . . . . . . . . . . . . . 270

Formula Index . . . . . . . . . . . . . . . . . . . . . . . . 271

Trivial Name Index . . . . . . . . . . . . . . . . . . . . . . 275

References . . . . . . . . . . . . . . . . . . . . . . . . . . 283

# List of Contributors

MURRAY, Dr. R. D. H., Chemistry Department, The University, Glasgow G12 8QQ, Scotland.

ROBINSON, Prof. Dr. J. A., Organisch-Chemisches Institut der Universität Zürich, CH-8057 Zürich, Switzerland.

# Chemical and Biochemical Aspects of Polyether-Ionophore Antibiotic Biosynthesis

J. A. ROBINSON, Organisch-Chemisches Institut der Universität Zürich, Zürich, Switzerland

With 64 Figures

## Contents

1. Introduction; What is a Polyether Antibiotic? . . . . . . . . . . . . . 1

2. A Polyketide Origin for the Polyether Antibiotics . . . . . . . . . . 6

3. The Mechanisms of Heterocyclic Ring Formation . . . . . . . . . . 15

4. Model Studies Related to the Late Stages of Polyether Biosynthesis . . . . 30

5. Evidence for the Processive Strategy of Polyketide Chain Assembly . . . . 36

6. A Unified Model of Polyether Antibiotic Structure and Biogenesis . . . . 43

7. Biochemistry of Polyketide Biosynthesis and Fatty Acid Biosynthesis . . . 50
   7.1. Programming Polyketide Assembly . . . . . . . . . . . . . . . 51
   7.2. Stereochemical Aspects of Polyketide Synthase Action . . . . . . . 52
   7.3. The Structure and Function of Fatty Acid Synthases . . . . . . . 54
   7.4. The Structure and Function of Bacterial Polyketide Synthases . . . . 61

References . . . . . . . . . . . . . . . . . . . . . . . . . . . 72

## 1. Introduction; What is a Polyether Antibiotic?

The polyether antibiotics are a large group of structurally related polyketide natural products, mainly of bacterial origin, which efficiently complex Group I or II metal cations (*1*). One well known example is the commercially important coccidiostat and growth promoter, monensin A (1) (see Fig. 1). Embedded within its carbon backbone is a complex array of functional groups that allow tight and specific binding of $Na^+$

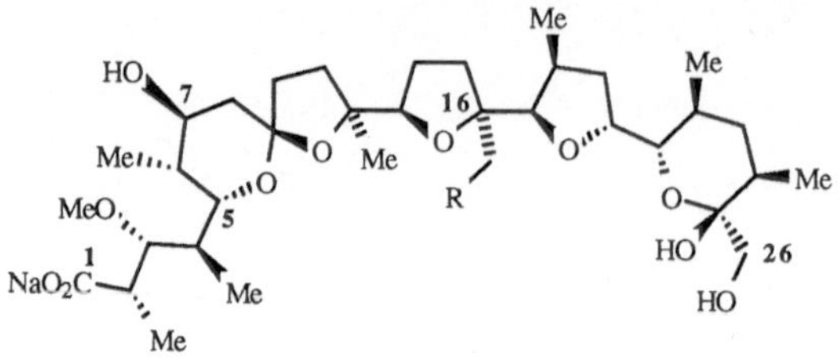

R=Me    monensin-A  (**1**)
R=H     monensin-B  (**2**)

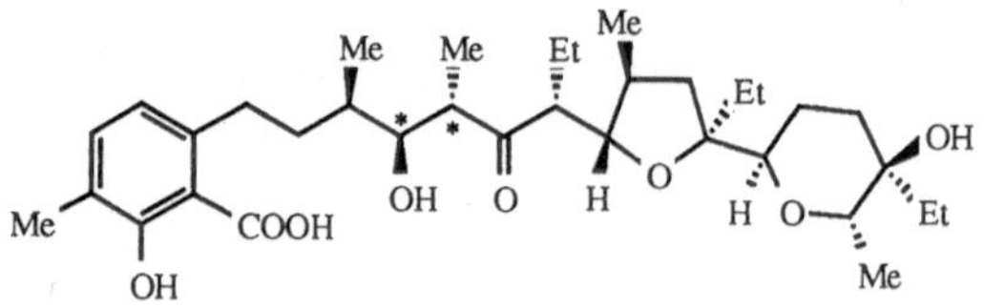

lasalocid A (**3**)

Fig. 1. Structures of monensin-A, -B and lasalocid A

ions. How this binding is achieved can be largely appreciated from a consideration of the monensin A-Na$^+$ X-ray crystal structure (*2*) (Fig. 2). The relative configuration of the molecule and the disposition of six oxygen atoms within or appended to the five heterocyclic rings allow the backbone to fold and create a cavity whose size exactly matches the ionic radius and ligand requirements of Na$^+$. In addition, a pair of hydrogen bonds is formed between the carboxylate oxygens at C1 and the two hydroxyl groups at C25 and C26, which tether the ends of the molecule together generating a shell which encompasses the metal ion. The exterior of the complex possesses a uniform hydrophobic surface which is ideal for diffusion through a lipid environment. The transport of metal ions across cell membranes mediated by the polyether antibiotics frequently leads to uncoupling of oxidative phosphorylation and ultimately to cell death (*1*). Unfortunately, this effect is seen also with mammalian cells, so the polyethers have not found clinical applications.

    A better appreciation of Nature's molecular design of the polyethers has come through chemical studies with monensin A. STILL and SMITH (*3*) pointed out how the array of substituents, and their relative configuration, in the acyclic segment of monensin at C1-C5 appears to rigidify the chain by avoidance of alternative conformers having relatively high

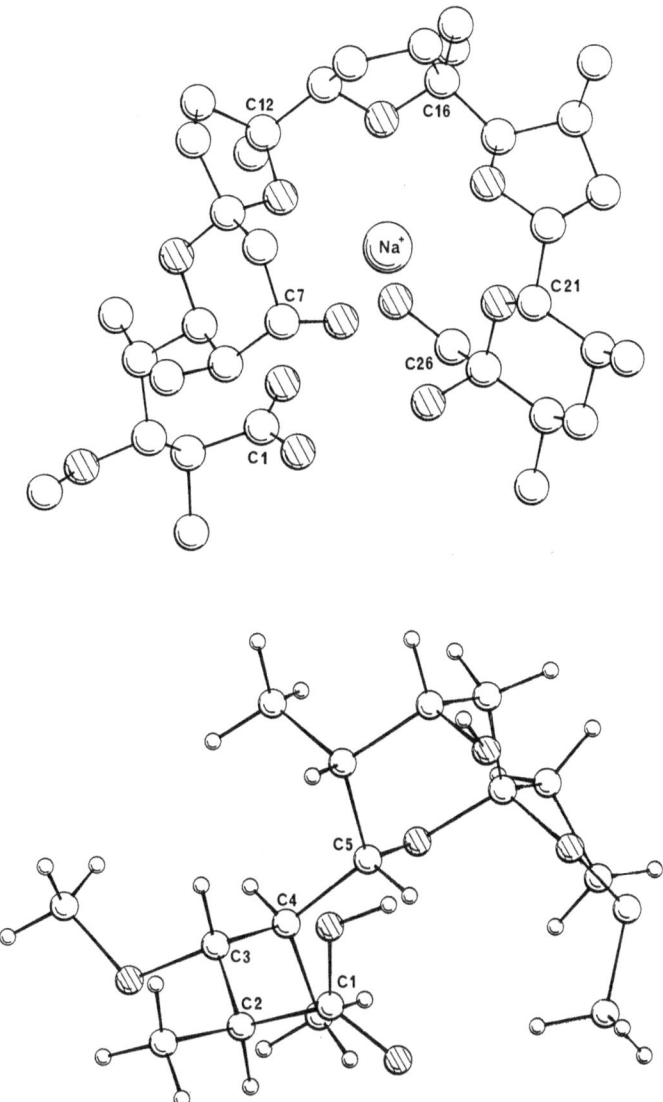

Fig. 2. X-ray crystal structure of Na$^+$-monensin A

energy + gauche/− gauche ( + g/− g) pentane interactions. The avoidance of + g/− g interactions leaves little opportunity for the C1-C5 fragment to adopt conformations other than that found in the crystal structure of Na-monensin A. The C2-C3 conformer shown (Fig. 2) would appear to be the least strained of the three possible staggered conformations since it places the planar C1 in nonbonded contact with C5, allows formation of the stabilising hydrogen bonds, and positions the anionic C1 carboxylate near to the bound metal cation. Such rigidifying features appear to be important in stabilising the binding conformation and thus enhancing metal ion binding. Support for this view has been obtained (3) through the synthesis of novel derivatives having a 3-isopropyl substituent, which possessed binding affinities for $Na^+$ in accord with those expected based on such conformational considerations.

Monensin A (1) is produced (4) together with smaller amounts of monensin B (2) by the microorganism *Streptomyces cinnamonensis*. There is apparently a lapse in specificity during the biosynthesis; monensin A differs from monensin B with respect to the alkyl substituent at C16. This seemingly minor structural change, however, has a surprisingly large effect upon the ion-complexing ability of the ionophore. Careful titration calorimetry experiments (5) have shown that changing the ethyl group at C16 to methyl affects both absolute binding strength and selectivity for metal ions {(1) + $Na^+$, $\Delta G$ (assoc.) = $-8.68$ Kcal/mol; and (2) + $Na^+$, $\Delta G$(assoc.) = $-7.98$ Kcal/mol, in MeOH at $25°C$}. Monensin A is selective for binding $Na^+$ over $K^+$ by $-1.41$ Kcal/mol, whereas for monensin B the selectivity is $-0.99$ Kcal/mol. Apparently, this increased selectivity of monensin A for $Na^+$ of about 40% relative to monensin B arises largely as a result of differences in the entropic contributions to the free energies of complexation (5).

Monensin A (1) possesses structural features that are shared by other members of the polyether family of ionophore antibiotics. Each contains a stereochemically complex array of ligands in a (usually) acyclic carbon chain incorporating rigidifying substructures, such as 5- and 6-membered cyclic ethers, and arrays of chiral centres which destabilise undesired rotomers and help to preorganise the ligating atoms so that metal ions may be bound effectively. Another illustration of this is seen in lasalocid A (3) (Fig. 1), one of the simplest polyethers. Two lasalocid molecules are found in the $Ca^{2+}$ and $Ba^{2+}$ complexes, with the association constant for $Ba^{2+}$ being over $10^6 M^{-1}$. Upon epimerising either or both of two different chiral centres (starred in Fig. 1) in the lasalocid backbone, a drop in the association constant is observed for the modified compound and $Ba^{2+}$ by more than two orders of magnitude relative to the natural complex (6). A computational study which examined lasalocid A and every single epimerised lasalocid diastereomer predicted a structure for the natural compound that was nearly identical to an X-ray derived structure of the Ba-lasalocid complex, whereas no other epimer was judged by this method to be capable of complexing ions as well as lasalocid A (*6*). Thus the available evidence suggests that the structural complexity of these metabolites has arisen largely, if not solely, in response to the need for optimal metal ion complexing properties. Since their biosynthesis requires dozens of enzymes encoded by tens of kilobases of genomic DNA it seems unlikely that they arose coincidentally with the correct molecular architecture. Rather it is most compelling to assume that evolutionary pressure created in the past the driving force for this structural optimisation, with a resulting benefit to the producing organism in its competition with other species (*7*).

However, for a molecule to act as a transmembrane ion carrier there must be a delicate balance between the binding energy of the ion in its cage and its hydration energy in solution (*8*). A very high association constant would mean that the ion could never escape from its encaged complex. Also, there must be a low enough activation energy to allow the cation to enter or leave its cage at a reasonably fast rate. This would be the case if the complexation rate approximates the diffusion controlled limit, and the association constant lies in the range $10^4$–$10^6$ M$^{-1}$, as it does for most polyether-metal ion complexes. The ion complexing abilities of typical polyether antibiotics most often lie between that of the spherand (**4**) ($\Delta G$ for Li$^+$ = $-$ 22 Kcal/M, i.e. $K_a = 10^{16}$ in CDCl$_3$ at 25°C), and the linear podand (**5**) (Fig. 3) which does not form stable complexes with Group I metal ions (*9*).

Whilst there is a clear relationship here between structure and function, the ability of a polyether to transport metal ions would, in an achiral environment, be independent of its absolute configuration. In

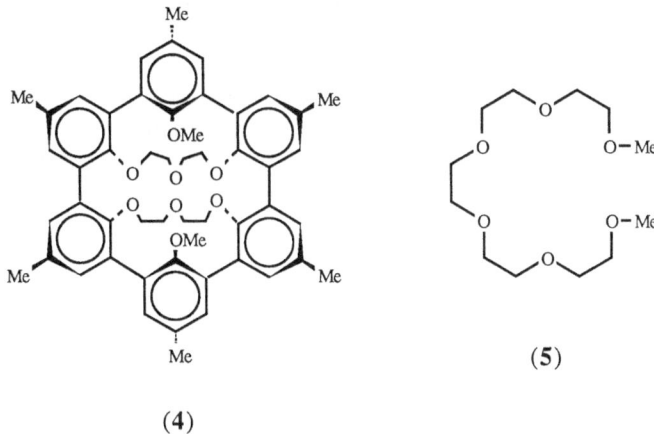

(**4**)                                    (**5**)

Fig. 3. A spherand (**4**) and a linear podand (**5**)

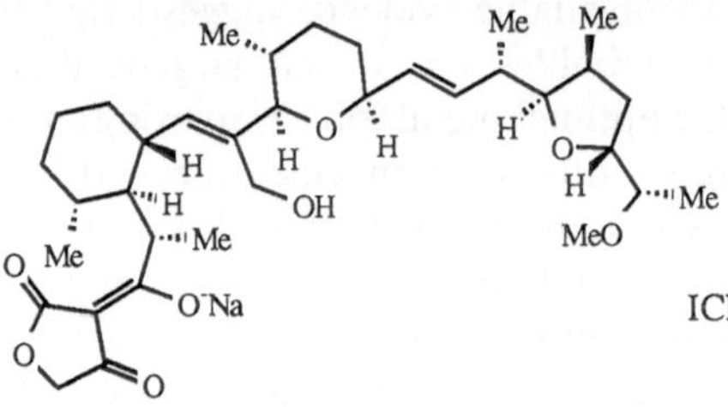

Fig. 4. Structures of ICI 139603, and tetronomycin

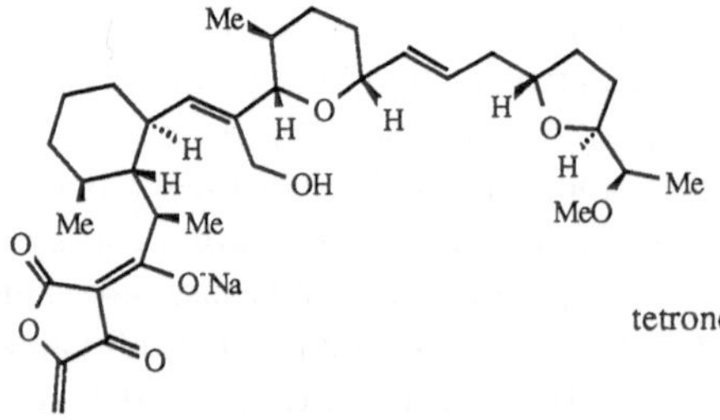

practice, close similarities in structure and absolute configuration (*10–12*) are often evident between different polyether antibiotics (*vide infra*), suggesting that their biosynthetic pathways share common evolutionary origins. There is, however, one interesting and striking exception. The polyethers ICI139603 (**6**) and tetronomycin (**7**) have closely similar structures (Fig. 4), but at each of the 10 common chiral centres the antibiotics possess opposite absolute configurations, as deduced reliably by X-ray crystallography (*13, 14*) and partial synthesis (*15, 16*).

Although the polyether antibiotics possess a high parenteral toxicity, they typically are potent coccidiostats[1] and they also act to improve the efficiency of feed utilisation in ruminants. These useful biological properties stimulated during the 1970's and 80's many attempts to isolate novel polyethers, so that now over eighty different structures in this class are known.

## 2. A Polyketide Origin for the Polyether Antibiotics

The polyketide hypothesis enunciated by BIRCH stands out as an incisive contribution to the field of polyketide biogenesis (*17, 18*). It

---

[1] Coccidia are parasitic protozoa of the subphylum Sporozoa, which exhibit a particular affinity to the epithelial cells in the digestive tracts of birds and mammals.

provided a valuable correlation between the structures of a large number of (largely) aromatic natural products and their probable modes of biosynthesis through the head-to-tail linkage of acetate units. The hypothesis was firmly grounded in mechanistic chemistry and showed how a carbon chain of β-ketone groups, retained from the successive condensation of acetate units (instead of being serially removed as in fatty acid biosynthesis), might undergo well precedented laboratory reactions, including aldol condensation, C-acylation, reduction, dehydration, methylation and oxidation, and thereby give rise to the large family of known polyketide metabolites. The biogenetic correlation was most apparent for certain plant phenolic compounds which have oxygen substituents attached β- to each other or to positions of ring closure, as a direct result of the β-positioning of ketone and methylene groups in the original chain. BIRCH also suggested that coenzyme-A esters other than acetyl-CoA might initiate chain assembly, while "propionate" units (methylmalonyl-CoA) could be incorporated in place of malonyl-CoA during chain extension (19).

Only a few years before, the first macrolide antibiotic had been reported by BROCKMANN and HENKEL (20), and the relevance of BIRCH's polyketide hypothesis to macrolide biosynthesis was quickly recognised. By 1957 at least six different macrolides had been discovered, and WOODWARD suggested that the macrolide rings could also arise by the stepwise condensation of acetate (or related) building blocks, with the oxygen atom of individual β-ketone groups retained as hydroxyl groups in the macrolide backbone (21). Since that time close to two hundred different macrolides have been isolated, largely from microorganisms of the genus Streptomyces, and many other important classes of antibiotics, including the polyethers, are now also known to be of polyketide origin. It is a remarkable aspect of natural products chemistry today that new, interesting, and therapeutically important polyketide metabolites continue to be discovered in Streptomycete screening programmes conducted largely within the pharmaceutical and agrochemical industries (e.g. avermectins, FK506, etc).

The first X-ray structure of a polyether, that of monensin A (1), was reported in 1967, and shortly thereafter biosynthetic experiments with radiochemically labelled precursors were carried out to identify the building blocks used for the biosynthesis of these metabolites in Streptomycetes (22). Although the specificity of any incorporations could only be ascertained through chemical degradation of the labelled antibiotics, in several cases convincing evidence was found for the assembly of polyether backbones from small fatty acid precursors, including acetate, propionate and butyrate. Later these conclusions were confirmed and extended through the application of stable isotope labelling techniques. The incorporation of $^{13}C$ labelled precursors can be detected straightforwardly by $^{13}C$-NMR spectroscopy, once the natural abundance $^{13}C$ spectrum of the antibiotic has been assigned (a task greatly facilitated by the emergence of 2D NMR methods). In the case of monensin A (1), five acetate units, seven propionates and one butyrate are specifically incorporated into the carbon backbone, thus proving its polyketide origin, as shown in Fig. 5 (23). The 3-O-methyl group is derived, as expected, from S-adenosylmethionine (23). In the case of lasalocid A (3) five acetate units, four propionates and three butyrates are specifically incorporated into the carbon backbone (22, 24) (see Fig. 5), and four of these precursors are involved in the construction of the aromatic ring.

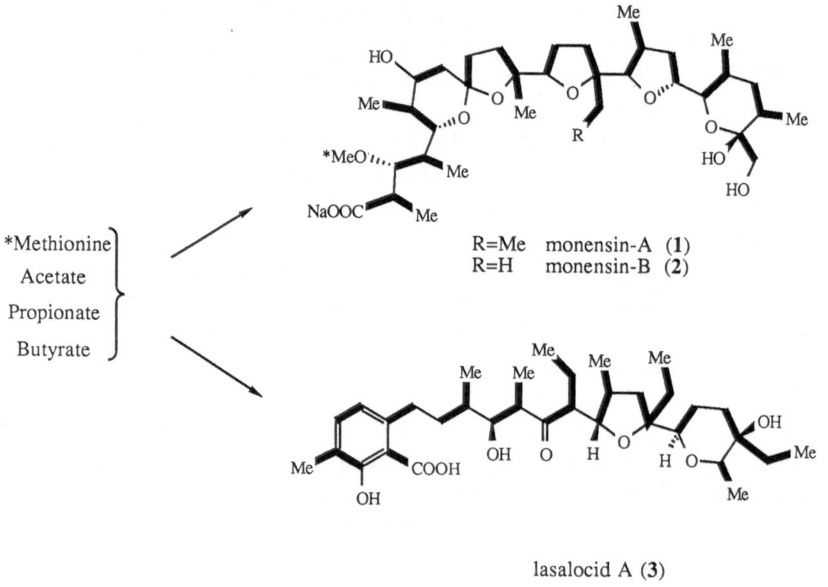

*Methionine
Acetate
Propionate
Butyrate

R=Me   monensin-A (1)
R=H    monensin-B (2)

lasalocid A (3)

Fig. 5. Incorporation of primary precursors into monensin A and lasalocid A

Other small carboxylic acids may occasionally play a role in polyether biosynthesis. The antibiotic calcimycin (8) (or A23187), for example, contains an α-ketopyrrole derived from proline, a centrally located polyketide spiroketal assembled from four propionate units and an acetate, and a benzoxazole whose benzenoid ring appears to be derived from shikimic acid (25, 26) (Fig. 6). The polyether antibiotic ICI139603 (6) contains a polyketide chain derived from seven acetate units, six propionates, combined with a $C_2$-unit of unknown origin (27) (Fig. 7).

*References, pp. 72–81*

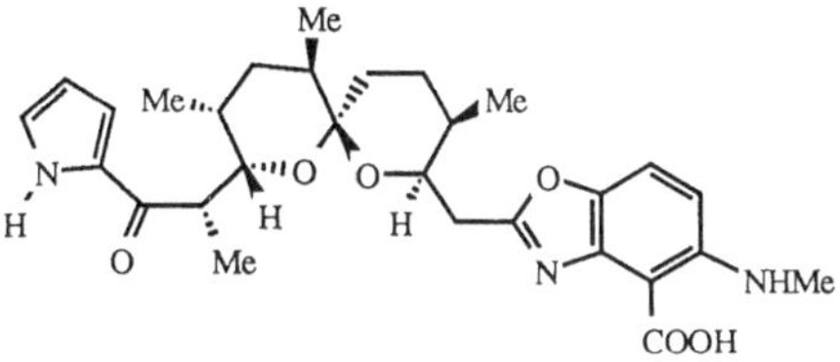

A23187 (**8**)

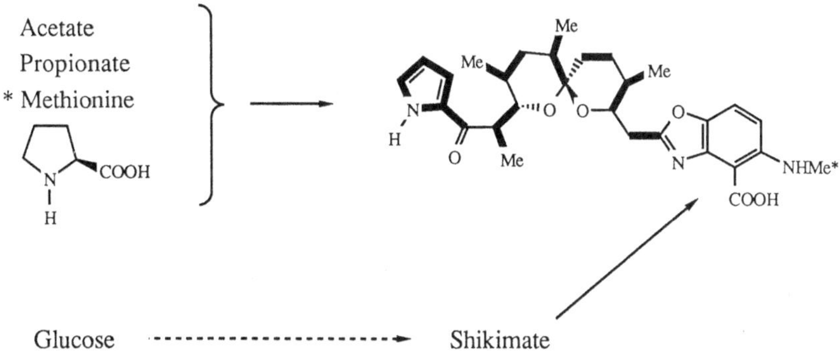

Acetate
Propionate
* Methionine

Glucose  - - - - - - - - - - - - - - - ▸  Shikimate

Fig. 6. Incorporation of primary precursors into A23187

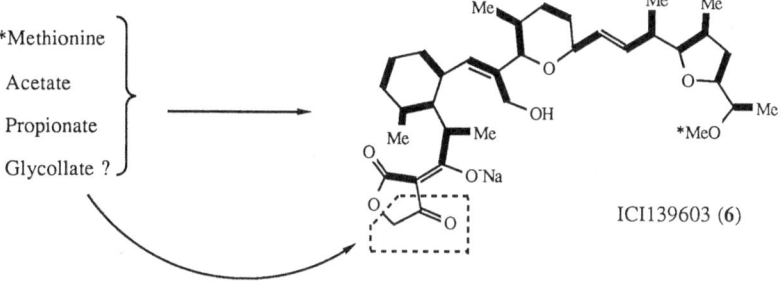

*Methionine
Acetate
Propionate
Glycollate ?

ICI139603 (**6**)

Fig. 7. Incorporation of primary precursors into ICI 139603

This $C_2$ unit might be glycollate, a conclusion strengthened by the efficient incorporation of $^{14}C$ labelled glycollate and glycerate into (**6**). Glycollate is also incorporated into the macrolide backbone of leucomycin $A_3$, and into the ansamacrolide geldanamycin. The macrotetrolide antibiotic nonactin (**9**) contains a $C_{10}$ subunit, nonactic acid (**10**), four molecules and both enantiomeric forms of which are required to form the macrotetrolide ring. Nonactic acid is built from two acetate units, one

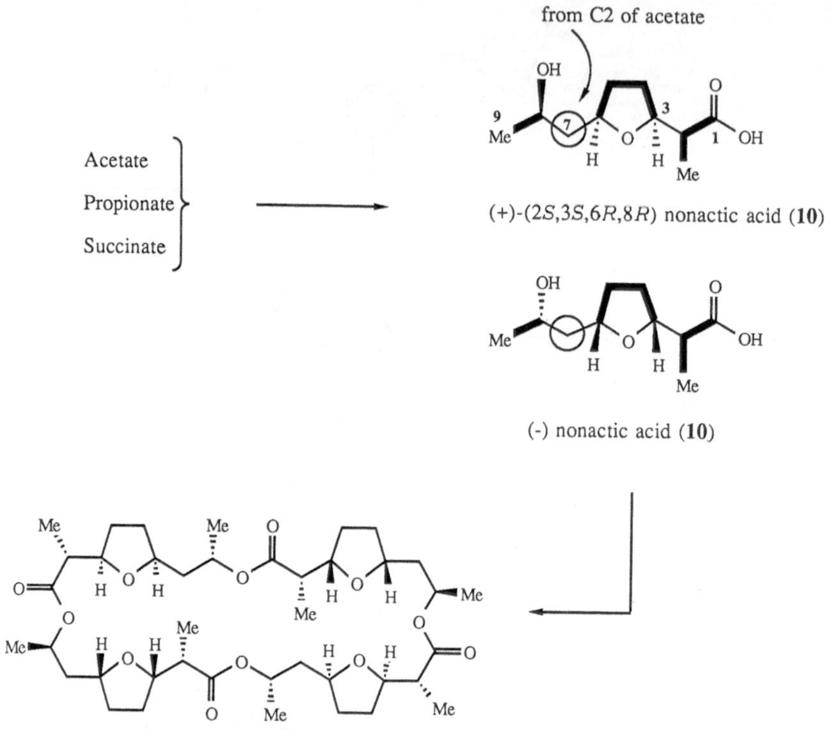

from C2 of acetate

$(+)$-$(2S,3S,6R,8R)$ nonactic acid (**10**)

Acetate  
Propionate  
Succinate  

$(-)$ nonactic acid (**10**)

nonactin (**9**)

Fig. 8. Structure of nonactin, and its biosynthetic origins from primary precursors

succinate and one propionate (Fig. 8), although only the C2 atom of one of the acetate units is incorporated during the assembly process, at C7. All four carbon atoms of the tetrahydrofuran ring originate intact from succinate (*28*). The macrodiolide boron-containing antibiotic aplasmomycin (**11**), isolated from *Streptomyces griseus*, is an unusual polyketide ionophore, since each half of the molecule is assembled from a glycerol-derived starter unit (at C17-C16-C15) and seven acetate extender units (*29*) (see Fig. 9). The polyketide is apparently modified by C-methylation during its assembly, since the branching methyl groups are derived from methionine rather than from intact propionate units. This is in contrast to the usual situation in macrolide and polyether biosynthesis, where methyl and ethyl branches are formed by incorporating propionate (i.e. methylmalonyl-CoA) or butyrate (i.e. ethylmalonyl-CoA) in place of

*References, pp. 72–81*

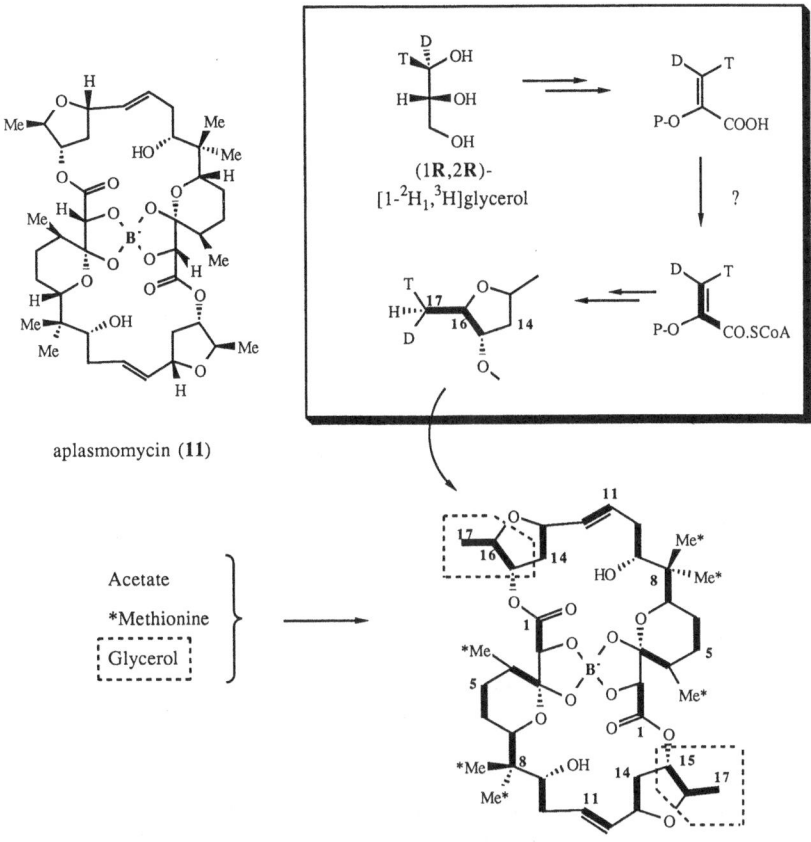

Fig. 9. Structure of aplasmomycin, the incorporation of primary precursors and chirally labelled glycerol

acetate (malonyl-CoA) during chain assembly. The origin of the starter unit appears to be phosphoglyceric acid or phosphoenolpyruvate (*29*). Stereochemical studies have revealed that when (1*R*, 2*R*)-[1-$^2$H$_1$,$^3$H]glycerol is incorporated into the antibiotic a chiral methyl group specifically of (*S*) absolute configuration is formed at C17, whereas the methionine derived methyls are introduced with inversion.

Whilst the vast majority of known polyether antibiotic are produced by microorganisms of the genus *Streptomyces* a smaller number have been isolated recently from marine sources, including sponges and dino-flagellates. Amongst these are brevetoxin-A (**12**), brevetoxin B (**13**), okadaic acid (14), acanthifolicin (15), pectenotoxin-1 (16), and nor-halicondrin A (17) (Fig. 10), all of which are highly cytotoxic. Although the biosynthesis of these metabolites has not been studied as extensively as that of the microbial polyethers, recent researches on brevetoxin B (13) indicate a non-normal polyketide origin (30, 31) for the carbon back-bone. Of the $^{13}$C-labelled precursors tested (acetate, propionate, succin-ate, mevalonolactone, and methionine), only $^{13}$C-labelled acetate and [$^{13}$C-Me]methionine gave levels of incorporation sufficient for analysis by $^{13}$C-NMR. A rather complex pattern of enrichments was observed (extents of enrichment were not reported) due to extensive in vivo meta-bolism of the precursors prior to incorporation. The results obtained, however, were interpreted as indicating a role for dicarboxylic acids from the citric acid cycle in the chain backbone building process, with the participation of $CO_2$ in the biosynthesis of C1 of (12) and (13). Unfortu-nately, the primary metabolism of these organisms has not been well characterised, and this makes it more difficult to draw firm conclusions about the identity of the direct precursors in the chain assembly process.

Often incorporation experiments with $^{13}$C labelled precursors, which are designed to identify primary biosynthetic precursors, can give addi-tional new information about primary metabolism in the antibiotic producing organism. For example, the incorporation of sodium [1-$^{13}$C]propionate into monensin A (1) leads to seven highly enriched centres, indicating intact incorporations into the backbone at the posi-tions shown in Fig. 11. When [1-$^{13}$C]n-butyrate is incorporated the same set of carbon centres is again strongly enriched, together with an extra one due to the expected intact incorporation of a single n-butyrate unit into the carbon backbone. On the other hand, when [1-$^{13}$C]iso-butyrate is fed, exactly the same pattern of $^{13}$C-enrichments is observed as seen with [1-$^{13}$C]n-butyrate. These observations could be explained by the in vivo conversion of isobutyrate into both n-butyrate and propionate, and of n-butyrate into propionate, prior to incorporation into the anti-biotic (32, 33). Similar results were also obtained in studies of tylosin and lasalocid biosynthesis (34, 35). More detailed labelling experiments (32) with other $^{13}$C and $^2$H enriched precursors and the monensin producing organism showed that the methylmalonyl-CoA mutase reaction could not be responsible for the formation of propionate units (i.e. methyl-malonyl-CoA) from n-butyrate (see Fig. 11), but rather suggested that an intramolecular "vicinal interchange" rearrangement of n- to iso-butyrate was occurring in vivo. Cell-free studies later indicated that a novel coenzyme-$B_{12}$ dependent mutase activity was responsible for catalysing the reversible interconversion of n-butyryl-CoA and iso-butyryl-CoA in Streptomyces (33). This enzyme appears to be important for channeling

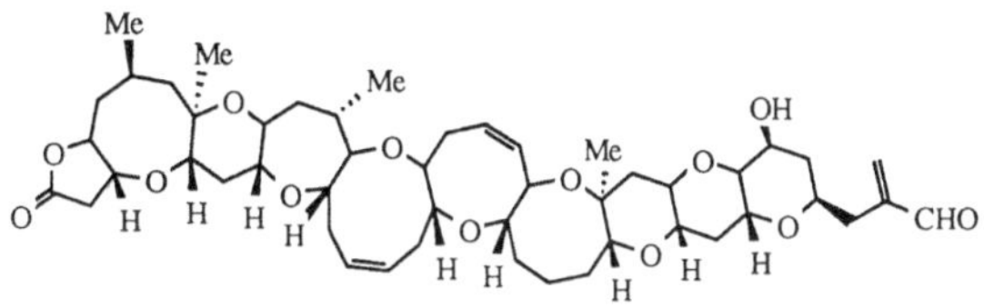

brevetoxin A (**12**)

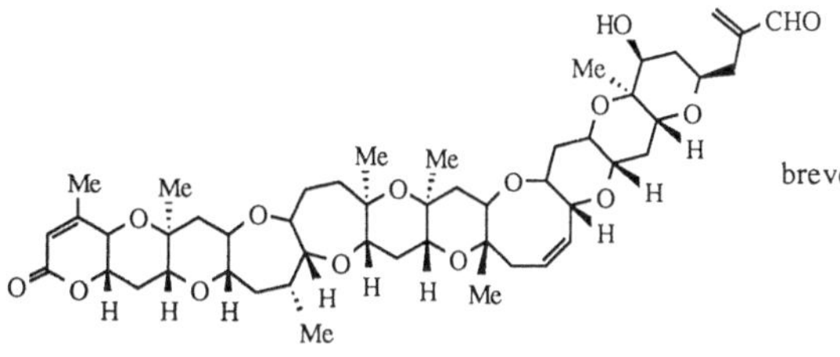

brevetoxin B (**13**)

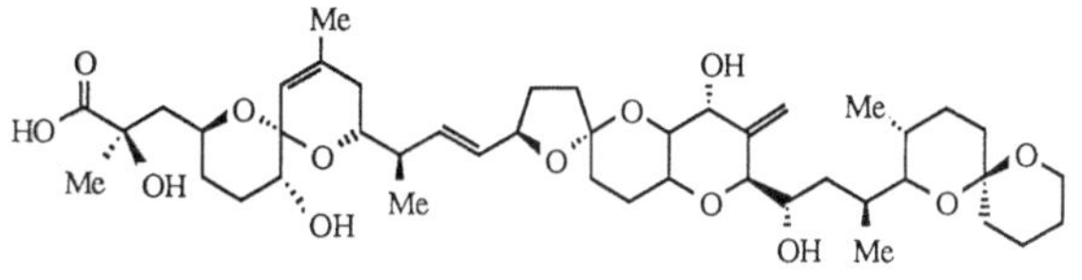

okadaic acid (**14**)

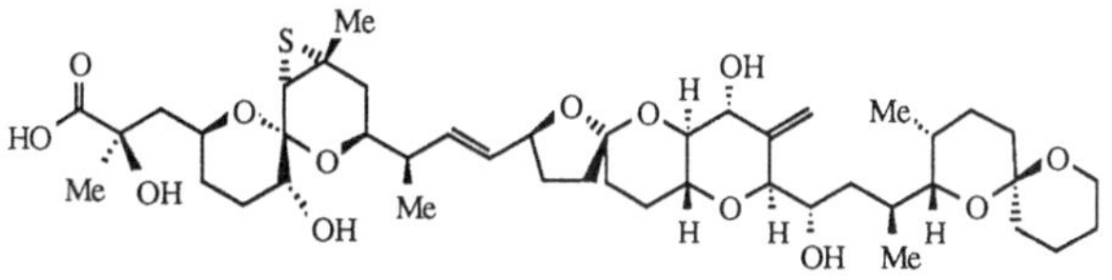

acanthifolicin (**15**)

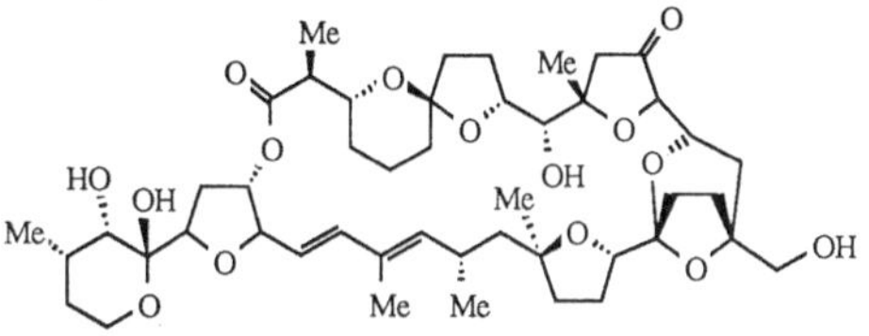

pectenotoxin-1 (**16**)

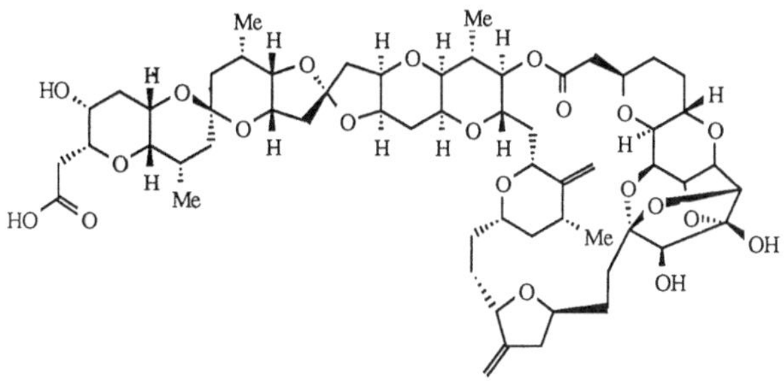

norhalichondrin A (**17**)

Fig. 10. Structure of some marine polyethers

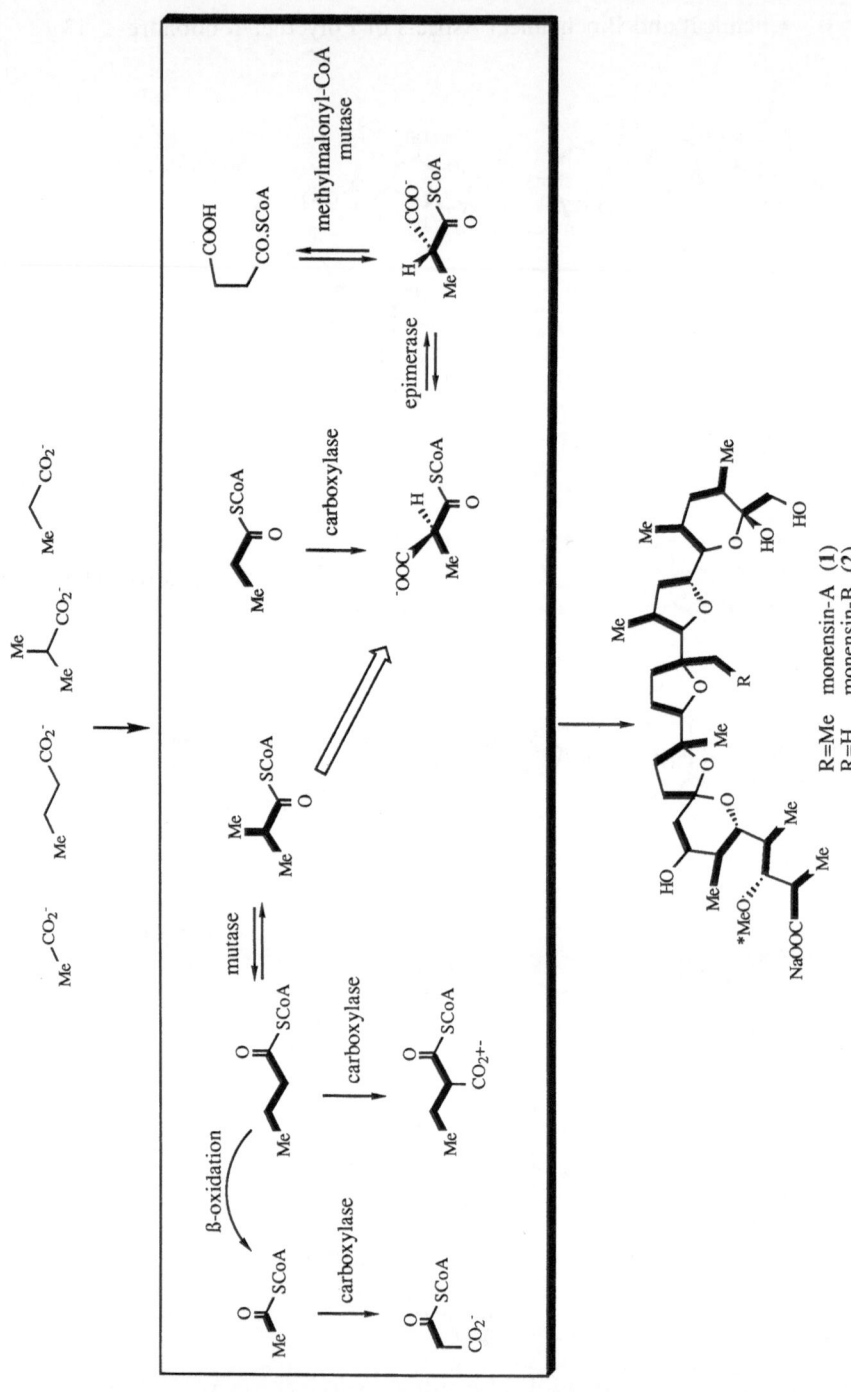

Fig. 11. Schematic view of primary metabolism furnishing building blocks for polyether biosynthesis in Streptomycetes

building blocks derived from the degradation of amino acids and fatty acids into antibiotic biosynthesis. The oxidation of isobutyryl-CoA to methylmalonyl-CoA appears to proceed in *Streptomyces* with oxidation of the pro-*S* methyl group to afford (*S*)-methylmalonyl-CoA (*32*) (see Fig. 11), by a route that is different from that seen in mammals and other bacteria. These aspects of the biochemistry of Streptomycetes are of interest since regulation of the biochemical pathways supplying precursors for antibiotic biosynthesis may be one of the control mechanisms influencing fermentation yields in the manufacture of these important commercial products.

Described below are two further aspects of polyether biosynthesis currently attracting great interest. One concerns the reaction mechanisms leading to heterocyclic ring formation in the late stages of polyether biosynthesis. The other focuses on the early biochemical events that culminate in the construction of a unique polyketide chain from activated forms of the simple fatty acid building blocks, as catalysed by a so-called polyketide synthase (PKS) multienzyme complex.

## 3. The Mechanisms of Heterocyclic Ring Formation

An important discovery, which strongly influenced early ideas about the mechanisms of heterocyclic ring formation in polyether biosynthesis, was made by WESTLEY and coworkers at Hoffmann La-Roche (*22*). They isolated a minor metabolite called iso-lasalocid (**18**) from *S. lasaliensis*, which differed from lasalocid A (**3**) only in the C-ring (see Fig. 12). The co-occurrence of these two structurally related metabolites suggested that the diepoxide (**19**) might be a common biosynthetic precursor. Furthermore, since microorganisms such as *Pseudomonas oleovorans* were known to epoxidise alkenes, the diene (**20**) could be viewed as a further likely intermediate, whose formation from acetate, propionate and butyrate could then be rationalised by the polyketide hypothesis (*vide infra*). In this way heterocyclic ring formation in both lasalocid (**3**) and iso-lasalocid (**18**) biosynthesis can be explained by well precedented intramolecular cyclisations occasioned by the attack of a hydroxyl group on an epoxide in a stereoelectronically favourable $S_N2$ manner. The final cyclisation step should proceed by two routes, giving either lasalocid or iso-lasalocid. Under acid catalysis in free solution the favoured route would presumably be that leading to iso-lasalocid. However, with enzymic catalysis the substrate might be bound in a conformation that favours cyclisation to lasalocid (**3**). The carboxyl group might also act as a general acid catalyst (*22*), as shown in Fig. 12.

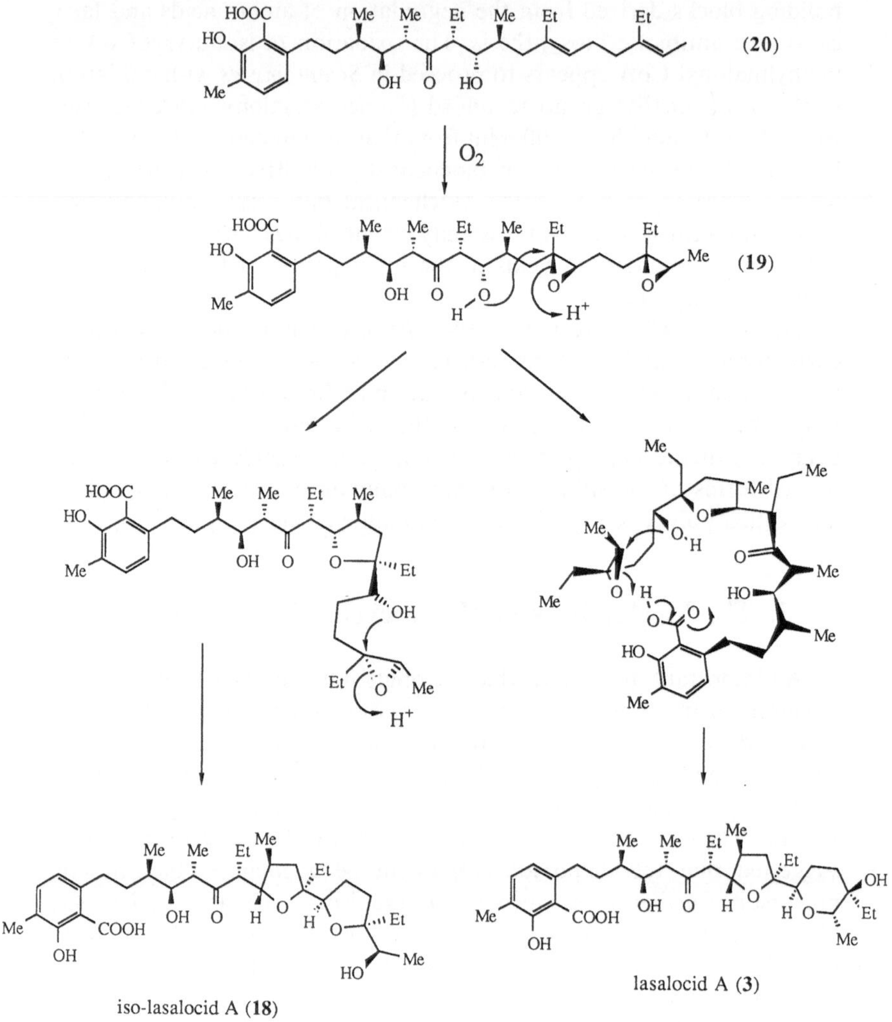

Fig. 12. Hypothetical pathways to lasalocid and iso-lasalocid, involving diene and diepoxide intermediates

Further interest in the polyepoxide concept was stimulated with the structure elucidation of metabolites isolated from *S. albus*, the producer of salinomycin (21) (22). Under appropriate fermentation conditions the major metabolite is not salinomycin (21) but rather 20-deoxy-17-*epi*-salinomycin (22), with 20-deoxysalinomycin (23) and salinomycin (21) being produced in minor quantities (Fig. 13). In the crystalline state

*References, pp. 72–81*

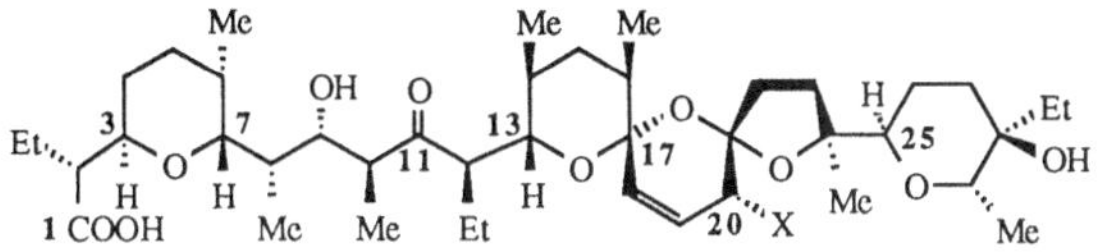

X = OH   salinomycin   (21)
X = H    20-deoxysalinomycin (23)

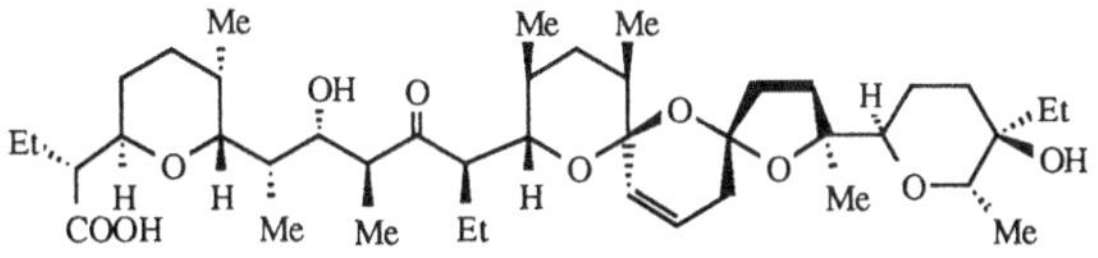

20-deoxy-17-epi-salinomycin  (22)

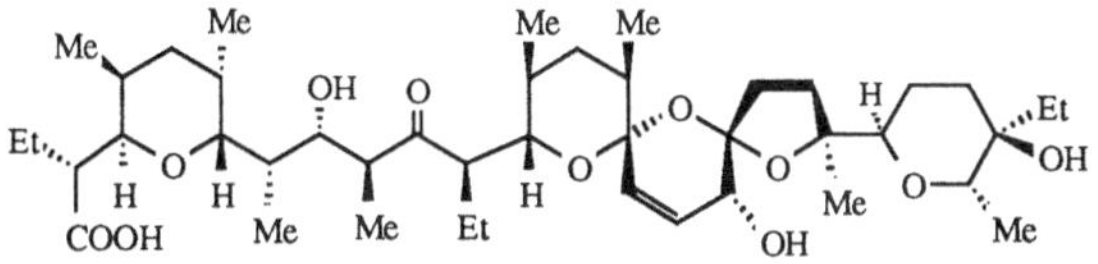

narasin  (27)

Fig. 13. Structures of salinomycin, narasin, and related naturally occurring derivatives

20-deoxy-17-*epi*-salinomycin (22) adopts an open conformation, whereas in contrast the *p*-iodophenacyl ester of salinomycin (21) adopts a U-shaped conformation in which the C20 hydroxyl group is hydrogen bonded with the C9 hydroxyl group. Again, these two different but closely related metabolites *might* arise from a common polyene inter-mediate (24) and closely related diepoxide intermediates (25 and 26), as shown in Fig. 14. However, it is unclear whether a single enzyme (or set of enzymes) could efficiently catalyse both modes of cyclisation. It cannot be excluded that these late cyclisation steps in the biosynthetic pathway might occur non-enzymically. However, an alternative scenario is sug-gested by observations from KISHI's group made during their syntheses of salinomycin (21) and narasin (27) (36). They observed that under acid catalysis the ketone (28) isomerised completely to (29) with the 20-deoxy-17-*epi*-salinomycin-like configuration, whereas the *epi*-keto alco-hol (30), equilibrated under the same conditions, gave a 1:4 ratio of (31) and (30) (Fig. 15). Furthermore, 17-*epi*-salinomycin (32) in methylene

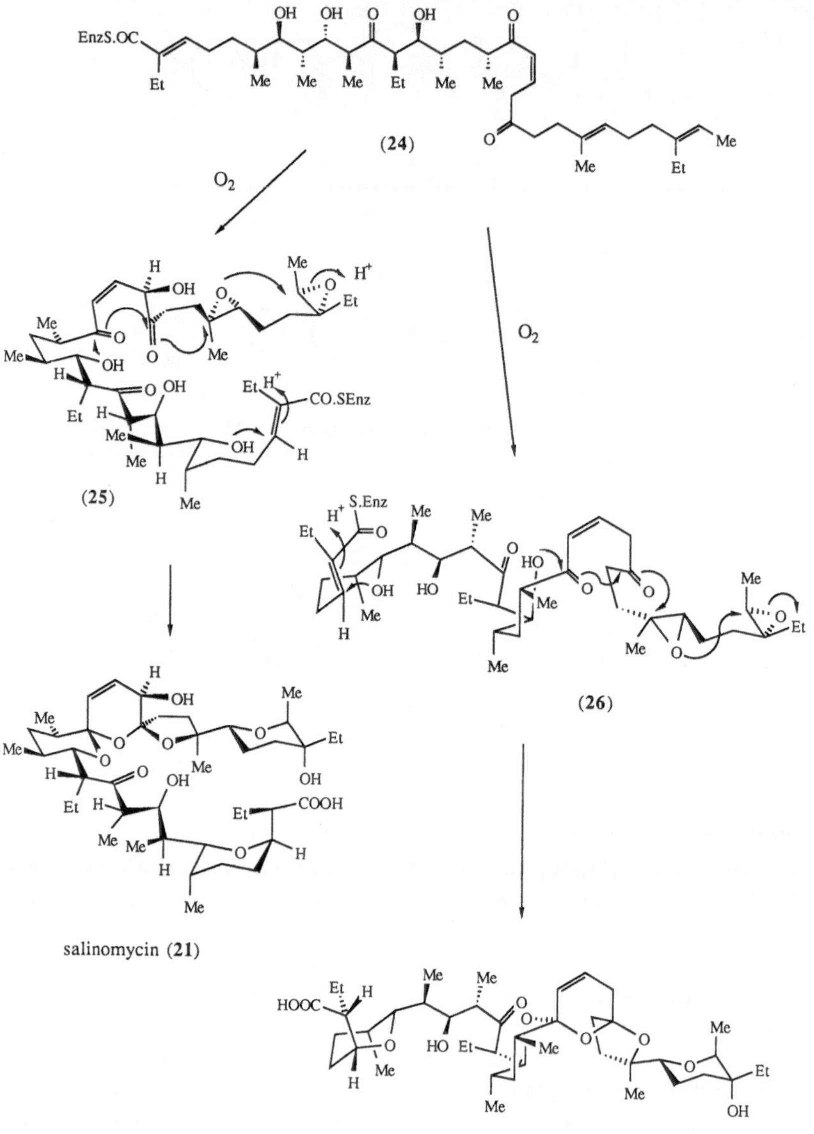

Fig. 14. Hypothetical pathways to salinomycin and 20-deoxy-17-*epi*-salinomycin via a common diene intermediate

*References, pp. 72–81*

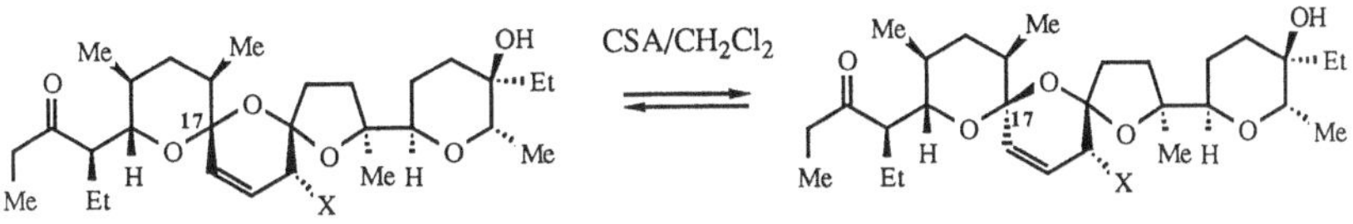

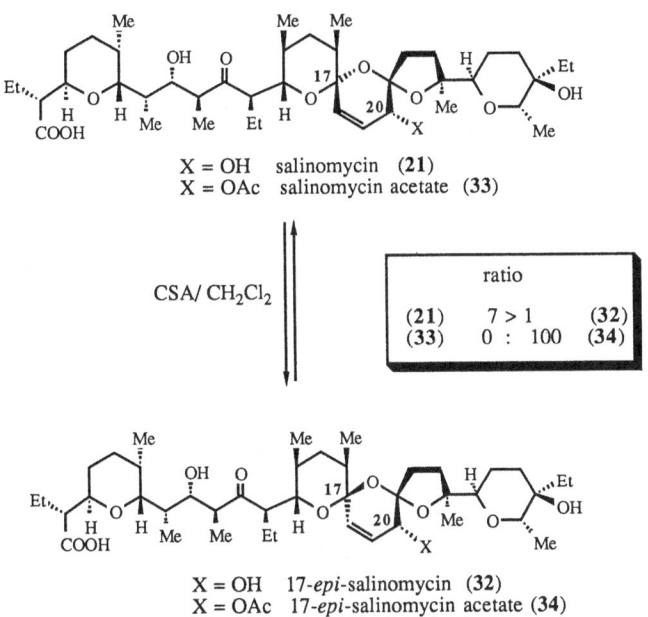

| | ratio | |
|---|---|---|
| (28) X = OAc | 0 : 100 | X = OAc (29) |
| (31) X = OH | 1 : 4 | X = OH (30) |

Fig. 15. Acid catalyzed epimerisation (36)

chloride with a small amount of trifluoracetic acid afforded almost exclusively salinomycin (21), whereas salinomycin acetate (33) upon acid treatment isomerised exclusively to 17-*epi*-salinomycin acetate (34) (Fig. 16). The dramatic difference in the equilibrium point between the ketone series (30, 31) and the antibiotic series (21, 32) may reflect the importance of hydrogen bond stabilisation involving the C20 hydroxy group and a *remote* position nearer to the carboxyl terminus (see (21) in Fig. 14).

X = OH   salinomycin   (21)
X = OAc   salinomycin acetate   (33)

CSA/ CH₂Cl₂

| | ratio | |
|---|---|---|
| (21) | 7 > 1 | (32) |
| (33) | 0 : 100 | (34) |

X = OH   17-*epi*-salinomycin   (32)
X = OAc   17-*epi*-salinomycin acetate   (34)

Fig. 16. Acid catalyzed epimerisation at C 17 in salinomycin and salinomycin acetate (36)

Such a hydrogen bond may also be important in compensating for a seemingly unfavourable dipole-dipole interaction at the bis-spiro centres in the salinomycin configuration (*36*). It is conceivable, therefore, that (**22**) might arise by spontaneous epimerisation of (**23**) during the isolation and extraction procedures.

The first direct support for the polyene-polyepoxide concept was obtained through the incorporation of carbon-13 and oxygen-18 doubly labelled precursors into the polyethers monensin A (**1**) and lasalocid A (**3**). The $^{18}$O-isotope effect on $^{13}$C NMR signals was first documented by VAN ETTEN (*37*) and VEDERAS (*38*) and their coworkers. Replacement of $^{16}$O by $^{18}$O causes an upfield shift of the $^{13}$C resonance of the directly attached carbon by about 0.015 to 0.05 ppm, depending mainly upon the C—O bond order. The value of $^{18}$O-labelled precursors for tracing the biosynthetic origins of oxygen atoms was first demonstrated by VEDERAS (*39*) in studies on the polyketide metabolite averufin (**35**) from *Aspergillus parasiticus*. As noted by these workers "this new non-degradative technique allows semiquantitative localization of $^{18}$O in highly oxygenated systems (by $^{13}$C NMR) where extensive degradations . . . would be necessary to determine structures of mass spectrometric fragments".

The incorporation of [1-$^{13}$C, $^{18}$O$_2$]propionate and [1-$^{13}$C, $^{18}$O$_2$] butyrate into lasalocid A (**3**), and analysis of the enriched antibiotic by 100 MHz $^{13}$C NMR, revealed (*40*) the intact incorporation of $^{18}$O at C3, C11, and C15 from labelled propionate, and at C1 and C13 from butyrate (at C1 via β-oxidation to acetate) (Fig. 17). Fortunately, very little wash-out of the $^{18}$O-label occurred during the incorporation *in vivo*, although this is not always to be expected. The only oxygenated centres not enriched were at C19-C23 and C22, and these are the oxygens that are predicted to arise from O$_2$ in the WESTLEY diene-diepoxide model of lasalocid biosynthesis (Fig. 12). To date a complementary experiment with $^{18}$O$_2$ to confirm the role of molecular oxygen in the biosynthesis has not been reported.

At the same time CANE and coworkers (*41*) reported similar experiments on the biosynthesis of monensin A (**1**). The results (see Fig. 18) demonstrated that 7 of the 11 oxygen atoms in monensin A are derived directly from the carboxylate oxygens of the precursors acetate and propionate, whereas the remaining ether oxygens, O(7), O(8), and O(9) as well as the hydroxyl group at C26, were derived from molecular oxygen (*42, 43*). This led to the attractive proposal shown in Fig. 19, where the biosynthesis leads to the all-*E* triene (**36**), possibly as the first enzyme-free intermediate. Following the epoxidation of each *E*- double bond, to give the (12*R*, 13*R*, 16*R*, 17*R*, 20*S*, 21*S*)triepoxide (**37**), an attack by the C5 hydroxyl at the C9 carbonyl carbon would initiate a cascade of ring closures to generate stereospecifically all five ether rings of monensin A.

*References, pp. 72–81*

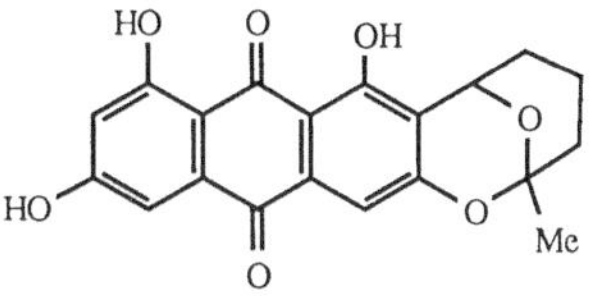

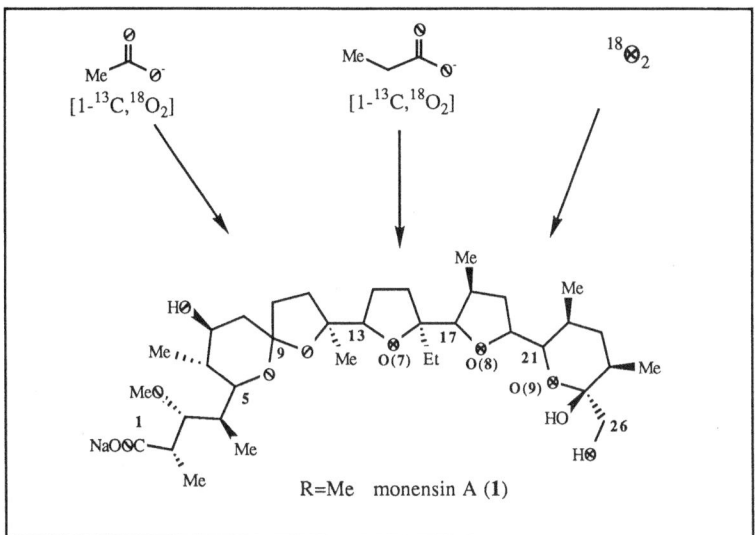

averufin (**35**)

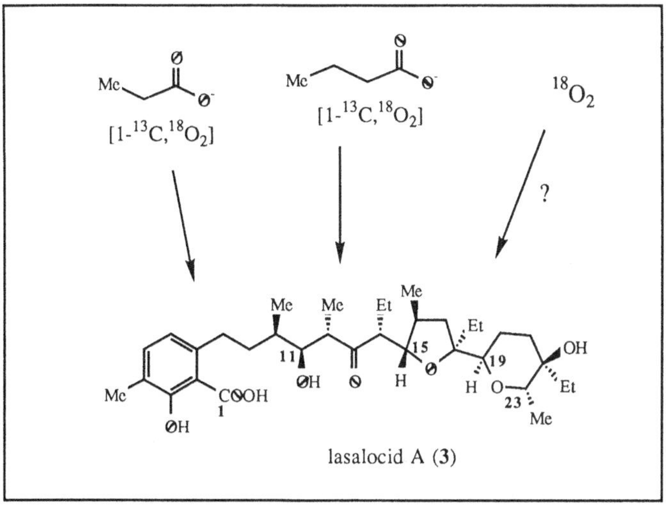

Fig. 17. Incorporation of $^{18}O/^{13}C$ labelled precursors into lasalocid

Fig. 18. Incorporation of $^{18}O/^{13}C$ labelled precursors into monensin A

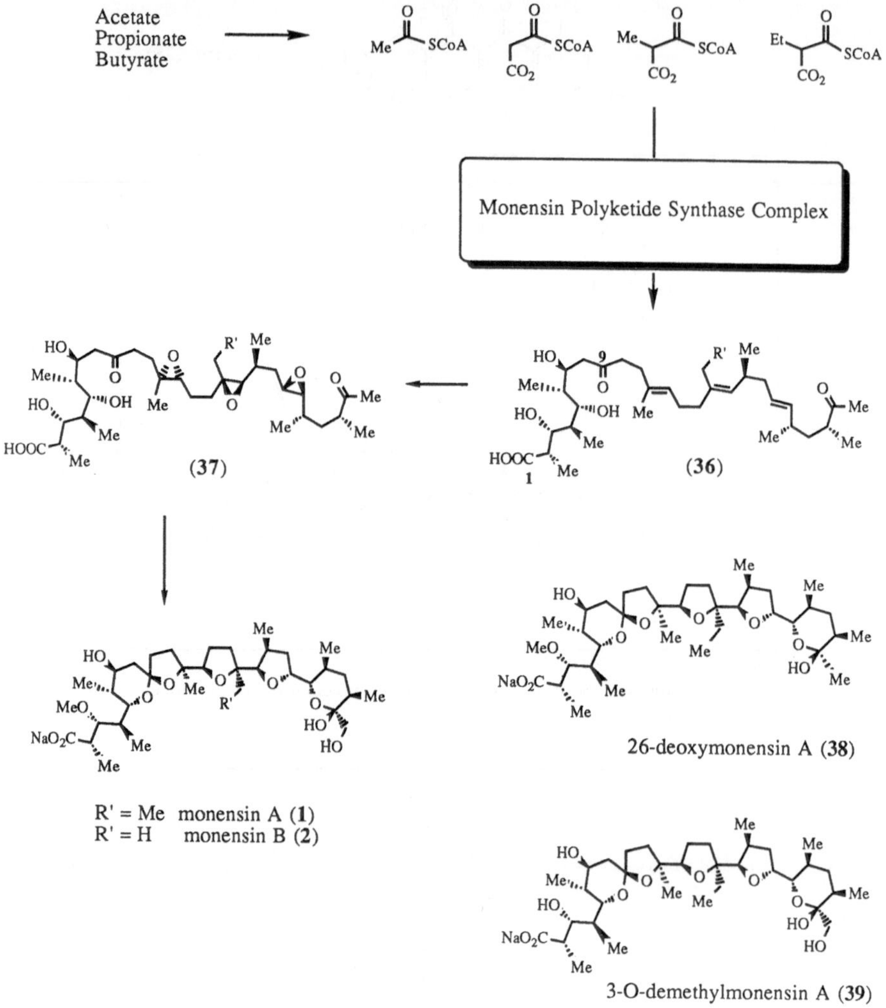

Fig. 19. A hypothetical biosynthetic pathway to monensin A, with the structures of 26-deoxymonensin and 3-O-demethylmonensin

Another important conclusion to emerge from these labelling studies on lasalocid A and monensin A concerns the mechanisms of stereocontrol at centres that retain the β-placed oxygen atoms from the precursors acetate and propionate during construction of the polyketide backbone. The fact that each of the β-placed oxygen atoms from C1 through to C9

*References, pp. 72–81*

in monensin A is incorporated intact from acetate or propionate indicates that the absolute configuration of the C(3), C(5), and C(7) centres is established directly through stereodivergent reduction of β-ketothioester intermediates, rather than by additional dehydration and rehydration steps. Since C(3) is *R*, and C(5) and C(7) are *S*, this would seem to require at least two different ketoreductases having opposite stereospecificities. Similar results have been obtained in studies of macrolide biosynthesis (erythromycin (**40**) (*44*), and tylosin (**41**) (Figs. 36 and 37) (*45*)) strengthening the idea of a common mode of stereocontrol during the assembly of these two classes of polyketide antibiotics.

Biosynthetic studies employing $^{18}$O-labelled precursors have also been reported for the polyethers narasin (**27**) (*46*), maduramicin (**42**) (*47*), lenoremycin (**43**) (*48*) and ICI139603 (**6**) (*49*), as well as the macrotetrolide nonactin (**9**) (*50*), and the macrolides avermectin (**44**) (*51*) and LL-F28249α (**45**) (*52*) (Fig. 20), all Streptomycete derived natural products. In the case of narasin (**27**) (*46*), the origins of the oxygen atoms, from either the precursors (acetate and propionate) or molecular oxygen, are in accord with its formation from the diene and diepoxide shown in Fig. 21. By analogy, these results also provide support for the hypothetical pathway to salinomycin (**21**) suggested by WESTLEY (cf. Fig. 14). The tetrahydropyran ring nearest the carboxyl terminus could be formed through attack by the C7-hydroxyl onto C3, most likely *via* a *syn*-Michael addition onto an *E*-α, β-unsaturated thiol ester. A related process has also been implicated in the nonactin (**9**) pathway (*50*) (see Figs. 22 and 8), since the tetrahydrofuran ring oxygen is incorporated intact with C6 from succinate, the C8 oxygen is derived from acetate, and the C1 carbonyl oxygen comes intact from propionate. Lenoremycin (**43**) and the closely related metabolites dianemycin, leuseramycin and moyukamycin are pentacyclic ethers containing a second tetrahydropyran-tetrahydrofuran spiroketal in place of the more commonly occurring pair of tetrahydrofuran rings typical of monensin. CANE and coworkers (*48*) showed that the polyepoxide concept can be extended to include this group of polyethers by postulating the intermediacy of the appropriate diepoxytriketone (**46**), shown in Fig. 23. The ionophore ICI139603 (**6**), on the other hand, contains an unusual cyclohexane ring in addition to tetrahydropyran and tetrahydrofuran rings. STAUNTON and coworkers (*49, 53*) suggested on the basis of their labelling results three possible mechanisms for formation of these rings, including one involving concerted formation of the two six membered rings, as shown in Fig. 24.

Although the polyene-polyepoxide concept provides in principle a chemically concise and stereospecific entry into the ether ring systems

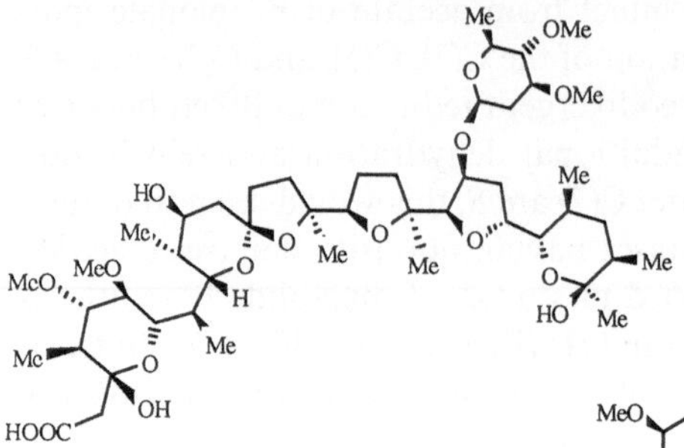

Fig. 20. Other macrolide and polyether antibiotics whose biosynthesis has been studied using $^{18}$O labelled precursors (see text for references)

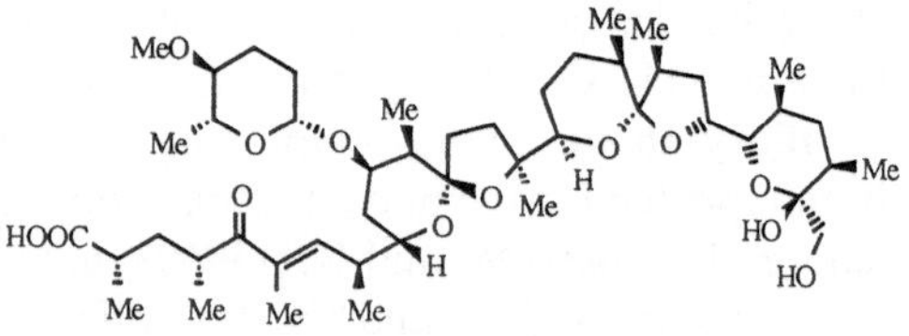

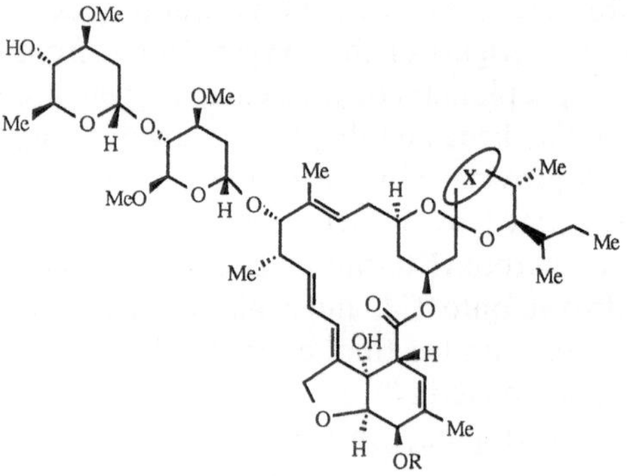

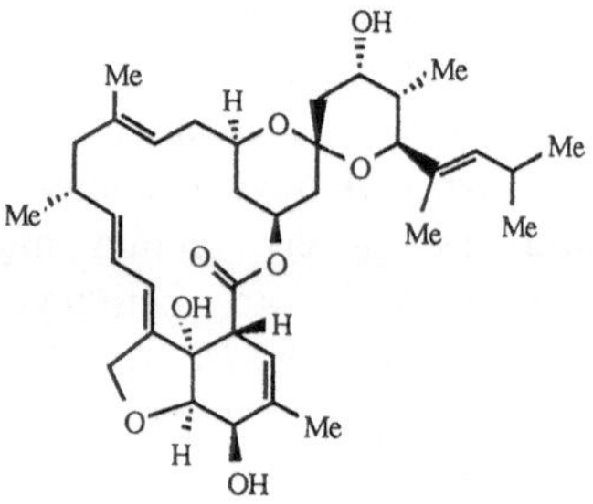

of the polyether antibiotics and its feasibility has been proven in model systems (*vide infra*), it should be noted that no *bona fide* intermediates on any pathway to any polyether have been described in the primary literature. In the case of monensin biosynthesis, 26-deoxymonensin (**38**) and 3-O-demethylmonensin (**39**) (see Fig. 19) have been isolated from

*References, pp. 72–81*

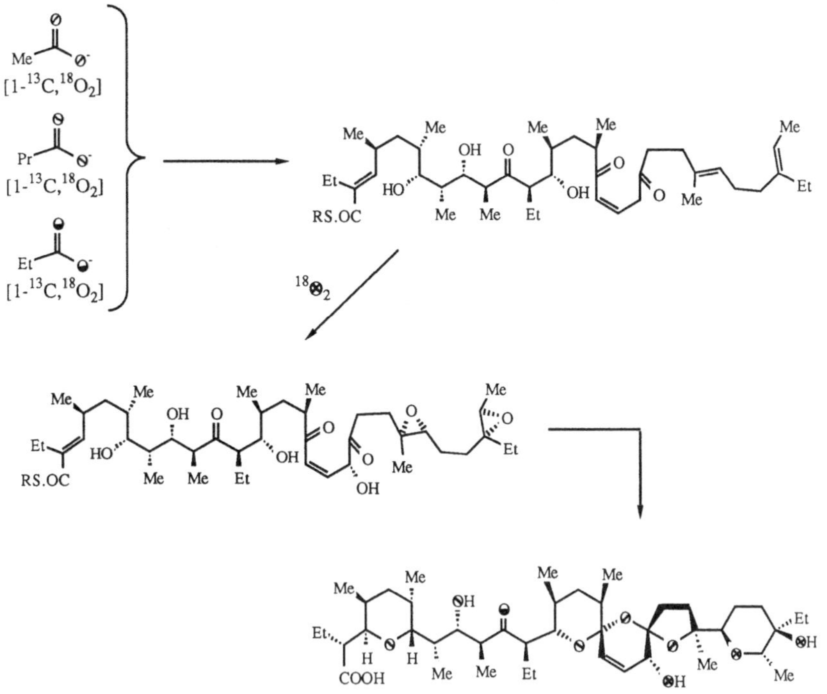

narasin (27)

Fig. 21. A hypothetical pathway to narasin, consistent with $^{18}$O-labelling results

a mutant (54) and from a wild type strain (55), respectively, but labelled forms of each were not incorporated efficiently into monensin in shake cultures of S. cinnamonensis (54), so their status in the pathway, as either intermediates or shunt products, remains unsettled. The timing of the methylation and C26-hydroxylation events is therefore uncertain, although they are apparently not essential for the formation of the ether rings. There is clearly a very substantial gap at this point in our knowledge of polyether ionophore antibiotic biosynthesis.

Finally, an intriguing suggestion has been advanced to explain the formation of the fused ring systems in brevetoxins A (12) and B (13) (30, 31), and related marine polyethers (56), which is very reminiscent of the polyene-polyepoxide model of microbial polyether biosynthesis. As shown in Fig. 25, after creation of an appropriate polyepoxide, a cascade of ring closures could be envisaged, leading to the fused 6/6, 6/7, 6/8, 7/9, 8/9 and 7/7 ring systems in these natural products. No oxygen-18

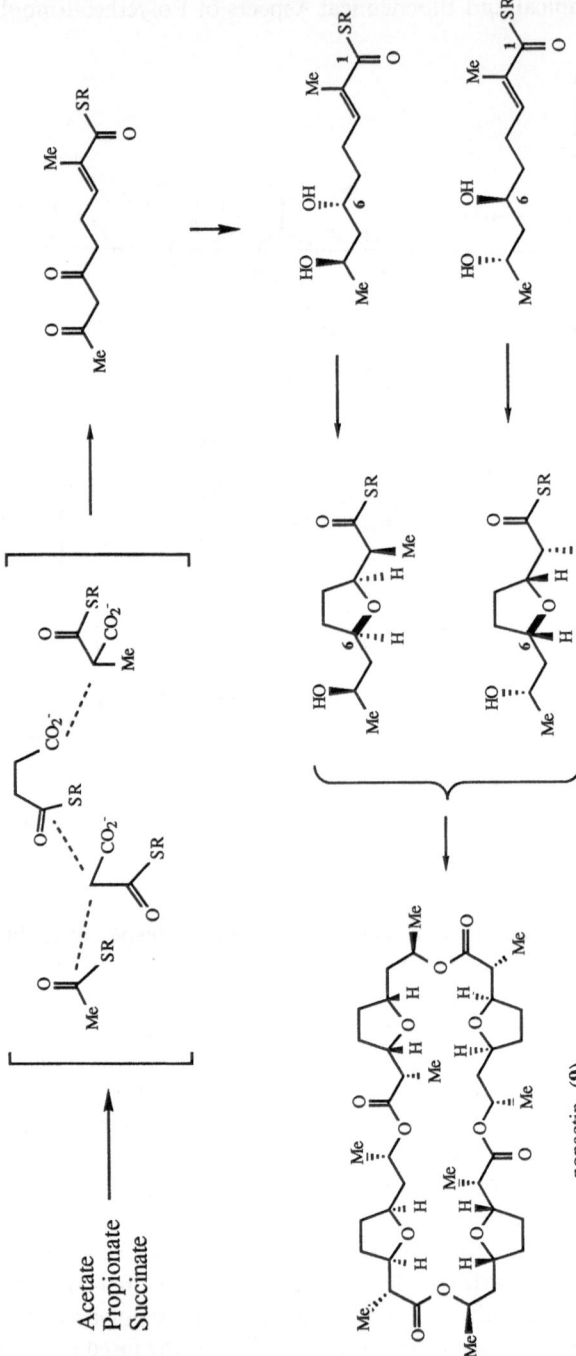

Fig. 22. A hypothetical pathway to nonactin, consistent with [18]O-labelling results

Acetate
Propionate
Succinate

nonactin (9)

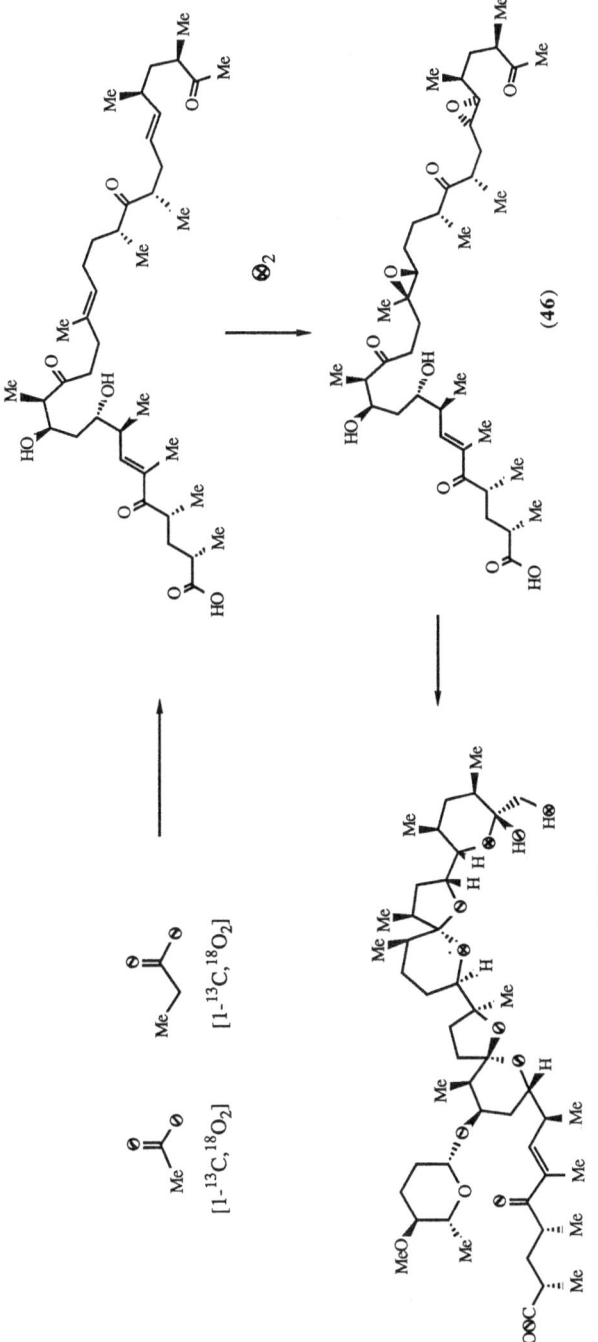

lenoremycin (**43**)

Fig. 23. A hypothetical pathway to lenoremycin, consistent with [18]O-labelling results

(**46**)

[1-[13]C, [18]O$_2$]

[1-[13]C, [18]O$_2$]

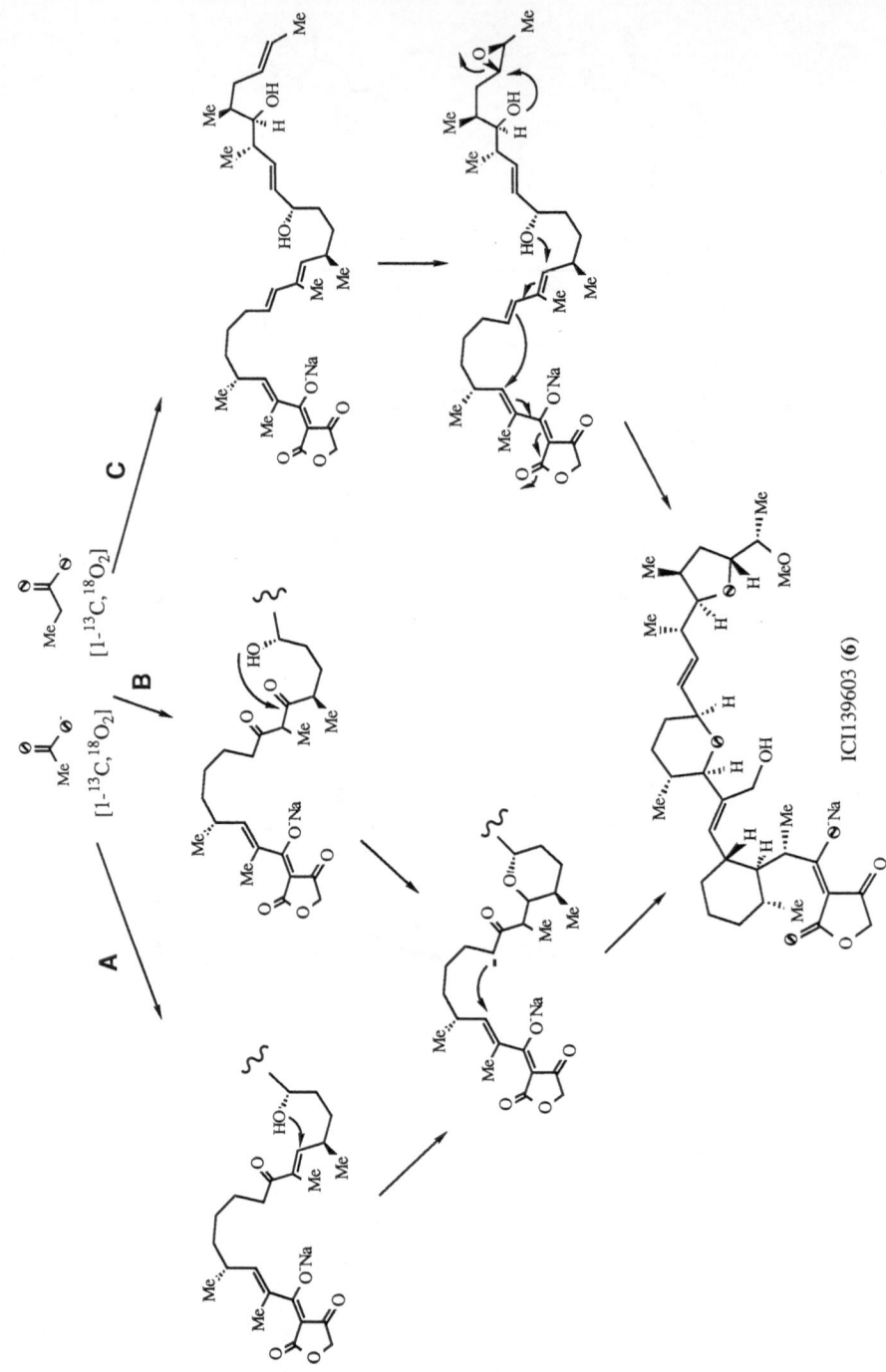

Fig. 24. Several possible mechanisms to account for ring formation in ICI 139603, all consistent with [18]O-labelling results

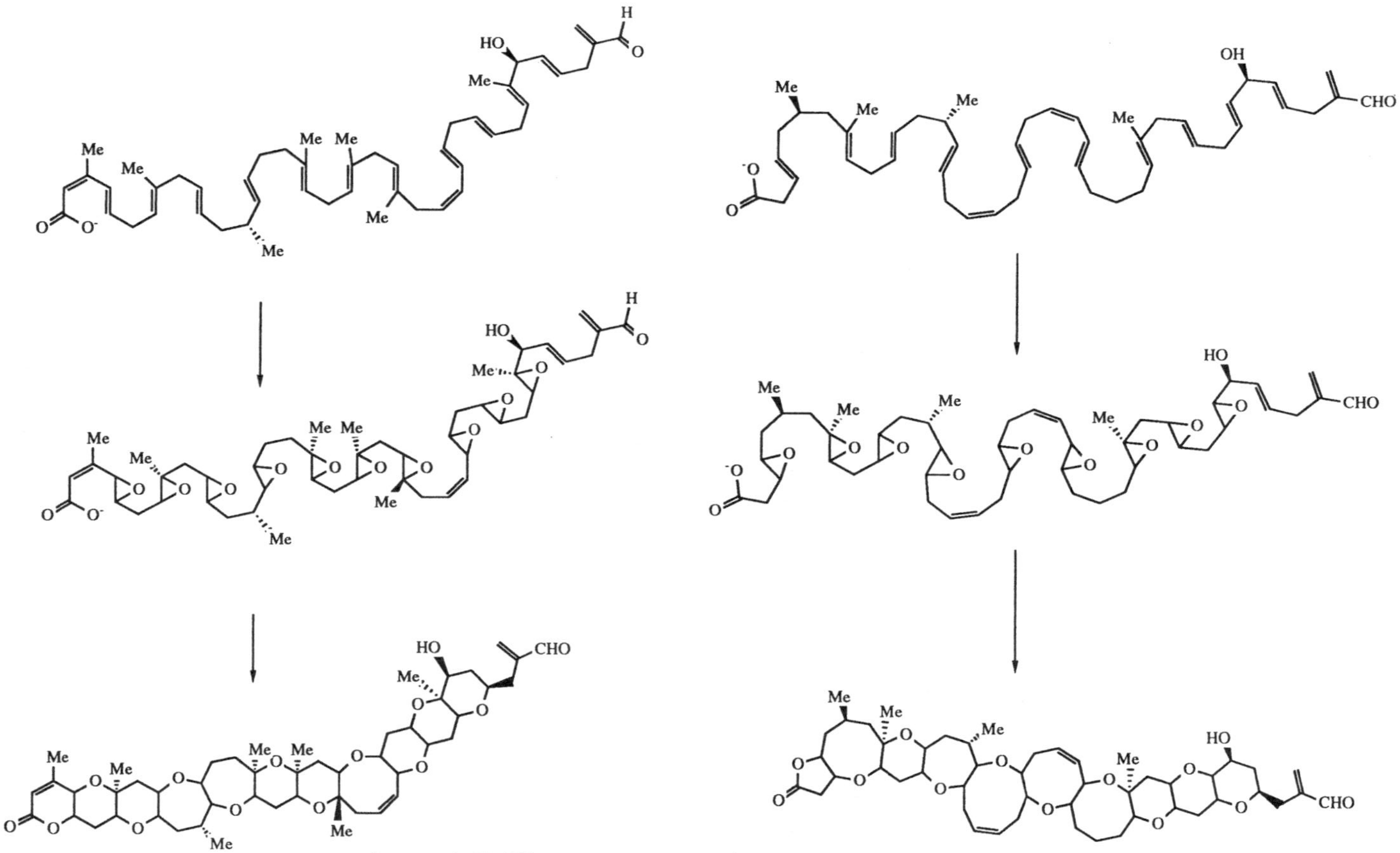

brevetoxin A (12)

brevetoxin B (13)

Fig. 25. Hypothetical biosynthetic routes to marine polyethers

labelling experiments have so far been reported in support of these proposals.

# 4. Model Studies Related to the Late Stages of Polyether Biosynthesis

It is not surprising that the polyene-polyepoxide concept of polyether biosynthesis has also stimulated interest in the development of putative biomimetic syntheses of these and related molecules. Several elegant model systems have now been reported which amply demonstrate that polyepoxide cyclisation cascades provide an efficient entry to multiply linked tetrahydrofuran and tetrahydropyran ring systems. In one of the first studies reported, STILL and ROMERO (57) synthesised the macrolide (47), whose conformational behaviour allowed a stereoselective conversion into the triepoxide (48) (see Fig. 26). The structure of (48) was confirmed by X-ray crystallography. Although the configuration of this triepoxide differs from that required for a synthesis of the central section of monensin B (2), (48) cyclised spontaneously to lactone (49) upon saponification and workup with acetic acid. In another study, by SCHREIBER and coworkers (58), the macrolide (50) was converted into diepoxide (51) stereoselectively, and this upon saponification and treatment with acid/acetone gave the lactone (52) (Fig. 27). In a related system, the diepoxides (53) and (56) were prepared in optically pure form

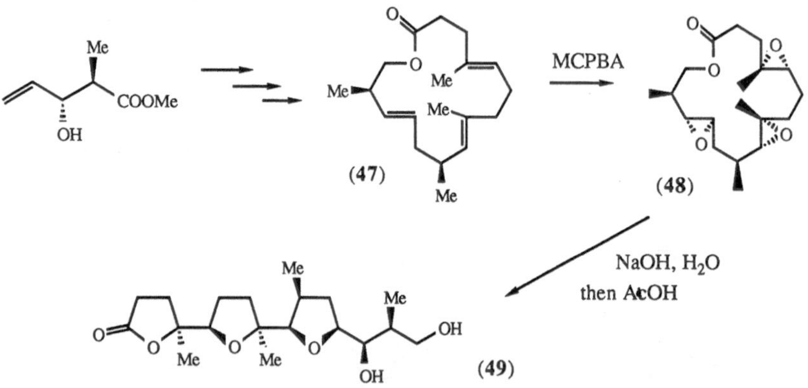

Fig. 26. A model polyene-polyepoxide cyclisation cascade (57)

References, pp. 72–81

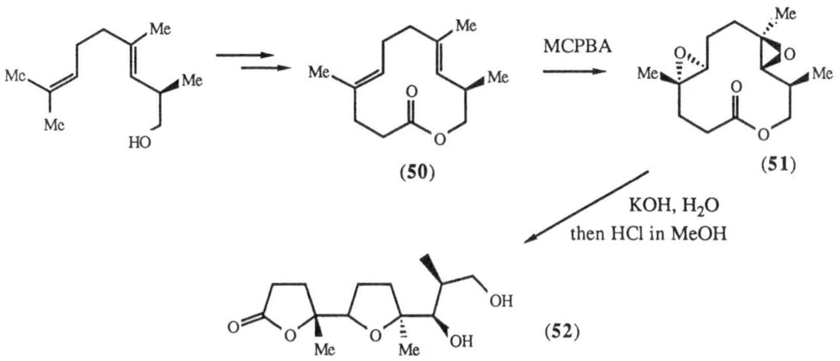

Fig. 27. Another model polyene-polyepoxide cyclisation cascade (*58*)

using the Sharpless epoxidation methodology, and each cyclised stereo-specifically to (**55**) and (**58**) in aqueous solution when treated with pig liver esterase (*50*) (Fig. 28). It is of interest that under these (near physiological) conditions the cyclisation cascade did not proceed in a concerted fashion; the lactones (**54**) and (**57**), detected by NMR, formed rapidly but cyclised more slowly to (**55**) and (**58**).

Several interesting observations have been made by HOYE and coworkers. For example, when the optically pure triepoxide (**59**) was exposed to aq. NaOH an almost racemic mixture of tetrols (**60**) and (**61**) was formed (*60*), by two cyclisation cascades proceeding at virtually equal rates, one starting from the "top-down" and the other from the "bottom-up" (see Fig. 29). Other cyclisation cascades which mimic closely that postulated to occur during monensin biosynthesis have been described by PATERSON and coworkers (*61, 62*). They synthesised the triepoxide (**62**) and diepoxide (**64**), and could show that, with acid catalysis, rearrangement occurs to give (**63**) and (**65**), respectively (see Fig. 30).

These studies show that polyepoxide cyclisation cascades provide an efficient vehicle for heterocyclic ring formation and provide valuable models for the processes thought to occur during polyether biosynthesis. It is also noteworthy that some strains of polyether producing *Streptomyces* can catalyse biotransformations which are reminiscent of epoxide-mediated cyclisations. For example, the nigericin (**66**) producing organism *S. albus* can bioconvert linalool into the linalool oxides shown in Fig. 31, in shake flask cultures (*63*). Although epoxide intermediates were not detected, the formation of these products appears to arise by an epoxidation-cyclisation pathway. Similar observations were made with

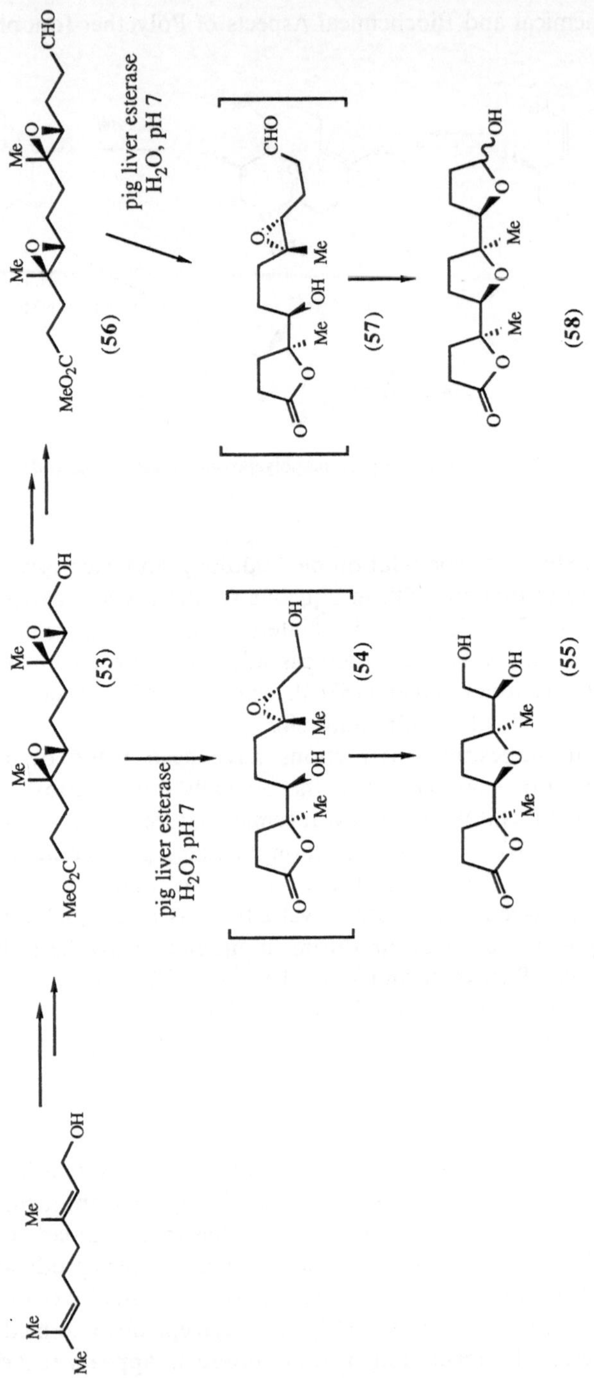

Fig. 28. A diepoxide cyclisation under (near) physiological conditions (59)

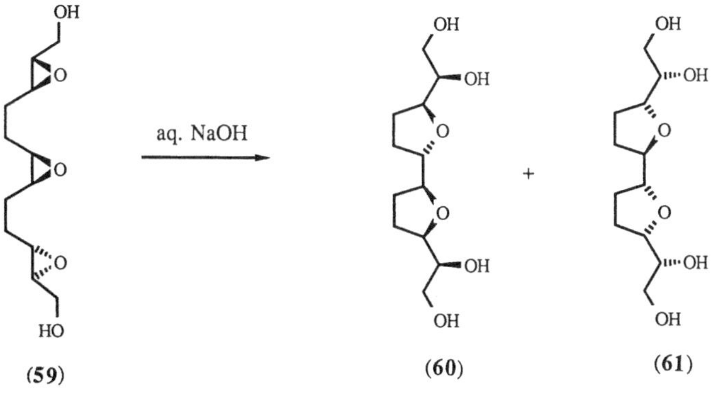

Fig. 29. A homochiral triepoxide leading to an almost racemic product (*60*)

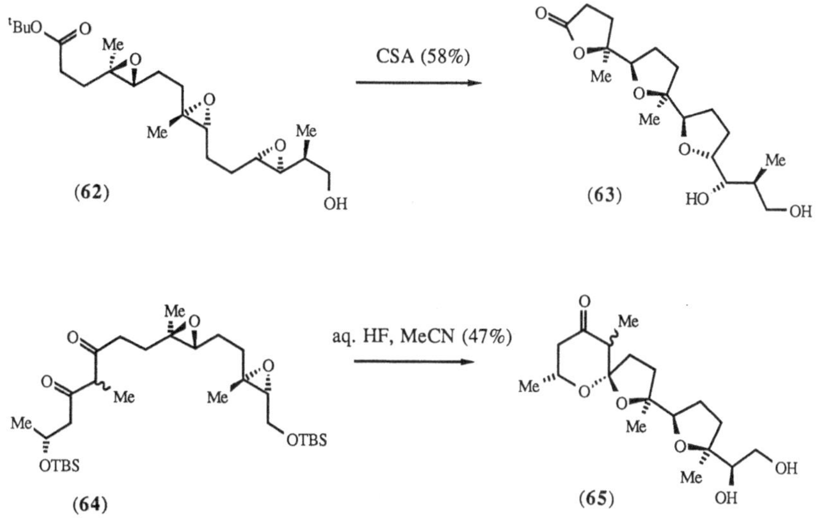

Fig. 30. Close models for the putative monensin cyclization cascade (see Fig. 19). (*61, 62*)

the monensin producing organism *S. cinnamonensis*, and in addition the biotransformation of *E*- and *Z*-nerolidol (*64*) gave rise to several new products (see Fig. 32), including one containing an intact epoxide moiety. However, in these experiments bio-oxidation of only the *Z*- and not the *E*-trisubstituted double bond was observed. While it is gratifying to find a polyether producing organism with the ability to catalyse epoxidations, it should be remembered that the normal *in vivo* function(s) for these

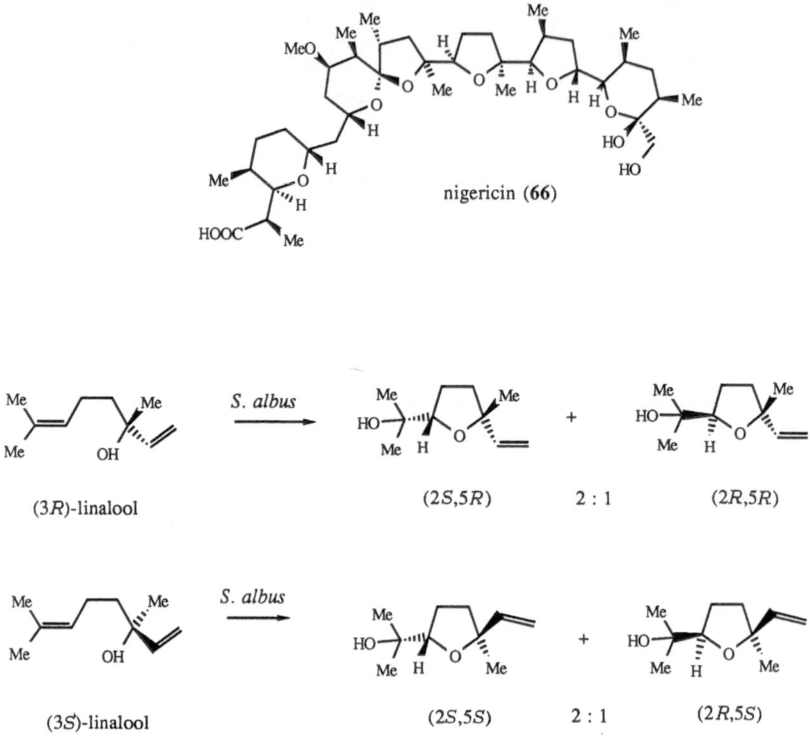

nigericin (66)

(3R)-linalool

(2S,5R)        2 : 1        (2R,5R)

(3S)-linalool

(2S,5S)        2 : 1        (2R,5S)

Fig. 31. Bioconversions catalyzed by S. albus, the nigericin producing organism (63)

enzymes is by no means certain, and they may have more to do with the catabolism of components of the fermentation medium than with cellular secondary metabolism.

Finally in this section, a quite different approach to the synthesis of fused tetrahydrofuran rings such as those found in monensin A, is introduced. KLEIN and ROJAN first reported that geranyl acetate and neryl acetate underwent stereoselective cyclisations in the presence of KMnO₄ to give a cis-tetrahydrofuran instead of the expected tetrol (65). Later, stereochemical studies by WALBA (66) and by BALDWIN (67) demonstrated that such reactions proceed with high stereoselectivity and that the configuration of the product is intimately related to that of the starting 1,5-diene, as shown in Fig. 33. A cis-tetrahydrofuran is formed and effectively two oxygen atoms are added in syn-fashion across one face of each double bond. WALBA and coworkers suggested a mechanism (shown in Fig. 34), involving initial [2+2] insertions of the olefinic

*References, pp. 72–81*

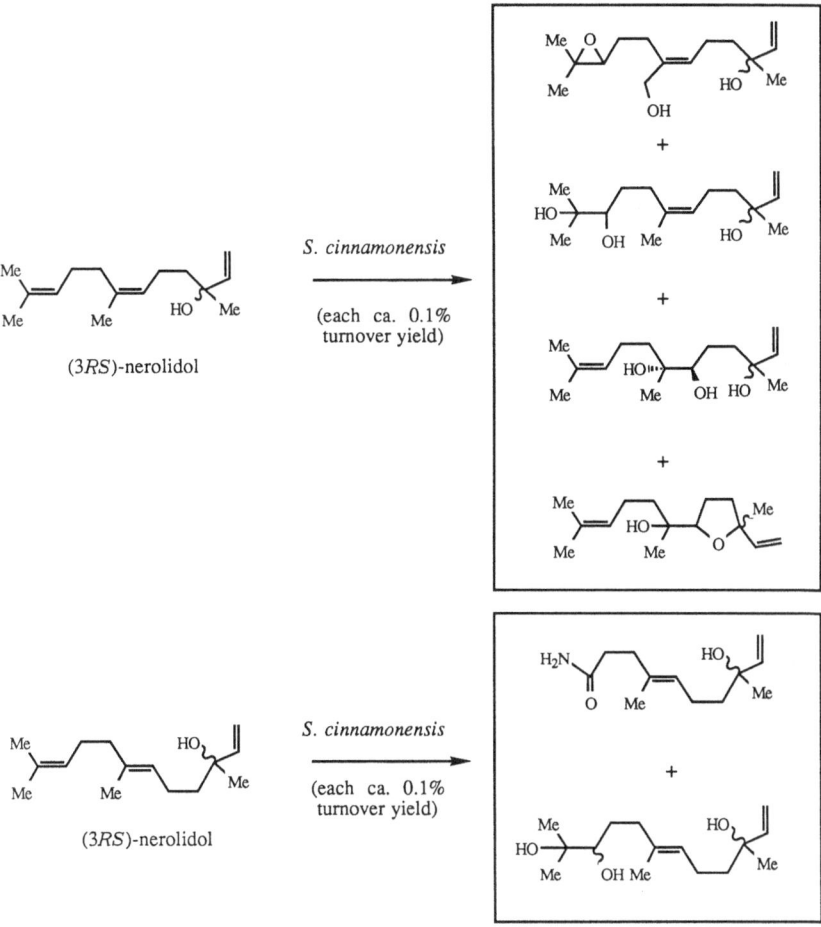

Fig. 32. Bioconversions catalyzed by *S. cinnamonensis*, the monensin producing organism
(*64*)

π-bonds into metal-oxo bonds of manganese followed by reductive
elimination with migration of carbon from manganese to oxygen with
retention of configuration (*66*). This methodology has been applied to
the synthesis of a fragment representing the BC-rings of monensin from
an achiral diene possessing two Z-double bonds (Fig. 35) (*68*), and to the
synthesis of an ionomycin fragment (*69*). More recently an asymmetric
variant was reported starting from chiral neroic acid imides (*70*).

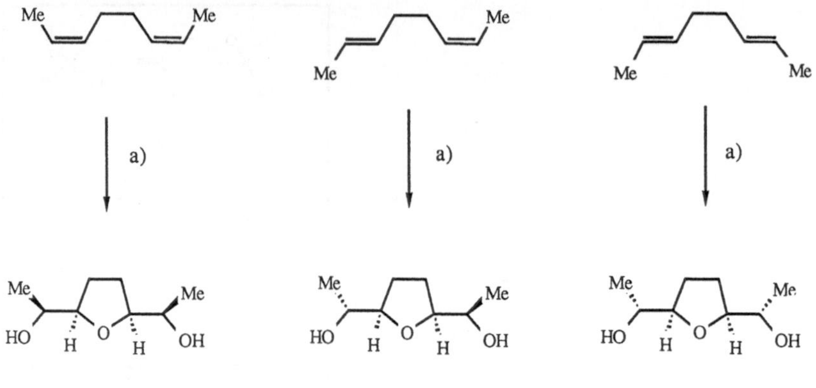

Reagent: a) KMnO$_4$, aq. acetone, -20°

Fig. 33. Permanganate-mediated oxidation of dienes

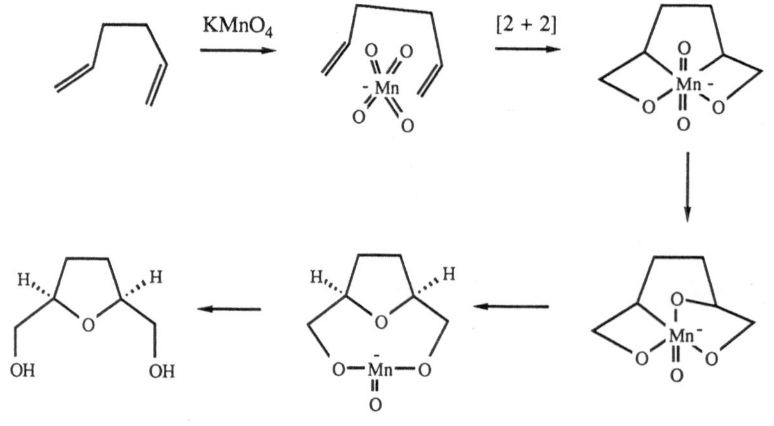

Fig. 34. A plausible mechanism for the reactions shown in Fig. 33

# 5. Evidence for the Processive Strategy of Polyketide Chain Assembly

Whereas attempts to isolate late stage intermediates in polyether antibiotic biosynthesis have so far been unsuccessful, macrolide antibiotic-producing organisms have typically been more forthcoming. From

*References, pp. 72–81*

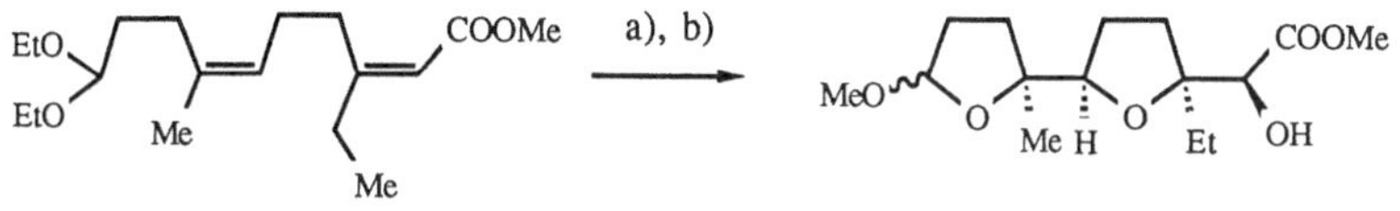

Reagents ; a) KMnO$_4$, aq. acetone ; b) HC(OMe)$_3$, C$_6$H$_6$, TsOH.

Fig. 35. Synthesis of a monensin BC-ring fragment (*68*)

these, or from macrolide non-producing mutants, it is usually possible to isolate the parent macrolide ring and all subsequent enzyme-free intermediates. During the biosynthesis of macrolide antibiotics, and presumably also of polyethers, it is after completion of carbon chain assembly that the backbone is oxygenated or glycosylated, steps which amplify the

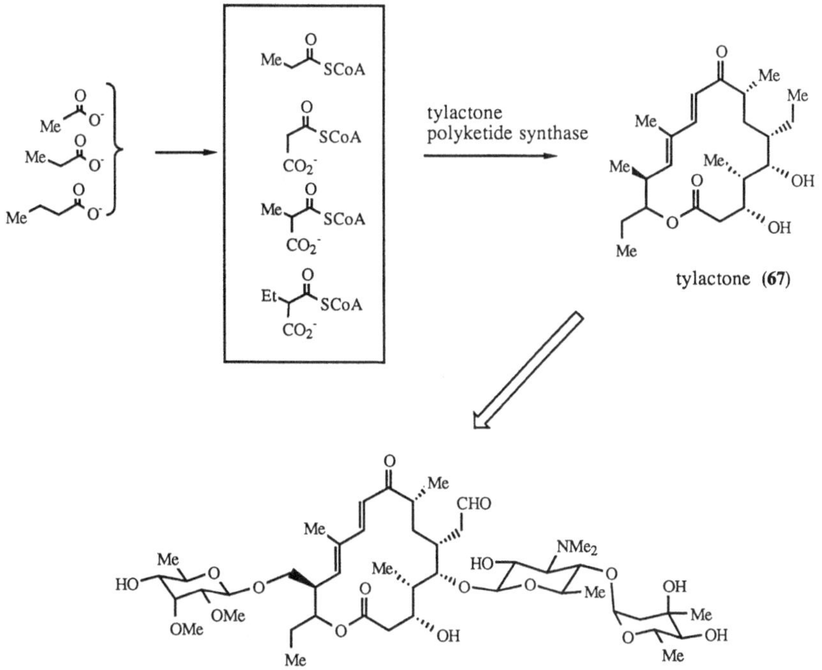

tylactone (**67**)

tylosin (**41**)

Fig. 36. Biosynthesis of tylosin showing only the first enzyme-free intermediate tylactone

scope for structural diversity within each class of antibiotic. The first free intermediate in tylosin (41) biosynthesis is tylactone (67), the likely end product of the tylosin polyketide synthase (PKS) complex (see Fig. 36), and each of the intermediates between tylactone and tylosin has been isolated and characterised (for a review see 71). In the case of erythromycin A (40) biosynthesis, the first detectable intermediate is 6-deoxyerythronolide-A (68), which is converted in five subsequent steps to erythromycin A (Fig. 37). It is generally assumed that macrolide and polyether backbone assembly occurs whilst the carbon chain remains covalently bound as a thioester to a carrier protein, in analogy to fatty acid biosynthesis. The triene (36) (Fig. 19) is then the logical end product of a series of condensation, reduction and dehydration events utilising substrates derived from acetate, propionate and butyrate and catalysed by the monensin PKS multienzyme complex, although conceivably the methylation at C(3)OH might still require an enzyme bound form of this

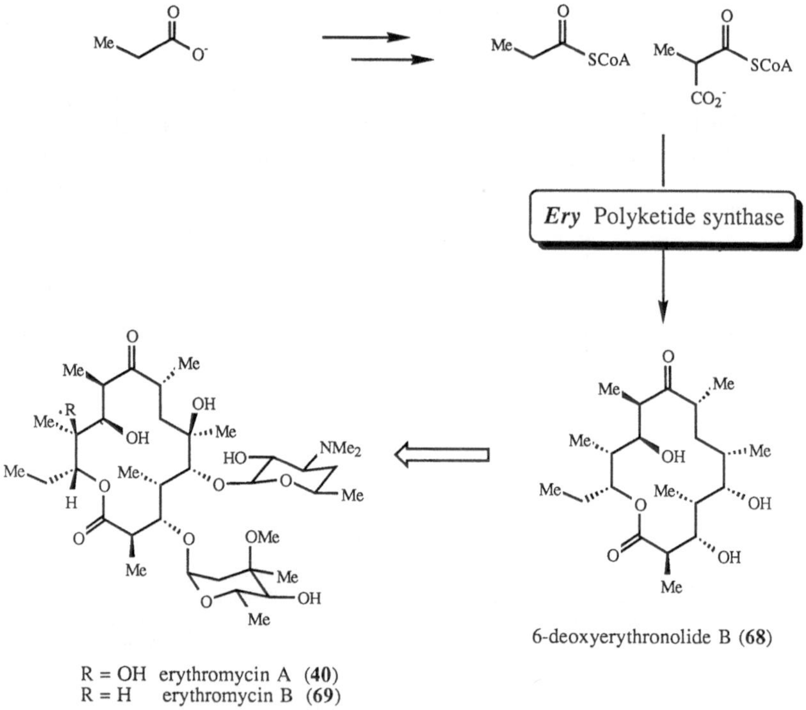

R = OH   erythromycin A   (40)
R = H    erythromycin B   (69)

6-deoxyerythronolide B (68)

Fig. 37. Biosynthesis of erythromycin A, showing the first enzyme-free intermediate, 6-deoxyerythronolide B

*References, pp. 72–81*

putative intermediate. A further important question concerns the order of these various events during chain assembly.

In macrolide biosynthesis evidence has accumulated recently in support of the so-called processive strategy of backbone assembly (72, 73). In this strategy the β-ketoacyl thioester, formed by decarboxylative condensation of an acyl chain with a malonyl, methylmalonyl, or ethylmalonyl extender unit, is modified (where necessary) before the next building block is incorporated. The modification can involve reduction only to a β-hydroxythioester, reduction and dehydration to give an olefin, or reduction, dehydration and further reduction to give a fully saturated chain (see Fig. 38). However, the same modification need not occur during each cycle of chain extension. The evidence supporting this model of macrolide assembly includes the incorporation of putative chain elongation intermediates into tylactone (**67**) (72), erythromycin

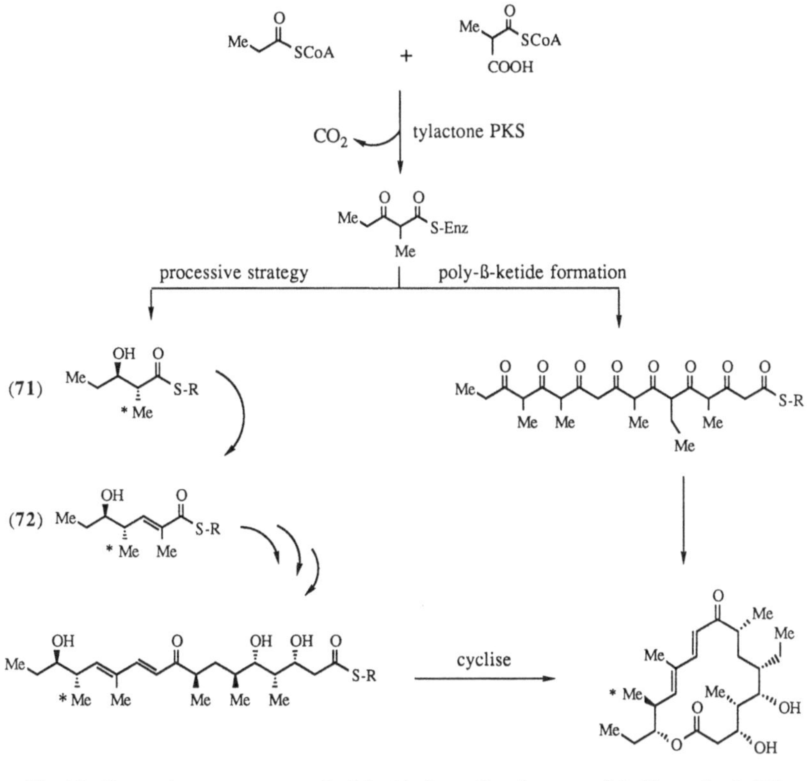

Fig. 38. Processive strategy *vs.* poly-β-ketide formation for macrolide biosynthesis (72)

B (69) (73) and nargenicin (70) (74), a metabolite of *Nocardia argenti-nensis*. For example, $^{13}$C-labelled forms of the β-hydroxythioester (71) and the enone (72), each activated as acetylcysteamine thioesters (R = —CH$_2$CH$_2$NHCOCH$_3$), were incorporated intact into tylactone (67) (Fig. 38) (72). The evidence for intact incorporation was based upon a selective enrichment of the $^{13}$C-resonance assigned to the methyls indicated by the asterisk in Fig. 38, although some of the labelled precursor was also degraded by β-oxidation prior to incorporation. These observations indicate that the PKS can accept the acyl portion of these precursors, presumably *via* a transthioesterification involving a nucleophile on one component of the PKS complex. Similarly the [2,3,-$^{13}$C$_2$]β-hydroxythioester (73) was incorporated into erythromycin B (69) (73) (see Fig. 39) in *S. erythraea*, and the same precursor was incorporated intact into nargenicin (70) (74) in *N. argentinensis*, again supporting a processsive strategy for the assembly of these macrolides. Whilst these conclusions are most probably valid, there is an inherent danger of

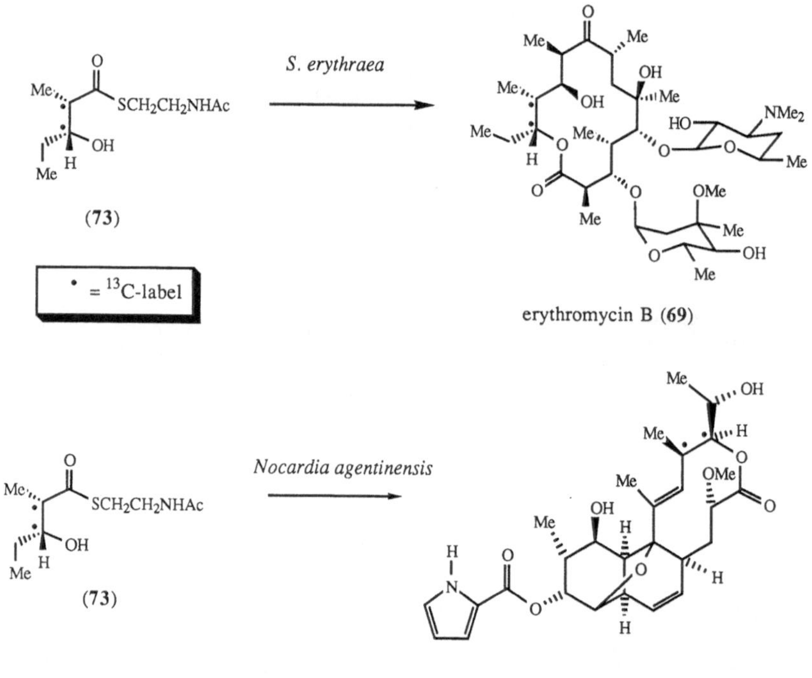

erythromycin B (69)

nargenicin (70)

Fig. 39. Incorporation experiments in support of the processive strategy of assembly (73, 74)

*References, pp. 72–81*

drawing a false conclusion in such whole cell feeding experiments, unless steps are taken to determine whether or not the acyl chain of the precursor is modified in some undesired way prior to its use in assembly by the respective PKS. It is not inconceivable that (71) or (73), for example, could be oxidised to a β-ketothioester prior to use in macrolide biosynthesis. Such an event would, of course, negate the validity of the experiment as a test of the processive strategy of assembly versus an alternative strategy leading to a poly-β-ketothioester.

A further indication that macrolide assembly follows a processive strategy was the isolation (75) and structure determination (76) of compounds considered to be derived from chain elongation intermediates in the biosynthesis of mycinamicin IV (74), in *Micromonospora griseorubida*. The compounds extracted from the culture broth included the carboxylic acids (75) and (76), and the ketone (77) (see Fig. 40). The ketone (77) could be derived from the β-ketoacid (78), by spontaneous decarboxylation.

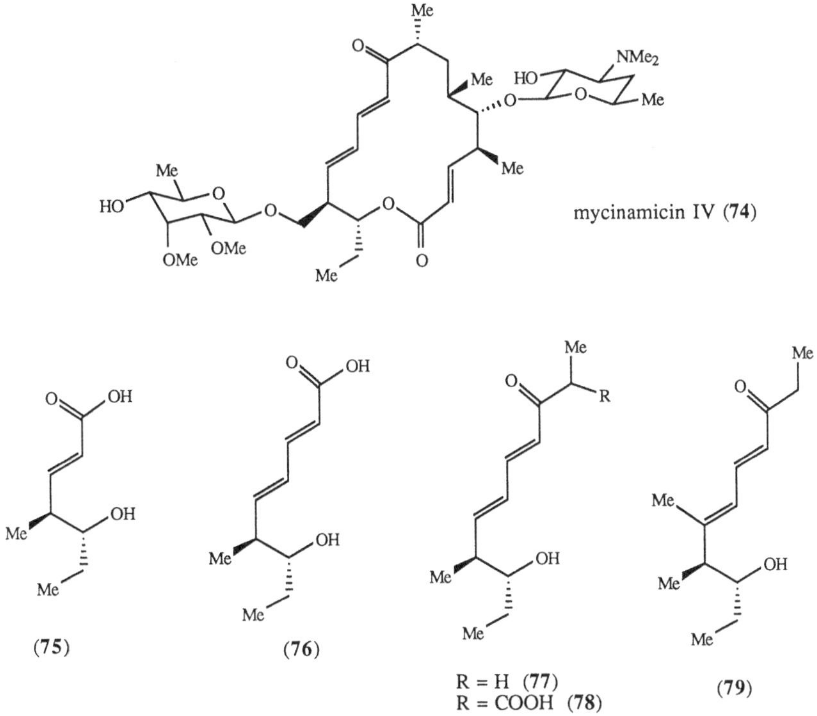

Fig. 40. Derivatives of putative chain elongation intermediates isolated from a mycinamicin producing organism (75, 76), and a tylactone producing organism (77)

A related ketone (79) had been isolated some years earlier (77) from a contaminated culture of the tylosin producer *S. fradiae* (cf. Fig. 38). The isolation of these compounds having the structure and absolute configuration expected for intermediates in chain assembly suggests strongly that they have been released fortuitously from the PKS complexes, and therefore that the processive strategy of assembly is followed in these cases. Additional support for this conclusion might be obtained by attempting incorporation experiments with suitably labelled forms of (75) and (76).

Incorporation experiments with putative partial chain elongation intermediates in monensin biosynthesis have also been attempted (or are

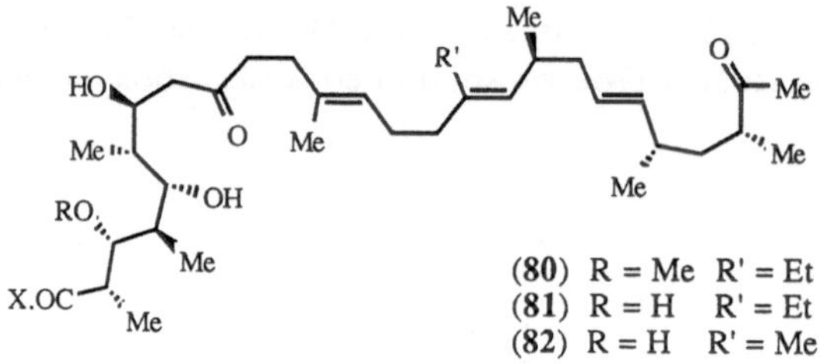

(80)  R = Me   R' = Et   X = OH
(81)  R = H    R' = Et   X = OH
(82)  R = H    R' = Me   X = SCH$_2$CH$_2$NHCOC$_7$H$_{15}$

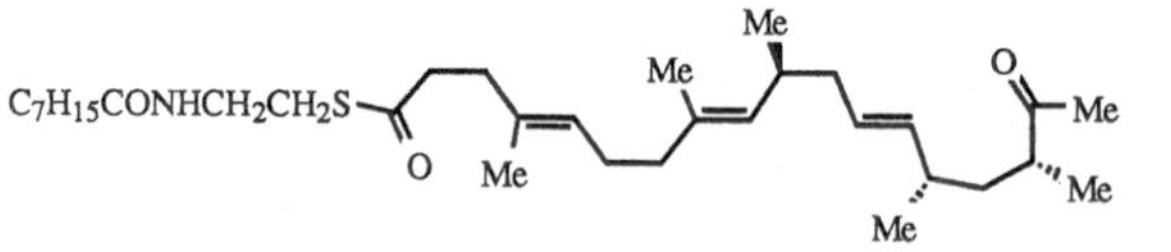

(83)

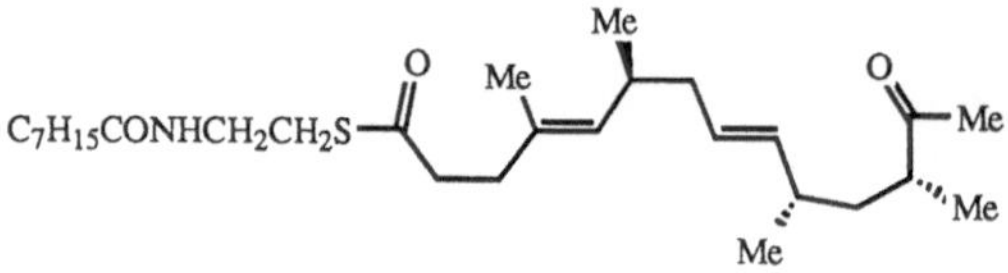

(84)

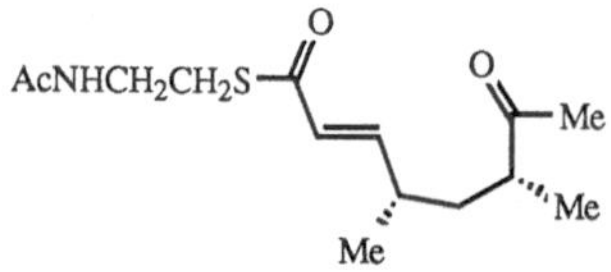

(85)

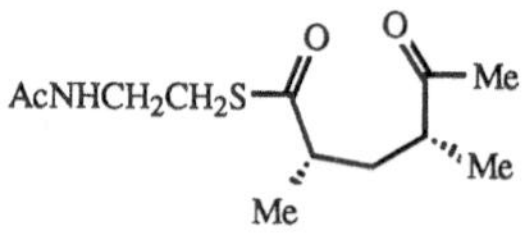

(86)

Fig. 41. Derivatives of putative chain elongation intermediates for feeding to the monensin producing organism

currently underway) with the monensin producing organism *S. cinna-monensis*. Thus the trienes (**80**) and (**81**) having a free carboxyl terminus (*78–80*) as well as the thioester-activated triene (**82**) (*81, 82*) have been synthesised as have likely chain elongation intermediates such as (**83**) and (**84**) (*82*), each activated as caprylcysteamine thioesters, and (**85**) and (**86**) (*83*) activated as acetylcysteamine thioesters (Fig. 41). In feeding experiments to *S. cinnamonensis* the trienes (**82**) and (**83**) were apparently unable to cross the intact cell membrane, so their future use may be limited to cell free studies of monensin biosynthesis, whereas the shorter precursor (**84**) was rapidly degraded by *S. cinnamonensis*, most likely through β-oxidation, without being incorporated significantly into monensin B (*82*). This latter problem of β-oxidation might be surmountable by using mutants of *S. cinnamonensis* blocked in fatty acid degradation in combination with inhibitors of β-oxidation (*84*). At the present time it seems very likely that a processive strategy of chain assembly operates during polyether biosynthesis, although successful incorporation experiments in support of this conclusion are not yet available.

# 6. A Unified Model of Polyether Antibiotic Structure and Biogenesis

In 1983 CANE, CELMER and WESTLEY (*10*) compared the structure and stereochemistry of a large number of polyether antibiotics and suggested a biogenetic basis for many of the structural regularities seen within this class of natural products in terms of the polyene-polyepoxide cyclisation concept. The important features of this model can be summarised as follows. The polyethers are divided into four main classes (*85, 86*) based on cation selectivity and chemical structure: 1) the monovalent polyethers including nonspiroketal, spiroketal and bis-spiroketal antibiotics (with or without glycosylation); 2) the divalent polyethers (e.g. lasalocid A (**3**), which bind divalent metals); 3) the pyrrole ethers (e.g. X14547A (**87**) and A23187 (**8**)) (Fig. 42); and 4) the acyltetronic acids (e.g. ICI139603 (**6**) and tetronomycin (**7**) (Fig. 4)). The structures of several monovalent polyethers, are shown in Fig. 43. If these structures are compared a striking stereochemical homology becomes apparent. In spite of a variety of substitution patterns and variations in the mix of acetate, propionate and butyrate building blocks used in the biosynthesis, the configuration of common alkyl or hydroxyl groups at any given position within the first twelve biogenetic subunits is constant throughout the entire series. A hybrid configurational map can therefore be drawn to represent the relative

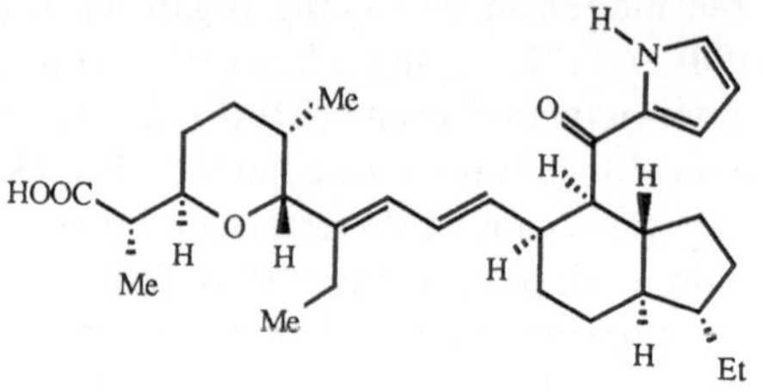

Fig. 42. Structures of two pyrrole-ether antibiotics

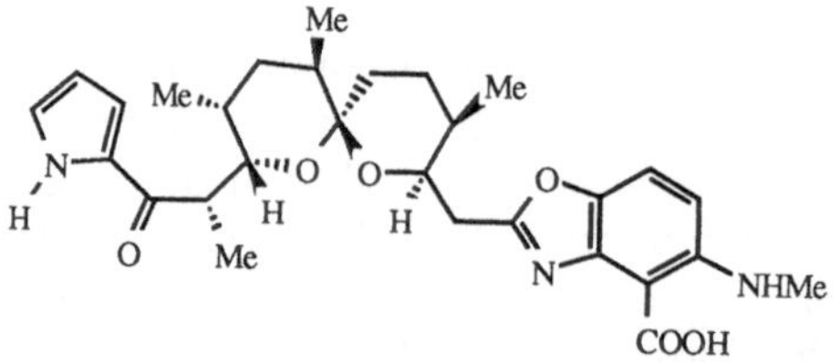

and absolute configuration of chiral centres at each carbon along the backbone (see Fig. 44). Since the first four biosynthetic building blocks are in all cases acetate, propionate, propionate and acetate: $A_4P_3P_2A_1$, the model in Fig. 44 was called the APPA prototypal polyether.

A further important assumption is that the polyene-polyepoxide model for monensin biosynthesis can be extended to account for the formation of all the other APPA polyethers as illustrated in Fig. 45. The structural and stereochemical regularity within the APPA polyethers would then be a reflection of a closer and more fundamental regularity in the putative triene intermediates. These trienes represent the presumed products of a series of condensation, reduction and dehydration events utilising substrates derived from acetate, propionate and butyrate, catalysed by the respective PKS complexes. The stereochemical regularity in the presumptive triene intermediates can also be represented by a prototypical model, as shown in Fig. 46A.

A second group of polyethers identified by CANE, CELMER and WES-TLEY possesses carbon chains that are assembled starting from $P_4A_3B_2A_1$, BABA or PAPA precursors. These four subunits appear in one of two closely related ring systems typified by the structures of lasalocid A (3) (5-ring + 6-ring) and isolasalocid (18) (5-ring + 5-ring) (see Fig. 47). Other members of this group contain additional structural features of interest, including the bis-spiroketal seen in narasin (27),

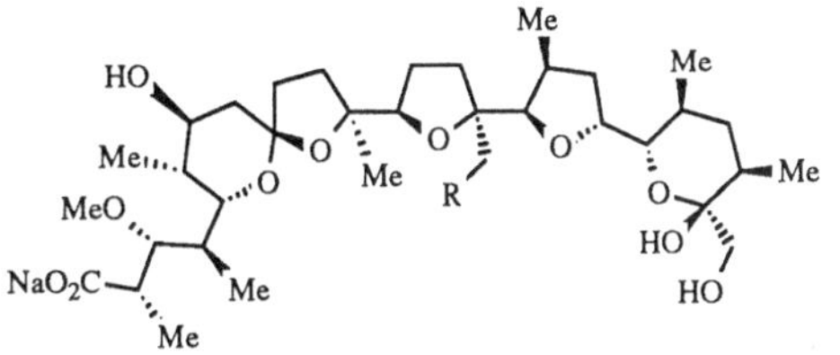

monensin A (1)

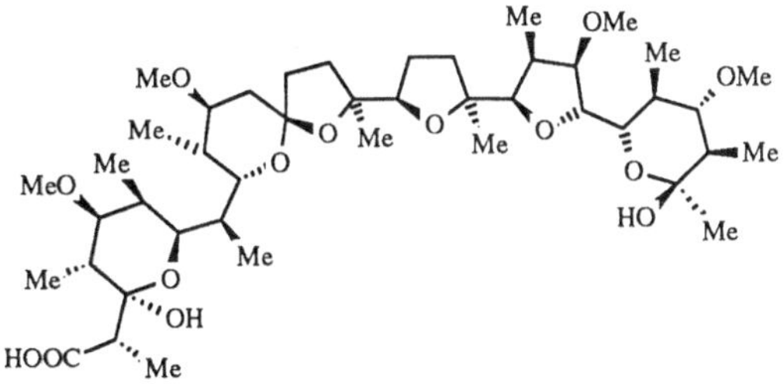

lonomycin A (88)

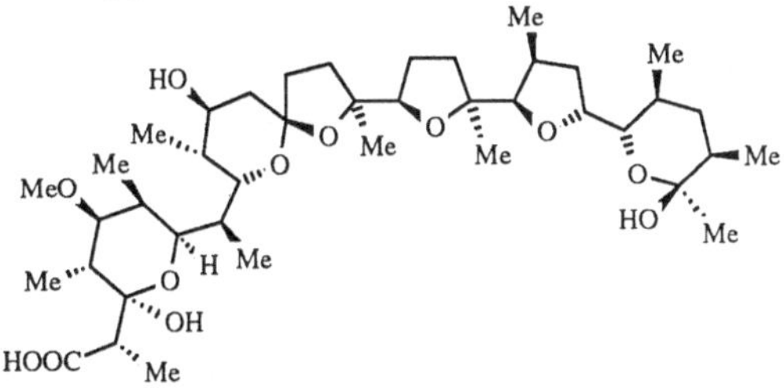

mutalomycin (89)

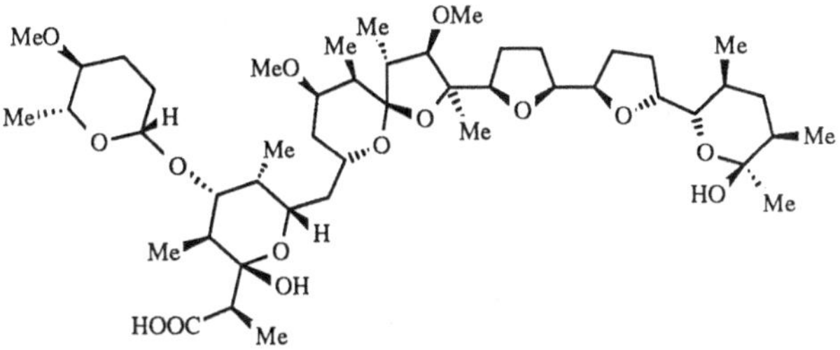

carriomycin (90)

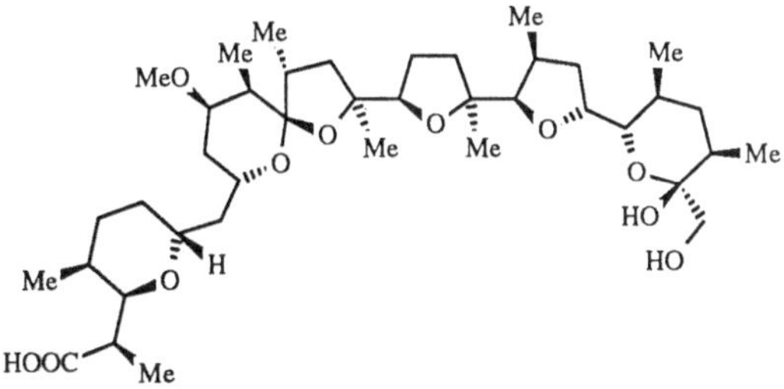

nigericin (66)

Fig. 43. Structures of several monovalent polyethers. Note the structural similarities

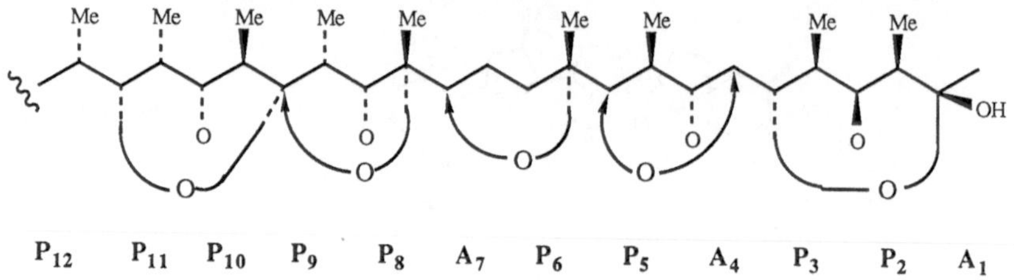

Fig. 44. A configurational model that correlates the relative and absolute configurations at common chiral centres in the monovalent polyethers (*10*)

salinomycin (**21**) and noboritomycin-A (**91**), and the aromatic ring system seen in noboritomycin-A (**91**), CP-44661 (**92**) and lasalocid A (**3**). The $^{18}$O-labelling results on lasalocid and narasin biosynthesis described above support the idea that the polyene-polyepoxide concept can be applied to this group of polyethers and suggest that the structural and stereochemical similarities apparent amongst them may once again reflect a close relationship at the level of the respective putative polyene intermediates (Fig. 48). A prototype configurational model which summarises the stereochemical regularities in the first twelve subunits of each hypothetical polyene intermediate, is illustrated in Fig. 46B. Lasalocid A (**3**) can be fitted to this model by assuming deletion of the central $P_8A_7A_6P_5$ subunits, while lysocellin (**93**) can be formally derived by deletion of $P_8A_7A_6A_5$ subunits from an analogous prototype (cf. Fig. 48).

The two stereochemical models shown in Fig. 46 successfully summarise the common relative and absolute configurations in more than thirty different polyether antibiotics, thereby emphasising the potential generality of a single biochemical mechanism, most likely involving the polyene-polyepoxide concept, for polyether biosythesis. Even where other polyethers do not fit either stereochemical prototype, putative polyepoxide biosynthetic intermediates can also often be readily visualised.

It is intriguing to note that the stereochemical prototypes for the APPA- and PAPA-type polyenes are closely related to each other as pointed out by Cane, Celmer and Westley (*10*). Subunits 5–12 of the APPA-model are essentially equivalent to subunits 1–5 and 8–11 in the PAPA model (Fig. 46). This process of comparing structures was taken a step further by O'Hagan (*11, 12*), who noted an analogy between the PAPA model and the backbones of certain macrolide antibiotics. Although the macrolides have a quite different mechanism of action, they also contain a long polyfunctionalised fatty acid backbone constructed

*References, pp. 72–81*

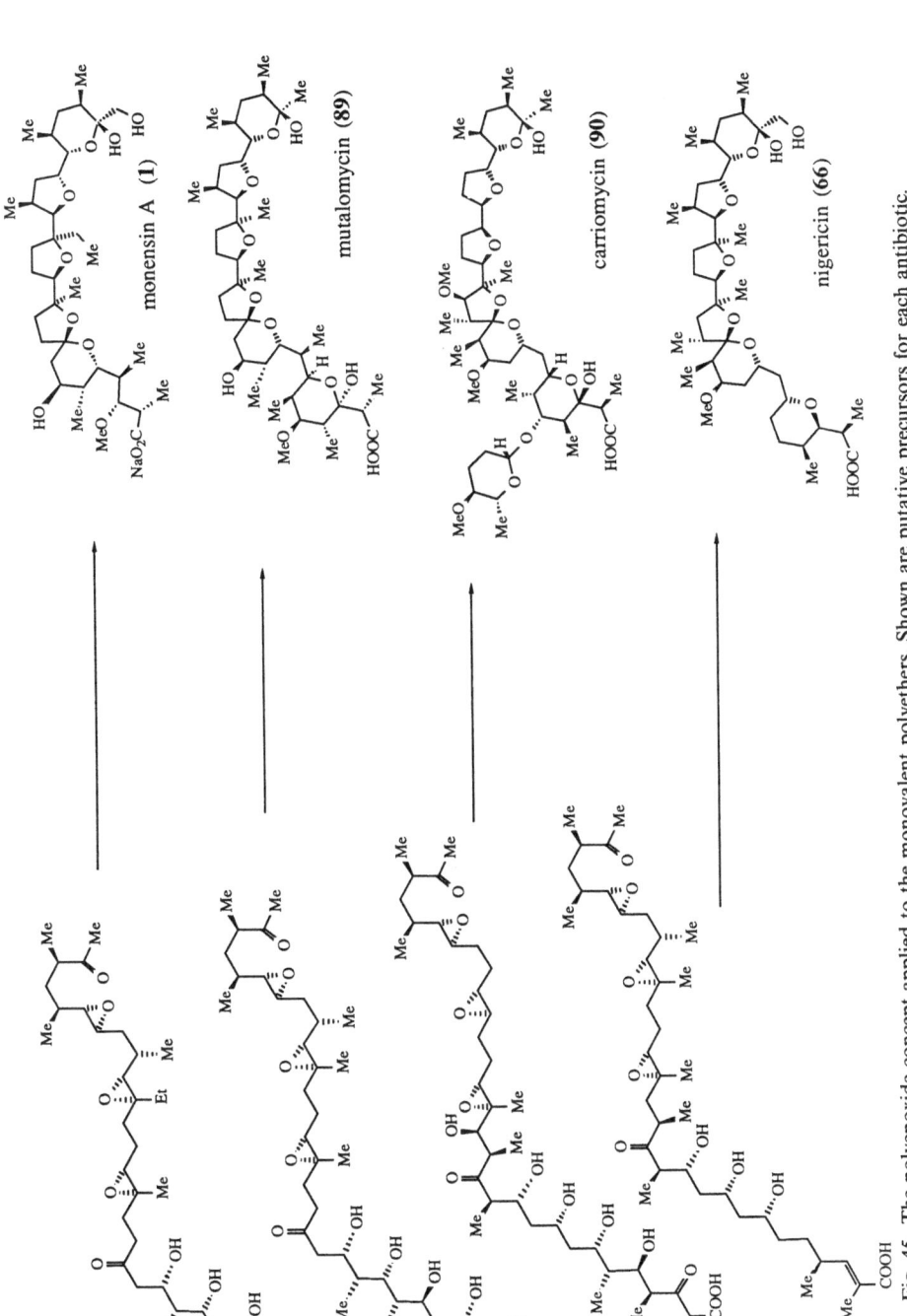

Fig. 45. The polyepoxide concept applied to the monovalent polyethers. Shown are putative precursors for each antibiotic.

47

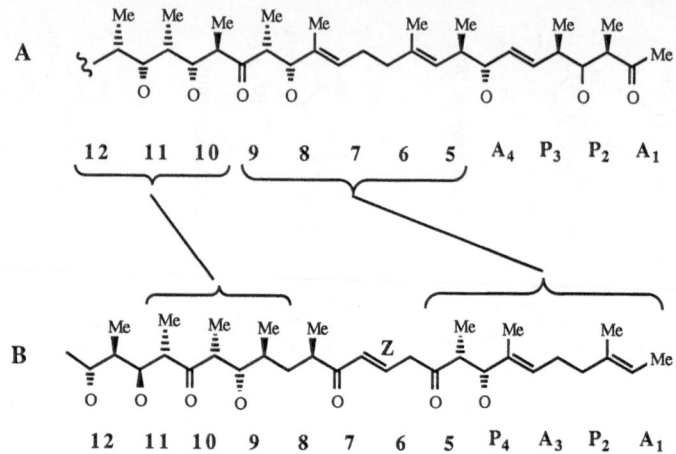

Fig. 46. Configurational models for the putative polyene intermediates: A) for the APPA family, and B) for the PAPA family. Regions of stereochemical homology are indicated

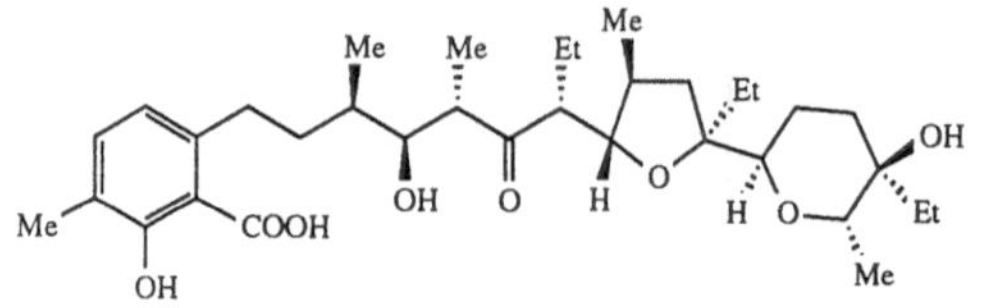

lasalocid A (3)

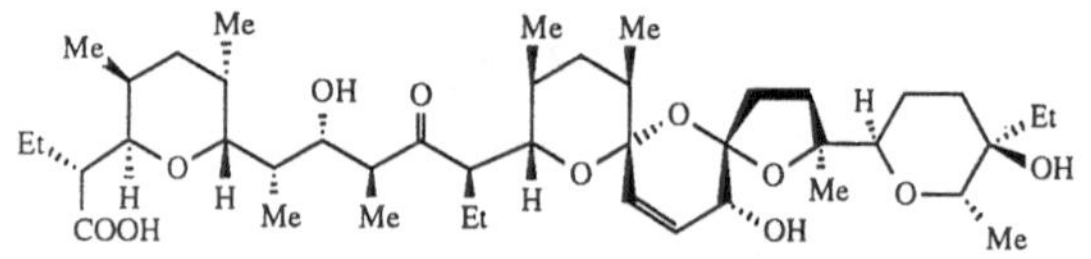

narasin (27)

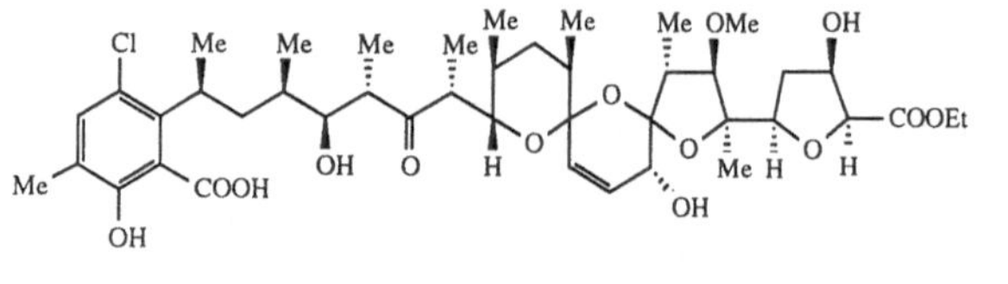

noboritomycin A (91)

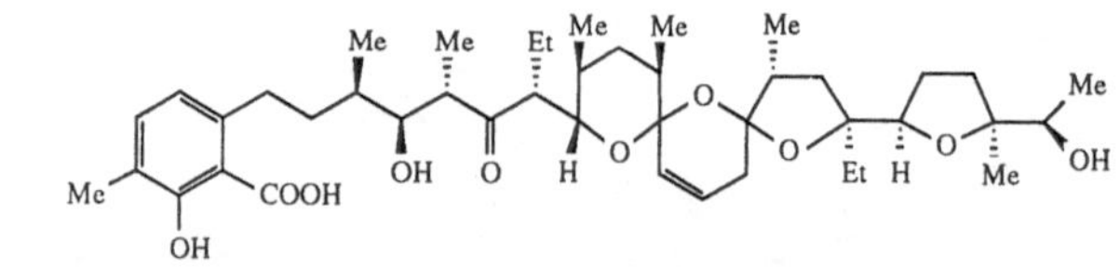

CP 44,661 (92)

Fig. 47. Structures of several polyethers of the PAPA (or PABA, BABA, BAPA) family

*References, pp. 72–81*          48

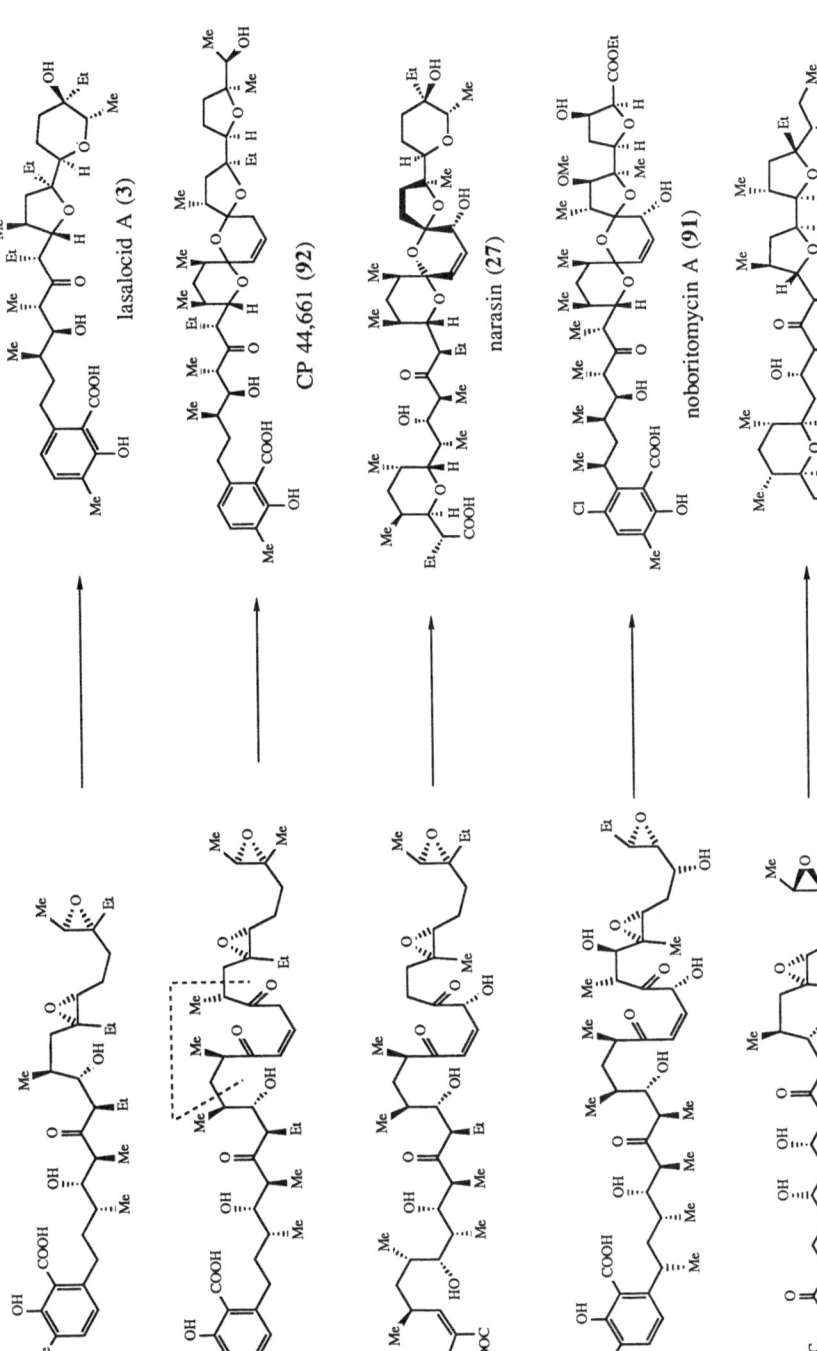

lasalocid A (3)

CP 44,661 (92)

narasin (27)

noboritomycin A (91)

lysocellin (93)

Fig. 48. The polyepoxide concept applied to the PAPA family of polyethers. Putative precursors for each antibiotic are shown

49

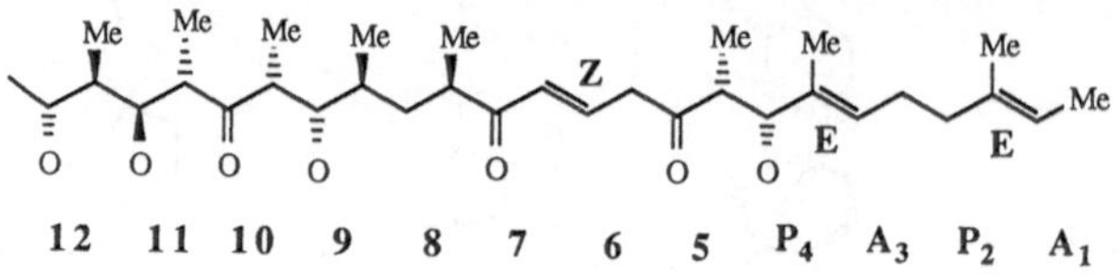

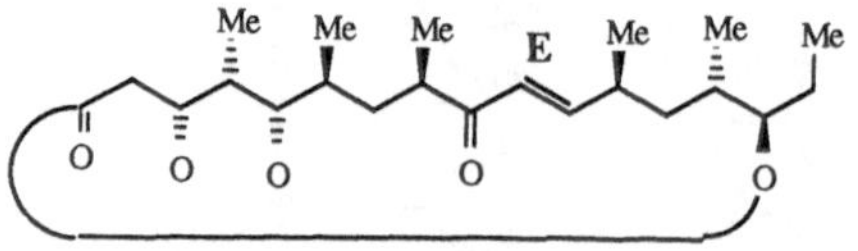

Fig. 49. Comparison of the PAPA configurational model, and a related model for 16-membered macrolide antibiotics

mainly from acetate, propionate and butyrate precursors (see Figs. 36 and 37). O'HAGAN proposed a stereochemical prototype for the 16-membered macrolides (including tylosin, leucomycin, spiramycin, mycinamicin etc.) that bears a similarity to subunits 4–10 of the PAPA-polyether model (see Fig. 49). This model for the 16-membered macrolides is closely related to the single stereochemical model representing the entire class of macrolides first constructed by CELMER (87, 88).

The structural relationships embodied in the stereochemical models of Figs. 44, 46, and 49 raise interesting biochemical and genetic questions about the mode of action, molecular organisation and evolution of the polyketide synthases involved in macrolide and polyether antibiotic biosynthesis. There is now a firm consensus that these proteins function in a manner very similar to fatty acid synthase, although it should be remembered that no Streptomycete PKS involved in polyether or macrolide biosynthesis has yet been detected in cell free extracts of the antibiotic producing organism. In the following sections the properties, structures and molecular organisation of these polyketide synthases are described in more detail.

# 7. Biochemistry of Polyketide Biosynthesis and Fatty Acid Biosynthesis

Fatty acid synthase is a remarkable and proficient machine for the synthesis of long chain fatty acids from simple acetate-derived building blocks (for an authoritative review see 89). Chemical, biochemical, and more recently molecular genetic studies have served to emphasise a fundamental similarity between fatty acid synthases and polyketide synthases, both in terms of their mechanisms of action and molecular organisation. However, this close relationship may not be readily apparent at first sight, since the synthesis of a typical polyketide must be more highly programmed than that of a fatty acid.

### 7.1. Programming Polyketide Assembly

During the assembly of the putative monensin triene intermediate (**36**) (Fig. 19), for example, the PKS following a processive strategy of assembly must select an acetate starter unit and then 12 further building blocks in a specific order from acetyl-, propionyl- and butyryl-residues, must incorporate the correct chemical functionality during each chain extension step, leaving a keto-, hydroxy-, enoyl- or saturated alkyl group, and must establish the correct relative and absolute configuration at each new chiral centre and double bond. Leaving aside the question of stereochemistry, this programming may be translated into a series of

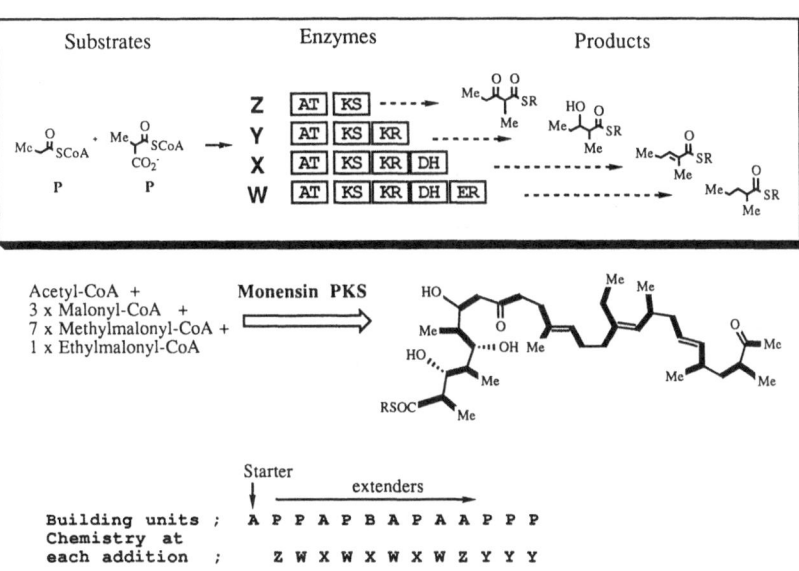

Fig. 50. The monensin PKS must select the desired building unit (*A* acetate, *P* propionate, *B* butyrate) and the correct chain extension chemistry (*AT* acyl transferase, *KS* β-ketoacyl synthase, *KR* β-ketoacyl reductase, *DH* dehydrase, *ER* enoyl reductase) at each round chain elongation

choices regarding the type of extender unit, and the type of chemistry following each subunit addition, using the notation shown in Fig. 50 (building blocks, $A$ = acetate, $P$ = propionate, $B$ = butyrate; chemistry at each addition, $Z$ = condensation, $Y$ = condensation + reduction, $X$ = condensation + reduction + dehydration, $W$ = condensation + reduction + dehydration + reduction). This generates a linear array of the minimum number of biochemical events required for polyketide chain assembly, excluding any extra acyl transferase steps which may be needed to transfer the growing polyketide chain between different proteins in the PKS complex, and a thioesterase step to release the final product as a free carboxylic acid. A similar shorthand notation can be written to describe the assembly of other putative polyether intermediates as well as the macrolide antibiotics. An important objective is to understand the basis of this biosynthetic programming, which allows a single PKS to assemble just one out of an enormous number of theoretically possible reduced polyketide chains; how is the order of biochemical events related to the structural organisation of the polyketide synthase complex at the protein and genetic levels? Before turning to a discussion of FAS and PKS structure, however, we will consider first one possible mechanism of stereocontrol during polyether and macrolide biosynthesis.

### 7.2. Stereochemical Aspects of PKS Action

One mechanism by which a PKS might exert control over the absolute configuration of chiral methine centres introduced during the incorporation of propionate derived building blocks was elaborated by HUTCHINSON and coworkers (90). The *in vivo* activated form of propionate used for antibiotic biosynthesis is methylmalonyl-CoA. If the carbon-carbon bond forming process occurs, in analogy to fatty acid biosynthesis, by decarboxylative-condensation with inversion of configuration (for a review see 91), catalysed by a condensing enzyme, then the newly created chiral methine centre would be $(S)$ or $(R)$, depending upon whether the enzyme uses the $(R)$ or $(S)$ isomer of methylmalonyl-CoA, respectively (see Fig. 51). Alternatively, the condensing enzyme(s) may only use $(S)$-methylmalonyl-CoA, but sometimes after condensation an epimerase inverts the new chiral methine centre to the $(S)$-absolute configuration. Feeding experiments that addressed this question involved the incorporation of $[2\text{-}^2H_2, 2\text{-}^{13}C]$propionate into lasalocid A (3) (92, 93) and erythromycin A (40) (94), and the incorporation of $[2\text{-}^2H_2]$-, $(R)$-$[2\text{-}^2H_1]$- and $(S)$-$[2\text{-}^2H_1]$-propionates into monensin A (1)

*References, pp. 72–81*

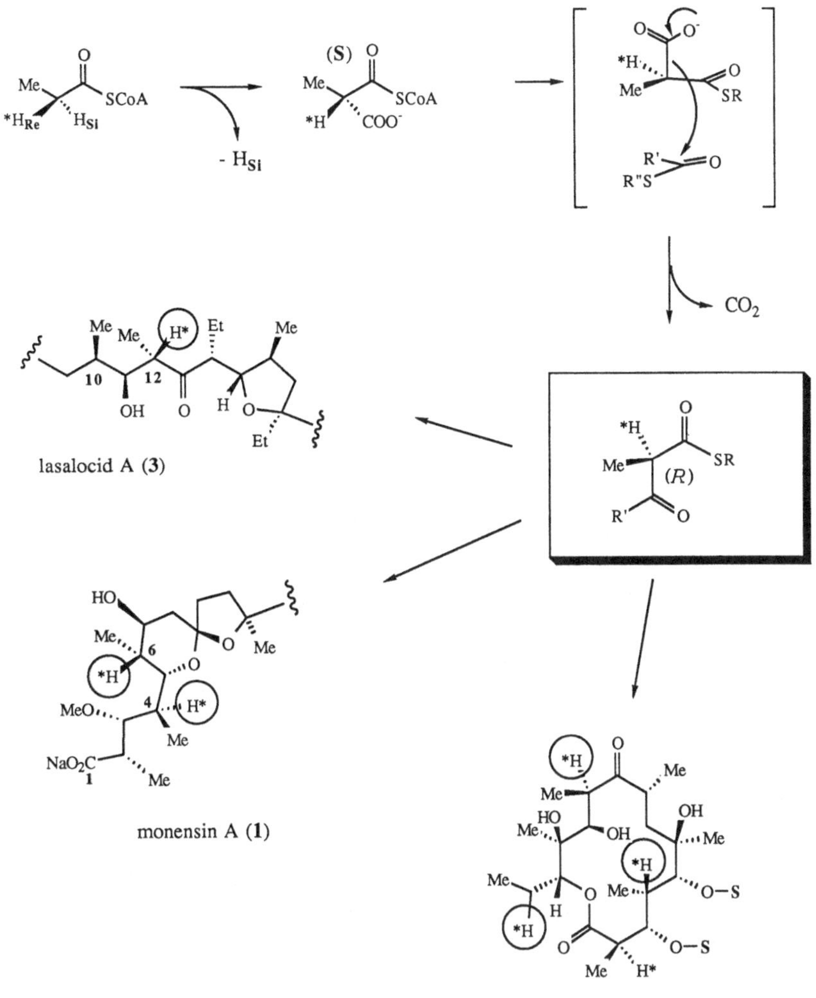

Fig. 51. Incorporation of methylmalonyl-CoA by decarboxylative-condensation with inversion, catalyzed by a condensing enzyme in the PKS complex

(95, 96). Each of these labelled propionates should be metabolised *in vivo* into labelled (S)-methylmalonyl-CoA's, by conversion into propionyl-CoA, and carboxylation on propionyl-CoA carboxylase or transcarboxylase. A propionyl-CoA carboxylase specific for (S)-methyl-malonyl-CoA has been isolated (97) from the erythromycin producer S.

*erythraea*, whereas there is no biochemical precedence for the formation of (R)-methylmalonyl-CoA by direct carboxylation of propionyl-CoA. If the (S)-methylmalonyl-CoA is then used directly in polyketide assembly the proton (or deuteron) at C2 will be incorporated intact at the new methine (R)-chiral centre (see Fig. 51). On the other hand, if the (S)-methylmalonyl-CoA is first converted into the (R)-isomer by epimerisation prior to use in antibiotic biosynthesis, deuterium label at C2 would be lost to the medium. In practice, wherever retention of deuterium label in these antibiotics was observed, its location was always at chiral methine sites whose absolute configurations were consistent with direct incorporation of (S)-methylmalonyl-CoA into the backbone by decarboxylative condensation with inversion. Furthermore, during the incorporation of $[2-^2H_2]-$, $(R)-[2-^2H_1]-$ and $(S)-[2-^2H_1]$-propionates into monensin A, deuterium retention occurs only at C(4) and C(6) when starting from $[2-^2H_2]-$ and $(R)-[2-^2H_1]$-propionate; the deuterium in $(S)-[2-^2H_1]$-propionate is completely lost to the medium. This again is consistent with the known stereochemical courses of bacterial propionyl-CoA carboxylase and transcarboxylase (91), both of which promote loss of the 2-$H_{Si}$ atom from propionyl-CoA during formation of (S)-methylmalonyl-CoA. This indicates that at least the (S)-isomer of methylmalonyl-CoA is a substrate for these PKS complexes and that the stereochemical courses of the reactions catalysed by the condensing enzymes of fatty acid biosynthesis and polyketide biosynthesis are probably identical. Unfortunately, it has not been possible by this method to determine whether (R)-methylmalonyl-CoA is also utilised by these PKS's during chain elongation, because of the difficulties in generating the relevant $(R)-[2-^2H_1]$-methylmalonyl-CoA *in vivo*. The question of whether the PKS can select either (R) or (S)-methylmalonyl-CoA to control the stereochemistry at new chiral methine centres therefore presently remains unresolved.

## 7.3. The Structure and Function of Fatty Acid Synthases

An observation which strengthens the biochemical analogy between fatty acid biosynthesis and polyketide biosynthesis is the inhibition of both by the antibiotic cerulenin. Cerulenin (**94**), a metabolite of the fungus *Cephalosporium caerulens*, specifically and irreversibly inhibits the condensing enzyme of FAS, by alkylating an essential cysteine residue in the active site (98–101) (see Fig. 52). Cerulenin also inhibits the biosynthesis of many polyketide metabolites, including polyethers, macrolides, tetracycline, and isochromanequinones such as nanaomycin and actinor-

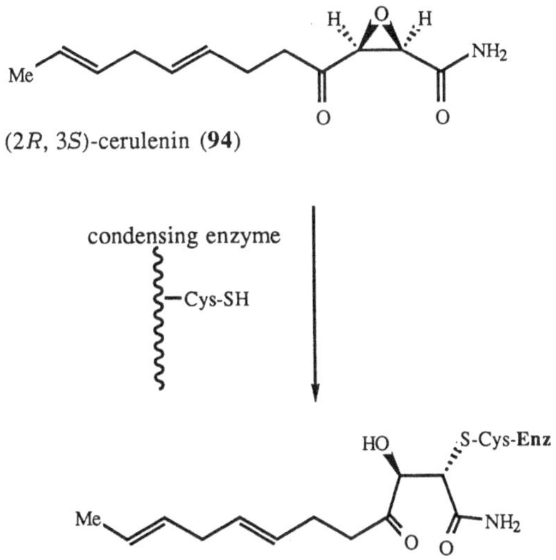

Fig. 52. Inhibition of the FAS condensing enzyme by cerulenin

hodin. This effect on macrolide biosynthesis has been exploited by OMURA and coworkers for the production of several hybrid macrolide antibiotics (102–104), as well as in studies of tylosin biosynthesis (105). Presumably cerulenin acts in a similar way by inactivating a condensing enzyme in the PKS complex. LEADLAY and coworkers (106, 107) showed that [³H]tetrahydrocerulenin specifically labelled a protein of $M_r$ 40,000 in extracts of S. erythraea, although it was not clear whether or not this was the condensing enzyme of the FAS. This raises important questions, however, about the size and molecular architecture of the FAS and the PKS multienzyme complexes in polyketide antibiotic producing organisms. In discussing the enzymology of PKS's it is helpful to draw upon comparisons to FAS's, where the relationships between structure and function are becoming more clearly understood. Already a great deal is known about the molecular organisation of FAS complexes in various other organisms, information that is worthwhile summarising at this point.

The most abundant fatty acid, palmitate, is synthesised de novo from acetyl-CoA, malonyl-CoA and NADPH in a series of steps that were elucidated largely from studies of fatty acid biosynthesis in cell-free extracts of E. coli (108, 109). These are illustrated in outline in Fig. 53. The acetyl starter unit and malonyl extender unit are first loaded onto the FAS by separate acyl transferases and then condensed with loss of

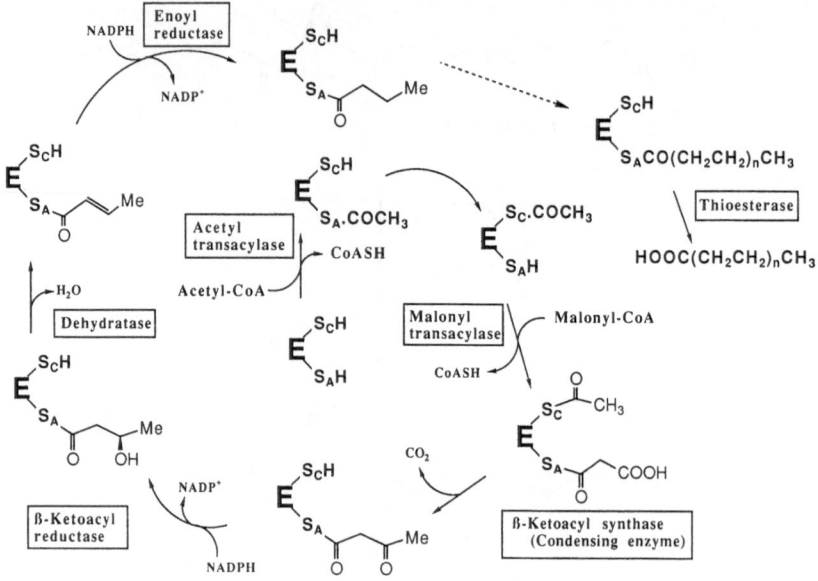

Fig. 53. The fatty acid synthase reaction cycle

$CO_2$ to afford a β-ketoacyl thioester. The simultaneous loss of $CO_2$ shifts the chemical equilibrium in favour of synthesis. The ketone group is then removed in three steps: by reduction to the (3R)-β-hydroxythioester, syn-elimination of water to give the E-unsaturated thioester, and reduction of the double bond to leave the fully saturated $C_4$ unit. There follow further rounds of condensation, reduction, dehydration and reduction, each requiring a new malonyl-CoA extender unit, until the chain reaches its final length. A thioesterase then releases the free fatty acid, as in bacteria and mammals, or the acyl chain is transferred to CoASH to form palmitoyl-CoA, as in yeast, or it may be utilized directly in the synthesis of phosphatidic acids, as in E. coli (89). With the exception of some highly specialised tissues (e.g. the lactating mammary gland) shorter chain biproducts are not produced.

The component activities of the type-II fatty acid synthases from plants and bacteria can be readily separated into discrete globular proteins, which together catalyse the consecutive steps of the fatty acid synthase cycle (see Fig. 53) (110). There is no evidence so far that these proteins form an aggregate within cells, and this is certainly not necessary to observe full fatty acid synthase activity in vitro. The following proteins comprise the *E. coli* FAS responsible for palmitic acid biosynthesis (*108*): 1) the acyl carrier protein (ACP), a relatively small protein of around 80 amino acids, containing the phosphopantetheine cofactor (Fig. 54). It plays a special role since it must bind the substrate at most steps of the cycle, and it must interact with all the component enzymes in the fatty acid synthase complex. The ACP's in Type-II FAS's from many sources have been sequenced and the 3D-structure of the *E. coli* (*111–115*) and spinach (*116*) ACP's have been investigated by NMR spectroscopy.; 2 and 3) the acetyl and malonyl transferases, that load the substrates acetyl-CoA and malonyl-CoA onto ACP's, each via a mechanism involving discrete O-acyl(Ser)-enzyme intermediates. The acetyl group must be transferred from the acetyl-S-ACP to the cysteine-SH of the condensing enzyme; the acetyl group is not transferred directly from CoASH to the β-ketoacyl synthases-I or -II (*vide infra*) since both acetyl-S-ACP and malonyl-S-ACP are required substrates for the condensation reaction (see *89*).; 4) the condensing enzymes (β-ketoacyl synthases), three of which have been isolated from *E. coli*, that differ in their substrate specificities. The substrates for the synthase-I and synthase-II, both homodimeric proteins, are an acyl-ACP and malonyl-ACP, but they differ in their substrate specificity with respect to the acyl-ACP. Both enzymes are essentially inactive with a fully saturated acyl chain longer than $C_{14}$, whereas the type-II but not type-I synthase functions with a $\Delta^9$-$C_{16}$ unsaturated acyl-ACP; both synthases function with the shorter $\Delta^5$-$C_{12}$ and $\Delta^7$-$C_{14}$ unsaturated acyl chains. Genetic studies suggest that synthase II plays a major role in the thermal regulation of fatty acid synthesis in *E. coli*, whereas synthase I is essential for unsaturated fatty acid biosynthesis (*117*). A third synthase was

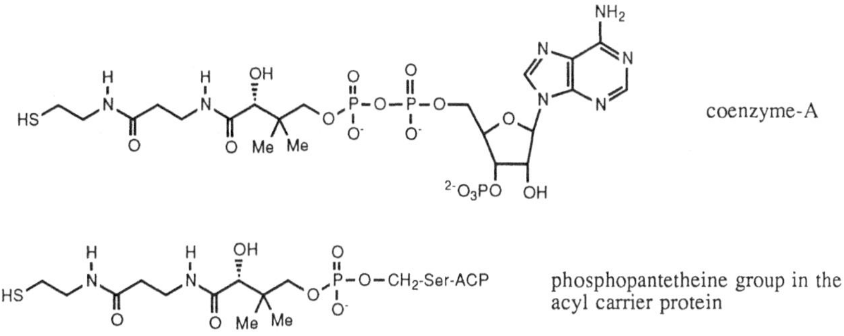

Fig. 54. Structures of coenzyme-A and the pantetheinyl group in the acyl carrier protein (ACP)

discovered recently (108) in E. coli, called acetoacetyl-ACP synthase. In contrast to the other two condensing enzymes, this type-III synthase condenses malonyl-ACP with acetyl-CoA rather than with acetyl-ACP (119); the synthases I and II can elongate acetyl-ACP but not acetyl-CoA. This enzyme appears to catalyse selectively the formation of acetoacetyl-ACP and may play an important role in the early stages of fatty acid biosynthesis in E. coli. Synthases-I and -II are inhibited by cerulenin (Fig. 52), whereas synthase-III is not. On the other hand, all three synthases are inhibited by the antibiotic thiolactomycin. The fabB gene encoding the E. coli β-ketoacyl synthase-I has been cloned and sequenced (120); 5) the β-ketoacyl-ACP reductase, which catalyses the NADPH dependent reduction of the β-ketoacyl-ACP to $(3R)$-β-hydroxyacyl-ACP; 6) the β-hydroxyacyl-ACP dehydrase, which catalyses syn-elimination with formation of E-2-enoyl-ACP; 7) the enoyl-ACP reductase. Two distinct enoyl reductases are known in E. coli having slightly different acyl chain length specificities, one NADH dependent and the other NADPH dependent; and finally 8) palmityl-ACP thioesterase, which catalyses hydrolysis of the acyl thioester when the acyl chain reaches $C_{16}$ in length.

In contrast to the dissociable subunits of the type-II FAS's, the type-I FAS's from yeast and mammals have a quite different architecture (89). In animal cells all the component activities of FAS are found on a single multifunctional polypeptide chain around of 2,500 amino acids in length. The native enzyme from chicken and rat is then a homodimer ($M_r$ 500,000). Upon dissociation the polypeptide chains retain all the component activities except that of the β-ketoacyl synthase. An elegant combination of protein biochemical (121–124) and molecular genetic experiments (125–128) has led to a detailed model for the architecture of the chicken FAS, in which the catalytic sites are arranged on a series of connected globular domains (see Fig. 55). Proteolytic cleavage of the chicken liver FAS leads initially to the release of three peptide fragments of $M_r$ 127,000, 107,000 and 33,000 corresponding to domains I, II and III in the intact protein. The smallest, domain III, contains the COOH terminus of the protein and the thioesterase. Domain I contains the $NH_2$ terminus, the β-ketoacyl synthase, and a single active Ser-OH used by both acetyl and malonyl transacylases. Domain II (the reduction domain) contains the dehydrase and enoyl and β-ketoacyl reductases as well as the acyl carrier site which connects the β-ketoacyl reductase to the thioesterase. This physical map based on proteolysis experiments is in accord with the predicted locations for the component activities in the polypeptide, based on an analysis of the entire sequence of the rat and chicken liver cDNA's. The rat FAS is transcribed into a single mRNA,

*References, pp. 72–81*

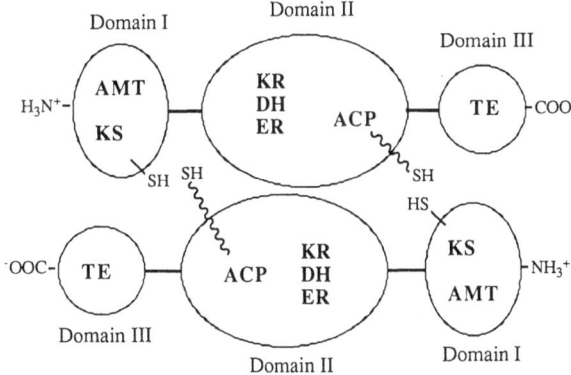

Fig. 55. Schematic model of the architecture of animal FAS, showing the domain structure, and head-to-tail arrangement of subunits

9156 nt long in rat FAS, including a 7515 nt coding sequence for the 2505 amino acid protein ($M_r$ 272,340). Using computer-assisted protein sequence comparisons, the order of catalytic activities within the multifunctional polypeptide was inferred to be (from N-terminus to C-terminus): condensing enzyme(KS)-acyl transferases(AMT)-dehydratase(DH)-enoyl reductase(ER)-ketoreductase(KR)-ACP-thioesterase(TE) (see Fig. 56). When compared, the deduced protein sequences of rat and chicken FAS's show a high level of sequence homology (*129*), consistent with a close evolutionary relationship. A portion of the chicken cDNA coding for the ACP and thioesterase domains was recently expressed (*130*) in *E. coli*. The r-protein possessed thioesterase activity, although

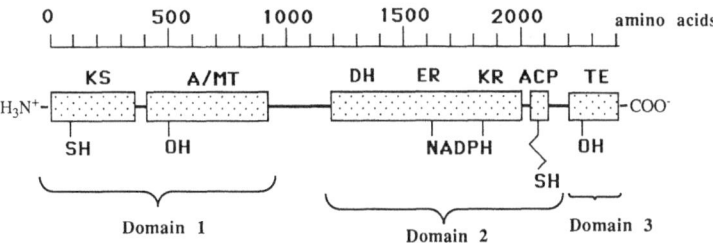

Fig. 56. The location of activities along the type-I mammalian fatty acid synthase polypeptide deduced from the cDNA sequence (*KS* β-ketoacyl synthase, *A/MT* acetyl/malonyl transferase, *KR* β-ketoacyl reductase, *DH* dehydrase, *ER* enoyl reductase, *ACP* acyl carrier protein domain, *TE* thioesterase). The organisation of activities into domains, as deduced by limited proteolysis is also shown (*89*)

the ACP site had undergone only partial pantothenylation by the endogenous *E. coli* ACP-synthase. These results augur well for more detailed structure-function studies of such multifunctional proteins by protein engineering techniques.

Clearly, for decarboxylative condensation to occur the malonyl extender unit on the ACP must be positioned next to the acyl chain attached to the active site cysteine on the β-ketoacyl synthase. Yet in the sequence of both chicken and rat FAS's, the ACP and β-ketoacyl synthase are located almost at opposite ends of the polypeptide.

Further important biochemical information about the spatial organisation of the three domains has come from studies with alkylating agents (89). For example, 1,3-dibromopropane reacts rapidly and specifically with the FAS, inactivating only the condensing enzyme, by alkylating intramolecularly *both* the cysteine-SH on one polypeptide and the pantetheinyl-SH in the ACP of the other chain. These and related observations with DTNB-inhibited FAS led to the proposal that the two chains are arranged head-to-tail so that the reactive thiols of the relevant ACP and condensing enzyme are juxtaposed as shown in Fig. 55. There are two full sites for fatty acid assembly. In this model each centre of palmitate synthesis derives the required component activities from both chains; one subunit contributes domain I while the second subunit contributes domains II and III. The following mechanism of FAS action then emerges: the cysteine-SH of the condensing enzyme is charged with an acetyl group and the ACP-SH with a malonyl group; condensation takes place, resulting in an acetoacetyl-group on the ACP and a free Cys-SH on the condensing enzyme; the acetoacetyl-group is then reduced, dehydrated and reduced by the enzymes in domain II of the other subunit, all of which are accessible to the ACP domain; the butyryl group is then transferred to the Cys-SH of the condensing enzyme, a new malonyl-group is loaded onto the ACP thiol and a second condensation occurs; the resulting β-ketohexanoyl residue is then reduced to a hexanoyl derivative. The process is repeated five more times until palmitoyl-S-ACP is formed, which is then hydrolysed by the neighbouring thioesterase in domain III.

The order of catalytic domains in the multifunctional yeast type-I FAS deduced from cDNA sequence data shows some interesting differences from that found in the animal FAS (131). The component activities are in this case distributed along two non-identical polypeptides. The active enzyme is an $\alpha_6\beta_6$ complex with a molecular weight of $2.4 \times 10^6$. These two proteins are encoded by two unlinked, intron-free genes, FAS1 and FAS2, which have both been cloned and sequenced (132–34). The α-subunit ($M_r$ 207,863) contains the acyl carrier site, the β-ketoacyl

*References, pp. 72–81*

synthase and β-ketoacyl reductase, whereas the β-subunit ($M_r$ 220,077) contains domains for the remaining activities, including an FAD binding site in the enoyl reductase domain. Also, there is no recognisable thioesterase domain, as expected since the end product is palmitoyl-CoA and not free palmitate. Thus the different end product of the FAS reflects a different combination of catalytic domains within the multifunctional protein, an observation that is of special interest in the context of the programming of polyketide synthase function.

### 7.4. The Structure and Function of Bacterial Polyketide Synthases

This extensive knowledge of FAS structure provides a good base for a discussion of PKS structure and function. Here also these multienzyme complexes can be grouped into two main classes, the type-I multifunctional proteins and the type-II systems comprising (apparently) dissociable component enzymes (131). One of the best studied PKS complexes is the type-I 6-methylsalicyclic acid synthase (MSAS) from the fungus *Penicillium patulum*. This enzyme generates 6-methylsalicyclic acid (95) from an acetyl-CoA starter unit, three malonyl-CoA extender units and one equivalent of NADPH (see Fig. 57). This is also one of the few PKS's to have been purified and characterised from a biochemical standpoint (135, 136), and recently its protein sequence was deduced from the cDNA sequence (137). The MSAS synthase has an apparent molecular mass of 0.8–1.1 MDa (136), with a subunit mass of about 188 KDa. This multienzyme complex should comprise acetyl and malonyl transferases to load the substrates, a β-ketoacyl synthase, a ketoreductase, a dehydratase but no enoyl reductase, at least one additional activity (cyclase + aromatase?) to from the aromatic ring, and a thioesterase to release the free acid. No detailed information is at hand so far on the order of all the chemical events catalysed by the complex, although a model for its mode of action was proposed by LYNEN (135). Recently, elegant stereochemical studies have provided new insights into the relative disposition of reactive groups in the active site of the enzyme (138, 139).

The gene for MSAS synthase includes a 5322 bp open reading frame coding for a protein of 1774 amino acids and 190,731 Da molecular mass. A computer assisted analysis of the deduced protein sequence revealed likely regions encoding a condensing enzyme and acetyl/malonyl transferases near the N-terminus of the polypeptide, where there are regions of high sequence similarity to the animal FAS condensing enzyme and acetyl/malonyl transferase domains, as well as to the FabB of *E. coli* β-ketoacylsynthase I (Fig. 58). Similarly, a strong candidate for the

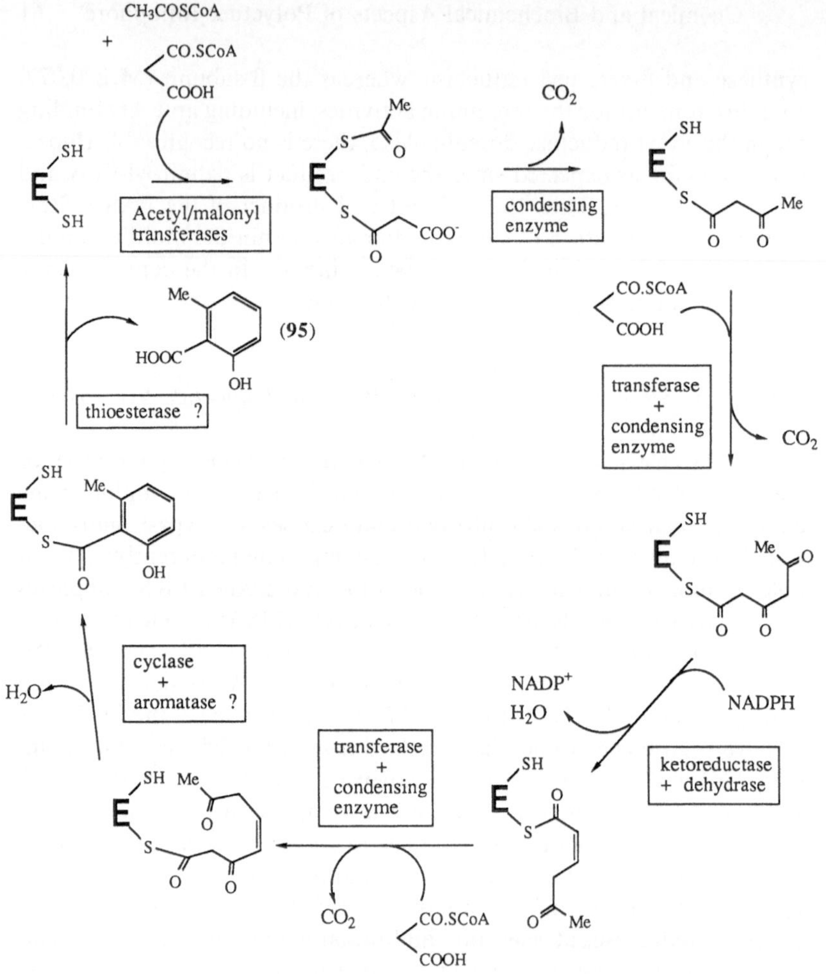

CH₃COSCoA — rendered as $CH_3COSCoA$

$CH_3COSCoA$

+

$CO.SCoA$
$COOH$

**E** SH / SH

Acetyl/malonyl transferases

**E** S—C(Me)=O / S—C(=O)—COO⁻

condensing enzyme → $CO_2$

**E** SH / S—C(=O)—CH₂—C(=O)—Me

$CO.SCoA$
$COOH$

transferase + condensing enzyme → $CO_2$

**E** SH / S—C(=O)—CH₂—C(=O)—CH₂—C(Me)=O

ketoreductase + dehydrase

$NADP^+$  $H_2O$ ← $NADPH$

**E** SH / S ...

transferase + condensing enzyme

$CO_2$  $CO.SCoA$ $COOH$

cyclase + aromatase ?

$H_2O$

**E** SH / S—C(=O)— (Me, OH aromatic)

thioesterase ?

(**95**) Me / HOOC / OH aromatic ring

Fig. 57. A hypothetical 6-methylsalicylic acid synthase reaction cycle

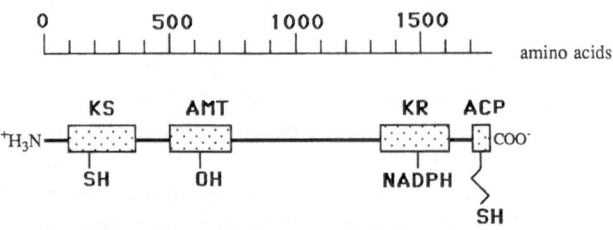

Fig. 58. The location of deduced activities along the 6-methylsalicylic acid synthase polypeptide determined from cDNA sequence (*KS* β-ketoacyl synthase. *AMT* acetyl/malonyl transferase, *KR* β-ketoacyl reductase, *ACP* acyl carrier protein domain) (*137*)

region comprising the β-ketoacyl reductase is found near the C-terminus, where a nucleotide binding site motif (Gly-Leu-Gly-Val-Leu-Gly) fits the consensus Gly-Xaa-Gly-Xaa-Xaa-Gly for the ketoreductase in the animal FAS (see Fig. 59). At the extreme C-terminus is a predicted ACP domain for the pantetheinyl prosthetic group. This leaves a large section in the middle of the protein for the remaining activities, where no specific functions have yet been assigned. A putative thioesterase domain has not been detected in the MSAS gene. *Penicillium patulum* was the first organism from which a FAS (*140*) and a PKS have been cloned and characterised. Perhaps somewhat surprisingly the MSAS sequence is more closely related to the heterologous animal FAS sequence than to its own homologous α/β subunit type-I FAS sequence.

**A    β-ketoacyl synthases**

```
       gra ORF1      G P V T M V S D G C T S G L D S V
       tcm ORF1      G P V T V V S T G C T S G L D A V
     whiE ORFIII     G P V Q T V S T G C T S G L D A V
       6-MSAS        G P S T A V D A A C A S S L V A I   (P. patulum)
       FAS fabB      G V N T S I S S A C A T S A H C I   (E. coli)
       FAS  rat      G P S I A L D T A C S S S L L A L
       FAS chicken   G P S L T I D T A C S S S L M A L
     ery ORFA (KS1)  G P A M T V D T A C S S G L T A L
     ery ORFA (KS2)  G P A V T V D T A C S S S L V A L
```

**B    ACP's**

```
       gra ORF3      E E L G Y D S L A L M E S A
       tcm ORF3      Q D L G Y D S I A L L E I S
     whiE ORFV       D T F G L D S L G L L G I V
       6-MSAS        A D L G V D S V M T V T L R
       rat FAS       A D L G L D S L M G V E V R
     chicken FAS     A D L G L D S L M G V E V R
     S. erythraea    E D L G M D S L D L V E X V
     E. coli FAS     E D L G A D S L D T V E L V
     ery ORFA (ACP1) R D L G F D S M T A V D L R
     ery ORFA (ACP2) T E L G F D S L T A V G L R
```

**C    NAD(P) binding domains**

```
       act ORF5           P V D V L V N N A G R P G G G A T A E L A D
       gra  ORF5          T V D I L V N N A G R S G G G A T A E I A D
       6-MSAS             L P R P E G T Y L I T G G L G V L G L E V A
   chick FAS1811 KR       S C P P T K S Y I I T G G L G G F G L E L A
    rat FAS1800 KR        F C P E H K S Y I I T G G L G G F G L E L A
   chick FAS1597 ER       K G E S V L I H S G S G G V G Q A A I A I A
    rat  FAS1587 ER       H G E T V L I H S G S G G V G Q A A I S I A
  ery ORFA1121 (KR1)      L E P L A G T V L V T G G I G A H L A R W L
  ery ORFA2558 (KR2)      S W E P A G T A L V T G G T G A L G G H V A
```

Fig. 59. Alignments of segments of deduced potein sequence found in various FAS's and PKS's (see text); A) Around the presumptive active site cysteine in β-ketoacyl synthases; B) around the active site serine of putative ACP's; and C) NAD(P)H binding domains in β-ketoacyl reductases and enoyl reductases

The extent of structural and biochemical knowledge of PKS's involved in polyether and macrolide antibiotic production is far less complete. However, genes encoding the biosynthetic pathways to several polyketide natural products have now been cloned from *Streptomycetes*, including those for actinorhodin (**96**) (*141*), granaticin (**97**) (*142*), oxytetracycline (**98**) (*143*), tetracenomycin (**99**) (*144*), daunorubicin (**100**) (*145*), erythromycin A (**40**) (*146–150*), tylosin (**41**) (*151, 152*), spiramycin (**101**) (*153*), carbomycin (**102**) (*154*), avermectin (**44**) (*155*), candicidin (**103**) (*156*), and a spore pigment of unknown structure encoded by the *whi*E locus (see *131*) (Fig. 60). Moreover, the DNA sequences of genes encoding the PKS's involved in actinorhodin (*act*) (*131, 157*), granaticin (*gra*) (*158*), tetracenomycin (*tcm*) (*159*), WhiE spore pigment (*whi*E) (*160*) and oxytetracycline (*otc*) (*131*) have been determined recently. The sequence information for these aromatic PKS's deduced so far spans almost 6 kb of DNA, and has revealed a common organisation of three characteristic open reading frames (ORF's = deduced protein coding regions) labelled ORF1, −2 and −3 (or ORF's III, IV and V) in Fig. 61, encoding deduced polypeptides of approximately 40 KDa, 40 KDa and 8 KDa respectively. The deduced protein sequences of the first two ORF's show high similarity with the FabB condensing enzyme of *E. coli*, although a potential active site Cys occurs only in the first ORF in each case. This fact, along with the possibility of translational coupling between the first two ORF's (because of overlapping 3′ stop/5′ start codons) suggests that the gene products are produced stoichiometrically to form a heterodimer rather than being two distinct condensing enzymes.

The ORF3 (and ORFV) in each cluster encodes a deduced protein showing high sequence homology to the type-II FAS ACP's, and the ACP domain of animal type-I FAS, with a characteristic motif centred upon a potential 4′-phosphopantetheine-binding serine (see also Fig. 59). It is notable that plant chalcone and resveratrole synthases, PKS's of plant origin involved in flavonoid biosynthesis, do not contain a covalently bound 4′-phosphopantetheine cofactor. Lying immediately downstream of ORF3 is a fourth ORF that shows relatedness between the clusters, although the sequence similarities extend only over part of the deduced protein sequences. The *act* and *gra* ORF4 deduced protein sequences are similar throughout the length of the protein, but each resembles *tcm* ORF4 only in the N-terminal half of the molecules. Since *act* and *gra* ORF4 DNA complement *act*VII mutants deduced to be defective in cyclisation and dehydration events, it has been suggested (*131*) that *act* and *gra* ORF4 deduced proteins comprise a bifunctional cyclase/dehydrase. In contrast, the deduced *tcm* ORF4 protein seems to

*References, pp. 72–81*

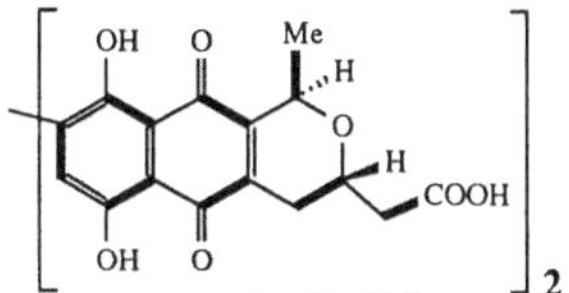

actinorhodin (96)

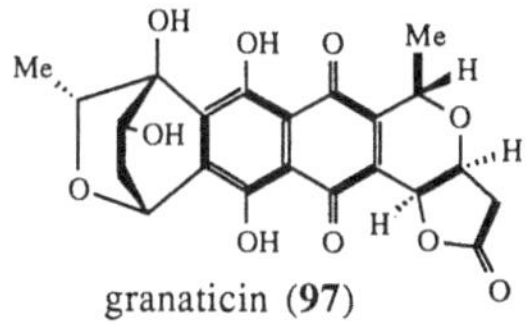

granaticin (97)

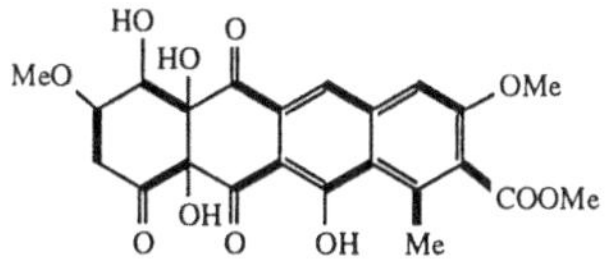

tetracenomycin (99)

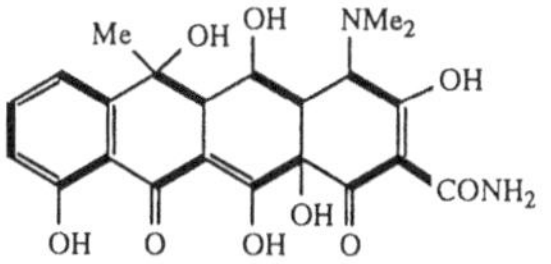

oxytetracycline (98)

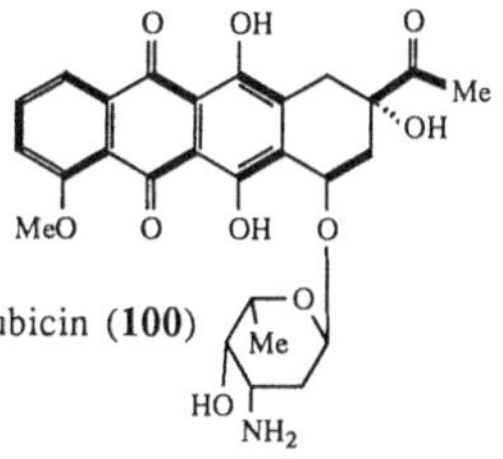

daunorubicin (100)

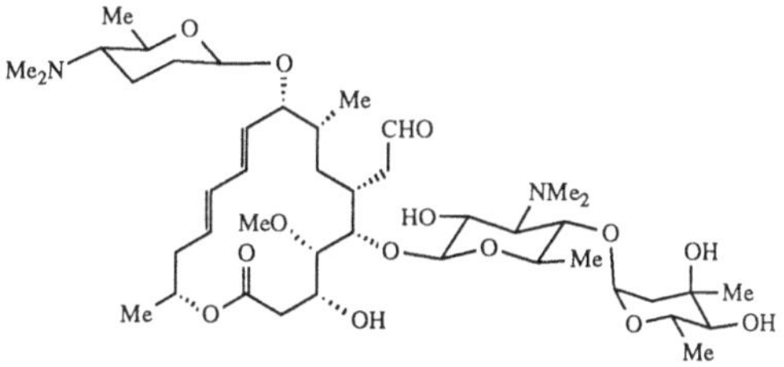

spiramycin (101)

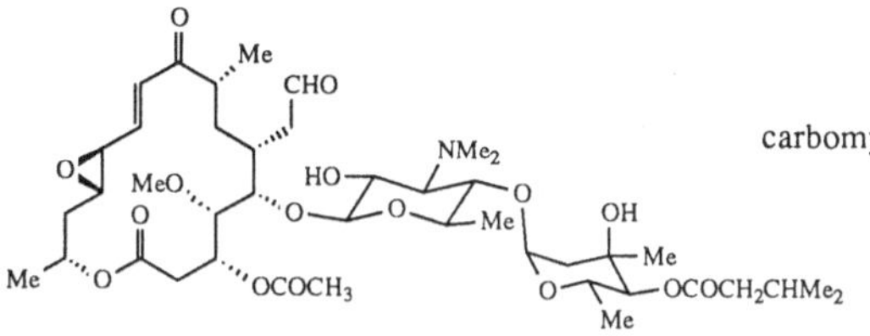

carbomycin (102)

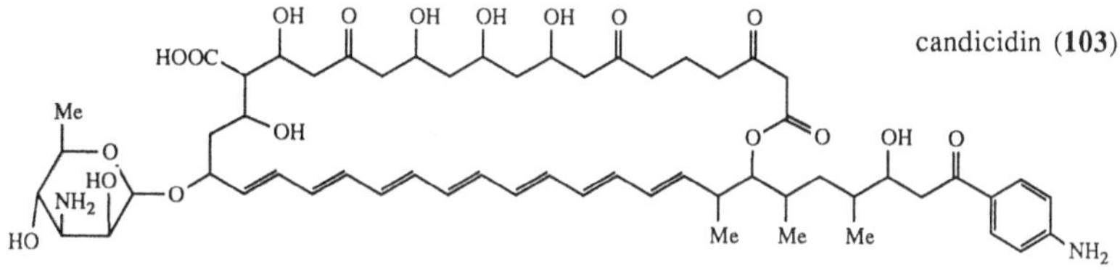

candicidin (103)

Fig. 60. Polyketide metabolites from Streptomycetes, whose biosynthetic genes have been cloned recently. See text for references

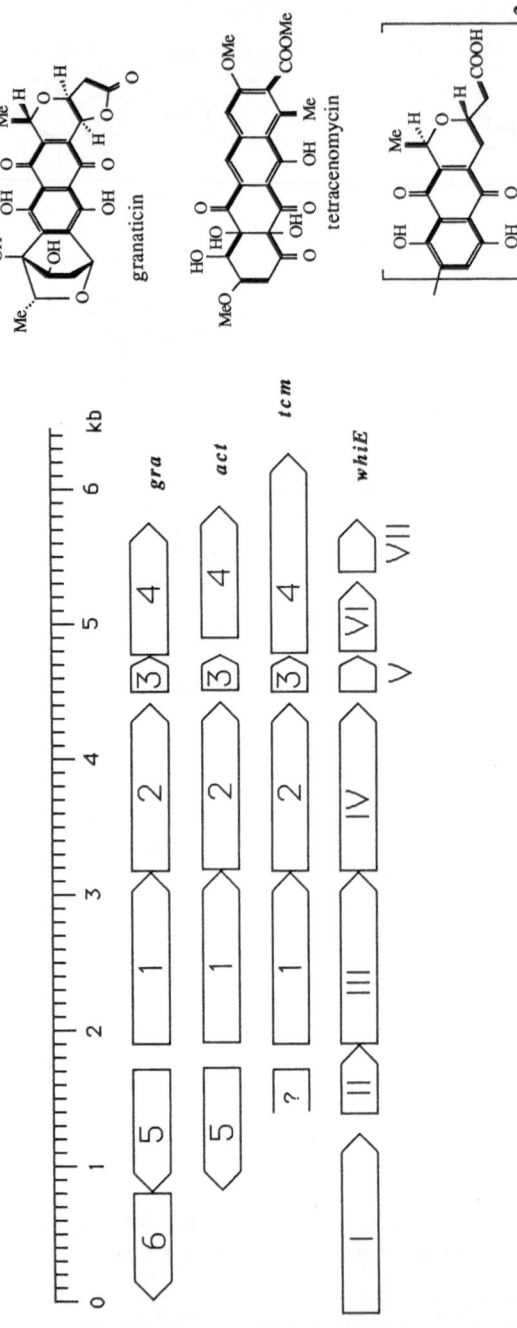

Fig. 61. The organisation of ORF's in the granaticin (*gra*), actinorhodin (*act*), tetracenomycin (*tcm*), and *whiE* polyketide synthase gene clusters. The arrows represent the length of each ORF in Kbp of DNA, and their relative positions and orientations of transcription. The deduced functions for these ORF's are; ORF1 + ORF2 (ORFIII + IV) = β-ketoacyl synthase; ORF3 (ORFV) = acyl carrier protein; ORF2 (ORFIII + IV) = β-ketoacyl synthase; ORF3 (ORFV) = acyl carrier protein; ORF4 = cyclase/dehydrase (*gra*, and *act*), cyclase/O-methyltransferase (*tcm*); ORF5 = ketoreductase; ORF6 = ketoreductase. See text for references

be a cyclase/O-methyltransferase because its C-terminal half resembles bovine hydroxyindole-O-methyltransferase and *tcm* ORF4 DNA complements *tcm* mutants lacking a specific O-methylation step (*131*).

The divergently transcribed ORF5 in the *act* and *gra* clusters most likely encodes a NAD-dependent ketoreductase (*157, 158*), and both deduced protein sequences contain the consensus nucleotide binding motif Gly-Xaa-Gly-Xaa-Xaa-Ala characteristic of nicotinamide coenzymes (see Fig. 59). The *tcm* cluster contains no sequence homologous to these putative oxidoreductases. This is consistent with the fact that no reductive step is needed in polyketide chain assembly; all the β-placed oxygens are retained!

In any event, an important conclusion from this work is that these PKS genes for aromatic polyketides appear to encode proteins most clearly resembling the type-II FAS from bacteria and plants rather than being large multifunctional polypeptides as seen in the type-I FAS and the 6-methylsalicylic acid synthase from *Penicillium patulum* (*137*). A great deal remains to be discovered about the specificity and mode of action of these (so far) deduced proteins.

Whilst these observations are of great interest and may be of direct relevance for polyethers containing an aromatic moiety, the PKS's involved in macrolide and polyether biosynthesis face a quite different problem in programming the polyketide chain assembly. In the case of macrolide antibiotics such as erythromycin A (**40**), tylosin (**41**), carbomycin (**102**) and spiromycin (**101**), regions within each biosynthetic cluster encoding the respective PKS's have been identified and these typically span tens of kilobases of genomic DNA. Perhaps the most fascinating (emerging) feature in these various systems is the existence of multiple sets of PKS genes. Thus a 30 kb segment in the erythromycin biosynthetic cluster, close to the originally cloned resistance gene, consists of repeated motifs whose sequence – so far incomplete – reveals proteins whose deduced functions mirror those found in the animal and *E. coli* FAS's and the other sequenced PKS systems. An analysis of this DNA is now beginning to furnish a remarkable view of macrolide PKS structure, which begins to provide an insight into how the synthesis of a macrolide might be programmed at the molecular level.

6-Deoxyerythronolide B (**68**), the first isolable intermediate on the pathway to erythromycin A, is assembled from a propionyl-CoA starter unit and six methylmalonyl-CoA extender units (see Fig. 37). The genes encoding erythromycin biosynthesis are clustered in the genome of *Saccharopolyspora erythraea* and in the middle of the cluster lies the resistance gene *ermE* (see Fig. 62). Located about 12 kb downstream from this resistance genes is a DNA segment capable of complementing

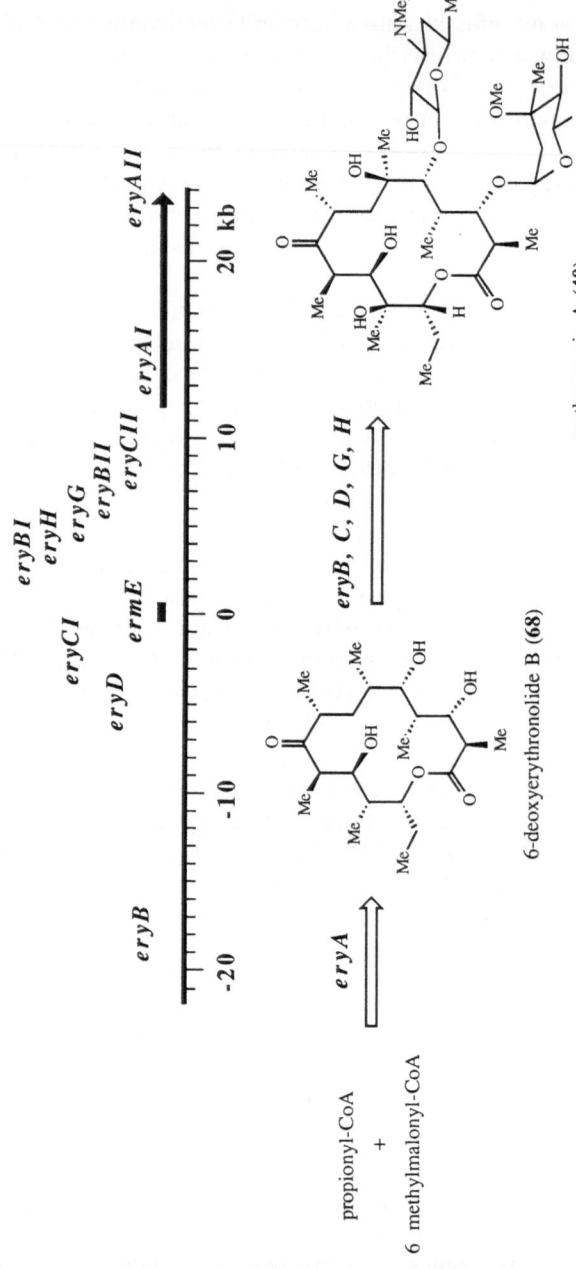

Fig. 62. The erythromycin biosynthetic gene cluster, showing relative locations of the resistance gene (*ermE*) and biosynthetic genes (*eryA-eryH*) (*150, 161*)

EryA mutants blocked in the biosynthesis of the erythronolide ring. This *eryAI* locus encodes the macrolide PKS. A further DNA segment homologous to *eryAI*, designated *eryAII*, was localised to a region about 35 kb downstream from *ermE*, and was also shown to encode genes for the macrolide PKS (*161*).

The *eryAI* and *eryAII* DNA has now been sequenced and this has provided crucial information not only about the size, but also about the probable functions of ORF's encoded in the ery PKS cluster. The sequence of one such ORF (ORFA) extending over 9.5 kb of DNA has been reported by LEADLAY's group (*162*). The deduced gene product is predicted to contain 3,178 amino acids. Upon comparison with available protein sequence databases nine separate portions of the deduced protein are found to be very similar to active site sequences found in the constituent catalytic activities of known FAS's and PKS's. The deduced activities for various regions of this predicted protein are shown in Fig. 63, and include in sequence AT(or "TE")-ACP-KR-AT-KS-ACP-KR-AT-KS. A comparison of putative active site residues in this ORF with those found in corresponding components of other FAS and PKS complexes is shown in Fig. 59. These results reinforce the earlier con-

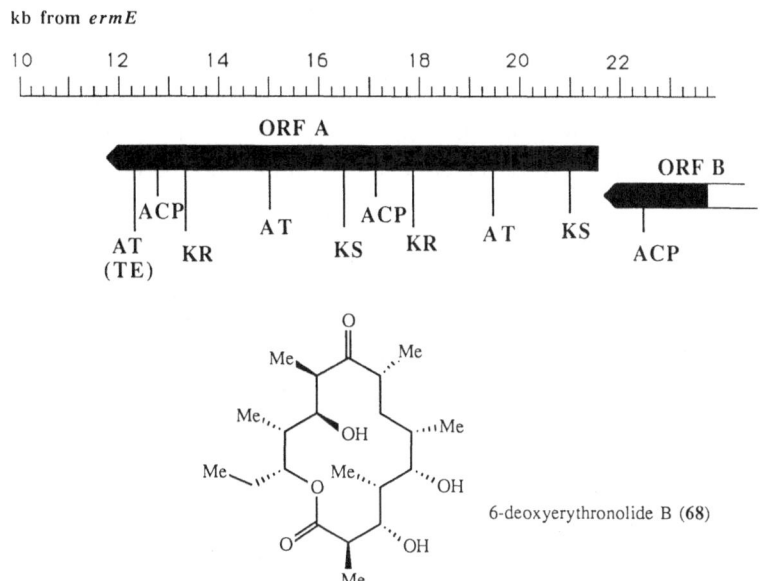

Fig. 63. A large open reading frame in the *ery* cluster (*eryAI*) encoding a multifunctional polypeptide, where the deduced activities are: *AT(TE)* acyl transferase (thioesterase), *ACP* acyl carrier protein domain, *KR* ketoreductase, *KS* β-ketoacyl synthase (*162*)

clusion that this portion of the *ery* cluster encodes a PKS and show also that the bacterial macrolide PKS's are not necessarily complexes of monofunctional proteins, a trend evident with the aromatic Streptomycete PKS's.

In a parallel investigation KATZ's group has also sequenced further downstream and found two additional large ORF's of comparable size to that described by LEADLAY. Each of these ORF's again appears to encode a single large multifunctional protein containing multiple copies of units comprising AT, ACP, KS, KR activities, and one ORF contains also one set of dehydrase and enoyl reductase activities (KATZ, L.; personal commun.). It seems, therefore, that the erythronolide PKS may comprise three large multienzyme complexes which possess all the activities required to catalyse six rounds of chain extension and modification.

With this information in hand it becomes possible to speculate as to how the individual steps in 6-deoxyerythronolide biosynthesis might be catalysed by the deduced activities encoded in the three *ery* ORF's. KATZ has proposed a "module hypothesis" in which each polypeptide carries the functions for two rounds of chain elongation and modification. The DNA sequence for each round is called "module". The deduced activities encoded in ORF A shown in Fig. 63 would comprise the last two modules required to add the final two propionate units and complete the synthesis of 6-deoxyerythronolide B. There are apparently two condensing enzymes (KS) and two ketoreductases (KR), and at the C-terminus of the deduced protein is a sequence showing similarity to known thioesterases (TE), which might therefore catalyse the final act of macrolide ring formation. There is, therefore, a correspondence between the catalytic steps required in these final two chain elongation events, and the deduced activities in the deduced protein, as shown in Fig. 64. This hypothesis implies that each of the deduced activities (KS, KR, DH and ER) is used once during the synthesis of each molecule of 6-deoxyerythronolide B, and fortunately there appear to be sufficient activities present; $6 \times KS$, $5 \times KR$, $1 \times DH$, $1 \times ER$. There are, however, also multiple copies of deduced ACP domains and AT activities. This leaves sufficient scope for differences in stereospecificity to arise amongst the ketoreductases, for example, or amongst the acyl transferases (($R$)- *vs.* ($S$)-methylmalonyl-CoA). There are no real clues yet as to how the growing acyl chain is transferred between active sites, and between the three different polypeptide chains. LEADLAY has suggested that the two halves of the ORFA gene product might fold back on each other so as to bring the predicted β-ketoacyl synthase active sites into close proximity to the ACP domains (*162*), but there is currently no firm structural information available for these *deduced* proteins.

*References, pp. 72–81*

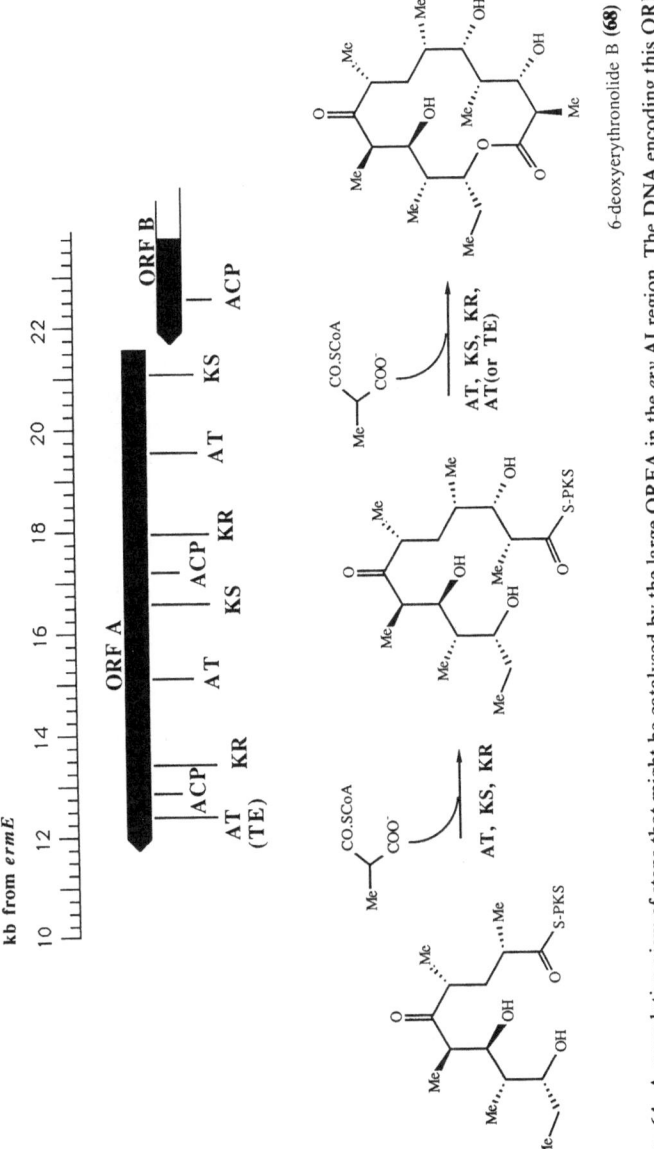

Fig. 64. A speculative view of steps that might be catalysed by the large ORFA in the *ery* AI region. The DNA encoding this ORF should represent the last two modules acting in the biosynthesis of 6-deoxyerythronolide B (**68**); *AT* acyl transferases for loading substrates, *KS* ketoacyl synthases for each decarboxylative condensation, *KR* ketoreductases for reducing each ketone to a β-hydroxythioester, and *AT* or *TE* for macrolide ring formation

The remarkable picture emerging from these studies is one of a colinearity between the biochemical steps in macrolide assembly and the genetic order of ORF's encoding the requisite multifunctional proteins in the genome of the producing organism. Whether this picture has been interpreted correctly will no doubt be tested by future biochemical experiments. One swallow, of course, does not make a summer. But if this correlation proves to be a general one amongst macrolide and polyether PKS complexes, then an important step forward will have been made in understanding how the programming of these polyketide pathways, such as that to the putative monensin triene intermediate shown in Fig. 19, is achieved at the biochemical and genetic levels.

The interesting prospect also arises of using rDNA methods to engineer new polyketide biosynthetic pathways. Protein engineering experiments might, for example, involve swapping modules or ORF's between macrolide or polyether pathways in order to make homologues of known antibiotics or bringing modules or ORF's into novel combinations so that the host microorganism is programmed to biosynthesise entirely new classes of natural products. Alternatively, modules or simply portions of modules might be deleted or disrupted to generate new end products or release intermediates in the assembly process. Although the prospects seem good, determining where the limits lie for the rational manipulation of polyketide biosynthesis will probably require a far more extensive knowledge of these processes at all levels.

# References

1. WESTLEY, J.W.: Polyether Antibiotics Naturally Occurring Acid Ionophores, Vol. 1, Biology. Ed. J.W. WESTLEY. New York: Marcel Dekker (1983).
2. DUESLER, E.N., and I.C. PAUL: X-ray structures of the polyether antibiotics. In: Polyether Antibiotics Naturally Occurring Acid Ionophones, Vol. 2, Chemistry. Ed. J.W. WESTLEY, New York: Marcel Dekker (1983).
3. SMITH, P.W., and W.C. STILL: The effect of substitution and stereochemistry on ion binding in the polyether ionophore monensin. J. Amer. Chem. Soc. 110, 7917 (1988).
4. HANEY, M.E., and M.M. HOEHN: Monensin, a new biologically active compound. I. Discovery and Isolation. Antimicrob. Agents Chemother. 349 (1967).
5. WALBA, D.M., and M. HERMSMEIER: Thermodynamics of complexation of monensin A and monensin B in methanol by titration calorimetry. J. Chem. Soc. Chem. Comm. 383 (1985).
6. STILL, W.C., P. HAUCK, and D. KEMPF: Stereochemical studies of lasalocid epimers. Ion-driven epimerisations. Tetrahedron Lett. 28, 2817 (1987).
7. WILLIAMS, D.H., M.J. STONE, P.R. HAUCK, and S.K. RAHMAN: Why are secondary metabolites (Natural Products) biosynthesised? J. Nat. Prod. Chem. 52, 1189 (1989).

8. DUNITZ, J.D., and M. DOBLER: Structural studies of ionophores and their complexes. In: Biological Aspects of Inorganic Chemistry. Ed. D. DOLPHIN. J. Wiley (1977), pp. 113–140.

9. Topics in Current Chemistry Vol. 98; Host Guest Complex Chemistry I. Ed. F. VÖGTLE. Berlin Heidelberg New York: Springer (1981).

10. CANE, D.E., W.D. CELMER, and J.W. WESTLEY: Unified stereochemical model of polyether antibiotic structure and biogenesis. J. Amer. Chem. Soc. 105, 3594 (1983).

11. O'HAGAN, D.: Structural and stereochemical homology between the macrolide and polyether antibiotics. Tetrahedron 44, 1691 (1988).

12. O'HAGAN, D.: Nat. Prod. Rep. 6, 205 (1989).

13. KELLER-JUSLEN, C., H.D. KING, M. KUHN, H.R. LOOSLI, W. PACHE, T.J. PETCHER, H.P. WEBER, and A. VON WARTBURG: Tetronomycin, a novel polyether of unusual structure. J. Antibiot. 35, 142 (1982).

14. DAVIS, D.H., E.W. SNAPE, P.J. SUTER, T.J. KING, and C.P. FALSHAW: Structure of antibiotic M139603; X-ray crystal structure of the 4-bromo-3,5-dinitrobenzoyl derivative. J. Chem. Soc., Chem. Comm. 1073 (1981).

15. HORI, K., K. NOMURA, S. MORI, and E. YOSHI: Synthesis of the acyltetronic acid fragment of tetronomycin. J. Chem. Soc., Chem. Comm. 712 (1989).

16. DOHERTY, A.M., and S.V. LEY: Synthetic studies towards the acyltetronic acid ionophore M139603. Tetrahedron Lett. 27, 105 (1986).

17. BIRCH, A.J., and F.W. DONOVAN: Studies in relation to biosynthesis I. Some possible routes to derivatives of orcinol and phloroglucinol. Austr. J. Chem. 6, 360 (1953).

18. BIRCH, A.J.: Biosynthetic relations of some natural phenolic and enolic compounds. Progr. Chem. Org. Nat. Prod. 14, 186–216 (1957).

19. BIRCH, A.J.: Biosynthesis of polyketides and related compounds. Science 156, 202 (1967).

20. BROCKMANN, H., and W. HENKEL: Pikromycin, ein bitter schmeckender Antibioticum aus Actinomyceten. Chem. Ber. 84, 284 (1951).

21. WOODWARD, R.B.: Struktur und Biogenese der Makrolide; Eine neue Klasse von Naturstoffen. Angew. Chem. 69, 50 (1957).

22. WESTLEY, J.W.: Polyether Antibiotics-Biosynthesis. In: Antibiotics Vol. IV, Biosynthesis, Ed. J.W. CORCORAN. Berlin Heidelberg New York: Springer (1981).

23. DAY, L.E., J.W. CHAMBERLIN, E.Z. GORDEE, S. CHEN, M. GORMAN, R.L. HAMILL, T. NESS, R.E. WEEKS, and R. STROSHANE: Biosynthesis of monensin. Antimicrob. Agents Chemother. 4, 410 (1973).

24. WESTLEY, J.W., D.L. PRUESS, and R.G. PRUESS: Incorporation of [1-$^{13}$C]butyrate into X-537A (lasalocid). J. Chem. Soc. Chem. Comm. 162 (1972).

25. DAVID, L., and S. EMADZADEH: Biosynthesis of the ionophorous antibiotic A23187. J. Antibiot. 35, 1616 (1982).

26. ZMIJEWSKI, M.J., R. WONG, J.W. PASCHAL, and D.E. DORMAN: The biosynthesis of antibiotic A23187. Tetrahedron 39, 1255 (1983).

27. BULSING, M.J., E.D. LAUE, F.J. LEEPER, J. STAUNTON, D.H. DAVIES, G.A.F. RITCHIE, A. DAVIES, A.B. DAVIES, and R.P. MABELIS: Biosynthesis of the polyketide antibiotic ICI139603 in Streptomyces longisporoflavus: Assignment of the $^{13}$C NMR spectrum by 2D methods and determination of the origin of the carbon atoms. J. Chem. Soc., Chem. Comm. 1301 (1984).

28. ASHWORTH, D.M., J.A. ROBINSON, and D.L. TURNER: Biosynthesis of the macrotetrolide antibiotics; The incorporation of carbon-13 and oxygen-18 labelled acetate, propionate and succinate. J. Chem. Soc. Perkin I Trans. 1719 (1988).

29. LEE, J.J., P.M. DEWICK, C.P. GORST-ALLMAN, F. SPREAFICO, C. KOWAL, C.-J. CHANG,

A.G. McInnes, J.A. Walter, P.J. Keller, and H.J. Floss. Further studies on the biosynthesis of the boron-containing antibiotic aplasmomycin. J. Amer. Chem. Soc. **109**, 5426 (1987).

30. Lee, M.S., G.-W. Qin, K. Nakanishi, and M.G. Zagorski: Biosynthetic studies of brevetoxins, potent neurotoxins produced by the dinoflagellate *Gymnodinium breve*. J. Amer. Chem. Soc. **111**, 6234 (1989).

31. Chou, H.-N., and Y. Shimizu: Biosynthesis of brevetoxins. Evidence for the mixed origin of the backbone carbon chain and the possible involvement of dicarboxylic acids. J. Amer. Chem. Soc. **109**, 2184 (1987).

32. Reynolds, K.A., D. O'Hagan, G. Gani, and J.A. Robinson: Butyrate metabolism in streptomycetes. Characterisation of an intramolecular vicinal interchange rearrangement linking isobutyrate and butyrate in *S. cinnamonensis*. J. Chem. Soc. Perkin I Trans. 3195 (1988).

33. Brendelberger, G., J. Retey, D.M. Ashworth, K. Reynolds, F. Willenbrock, and J.A. Robinson: The enzymic interconversion of isobutyryl and n-butyryl-carba(dethia)-coenzyme-A; A coenzyme $B_{12}$ dependent carbon skeleton rearrangement. Angew. Chem. Int. Ed. **27**, 1089 (1988).

34. Omura, S., K. Tsuzuki, Y. Tanaka, H. Sakakibara, M. Aizawa, and G. Lukacs: Valine as a precursor of n-butyrate unit in the biosynthesis of macrolide aglycone. J. Antibiot. **36**, 614 (1983).

35. Sherman, M.M., S. Yue, and C.R. Hutchinson: Biosynthesis of lasalocid A. Metabolic interrelationships of carboxylic acid precursors and polyether antibiotics. J. Antibiot. **39**, 1135 (1986).

36. Kishi, K., S. Hatakeyama, and M.D. Lewis: Total synthesis of polyether antibiotics narasin and salinomycin. In: Frontiers of Chemistry; 28th IUPAC Congress, Aug 1981. Ed. K.J. Laidler. Oxford: Pergamon Press (1982).

37. Risley, J.M., and R.L. van Etten: $^{18}$O-Isotope effect in $^{13}$C NMR spectroscopy 3. Additivity effects and steric effects. J. Amer. Chem. Soc. **102**, 6699 (1980).

38. Vederas, J.C.: Structural dependence of $^{18}$O-Isotope shifts in $^{13}$C NMR. J. Amer. Chem. Soc. **102**, 374 (1980).

39. Vederas, J.C., and T.T. Nakashima: Biosynthesis of averufin by *Aspergillus parasiticus*; Detection of $^{18}$O-label by $^{13}$C-NMR isotope shifts. J. Chem. Soc. Chem. Comm. 183 (1980).

40. Hutchinson, C.R., M.M. Sherman, J.C. Vederas, and T.T. Nakashima: Biosynthesis of macrolides 5. Regiochemistry of the labeling of lasalocid A by $^{13}$C, $^{18}$O-labeled precursors. J. Amer. Chem. Soc. **103**, 5953 (1981).

41. Cane, D.E., T.-C. Liang, and H. Hasler: Polyether biosynthesis. Origin of the oxygen atoms of monensin A. J. Amer. Chem. Soc. **103**, 5962 (1981).

42. Cane, D.E., T.-C. Liang, and H. Hasler: Polyether Biosynthesis 2. Origin of the oxygen atoms of monensin A. J. Amer. Chem. Soc. **104**, 7274 (1982).

43. Ajaz, A., and J.A. Robinson: The utilisation of oxygen atoms from molecular oxygen during the biosynthesis of monensin A. J. Chem. Soc., Chem. Comm. 679 (1983).

44. Cane, D.E., H. Hasler, and T.-C. Liang: Macrolide Biosynthesis. Origin of the oxygen atoms in the erythromycins. J. Amer. Chem. Soc. **103**, 5960 (1981).

45. O'Hagan, D., J.A. Robinson, and D.L. Turner: Biosynthesis of the macrolide antibiotic tylosin. Origin of the oxygen atoms in tylactone. J. Chem. Soc. Chem. Comm. 1337 (1983).

46. Spavold, Z., J.A. Robinson, and D.L. Turner: Biosynthesis of the polyether antibiotic narasin. Origins of the oxygen atoms and the mechanisms of ring formation. Tetrahedron Lett. **27**, 3299 (1986).

47. Tsou, H.-R., S. Rajan, R. Fiala, P.C. Mowery, M.W. Bullock, D.B. Borders, J.C. James, J.H. Martin, and G.O. Morton: Biosynthesis of the antibiotic maduramicin. Origin of the carbon and oxygen atoms as well as the $^{13}$C NMR assignments. J. Antibiot. **37**, 1651 (1984).

48. Cane, D.E., and B.R. Hubbard: Polyether biosynthesis. 3. Origin of the carbon skeleton and oxygen atoms of lenoremycin. J. Amer. Chem. Soc. **109**, 6533 (1987).

49. Demetriadou, A.K., E.D. Laue, J. Staunton, G.A.F. Ritchie, A. Davies, and A.B. Davies: Biosynthesis of the polyketide polyether antibiotic ICI139603 in *Streptomyces longisporoflavus* from $^{18}$O-labelled acetate and propionate. J. Chem. Soc., Chem. Comm. 408 (1985).

50. Ashworth, D.M., J.A. Robinson, and D.L. Turner: Biosynthesis of the macrotetrolide antibiotics; The incorporation of carbon-13 and oxygen-18 labelled acetate, propionate and succinate. J. Chem. Soc. Perkin Trans, I. 1719 (1988).

51. Cane, D.E., T.-C. Liang, L. Kaplan, M.K. Nallin, M.D. Schulman, O.D. Hensens, A.W. Douglas, and G. Albers-Schönberg: Biosynthetic origins of the carbon skeleton and oxygen atoms of the avermectins. J. Amer. Chem. Soc. **105**, 4110 (1983).

52. Tsou, H.-R., Z.H. Ahmed, R.R. Fiala, M.W. Bullock, G.T. Carter, J.J. Goodman, and D.B. Borders: Biosynthetic origin of the carbon skeleton and oxygen atoms of the LL-F28249α, a potent antiparasitic macrolide. J. Antibiot. **42**, 398 (1989).

53. Doddrell, D.M., E.D. Laue, F.J. Leeper, J. Staunton, A. Davies, A.B. Davies, and G.A.F. Ritchie: Biosynthesis of the polyether antibiotic ICI139603 in *Streptomyces longisporoflavus*: Investigation of deuterium retention after incorporation of $CD_3$$^{13}CO_2H$, $^{13}CD_3CO_2H$, and $CH_3CD_2$$^{13}CO_2H$ using $^2H$ NMR and edited $^{13}C$ NMR spectra. J. Chem. Soc., Chem. Comm. 1302 (1984).

54. Ashworth, D.M., D.S. Holmes, J.A. Robinson, H. Oikawa, and D.E. Cane: Selection of a specifically blocked mutant of *Streptomyces cinnamonensis*: Isolation and synthesis of 26-deoxymonensin A. J. Antibiot. **42**, 1088 (1989).

55. Pospisil, S., P. Sedmera, J. Vokoun, Z. Vanek, and M. Budesinsky: 3-O-demethylmonensins A and B produced by *S. cinnamonensis*. J. Antibiot. **40**, 555 (1987).

56. Prasad, A.V.K., and Y. Shimizu: The structure of hemibrevetoxin-B; A new type of toxin in the gulf of Mexico red tide organism. J. Amer. Chem. Soc. **111**, 6476 (1989).

57. Still, W.C., and A.G. Romero: Model of the polyepoxide cyclisation route to polyether antibiotics. J. Amer. Chem. Soc. **108**, 2105 (1986).

58. Schreiber, S.L., T. Sammakia, B. Hulin, and G. Schulte: Epoxidation of unsaturated macrolides: Stereochemical routes to ionophoric subunits. J. Amer. Chem. Soc. **108**, 2106 (1986).

59. Russell, S.T., J.A. Robinson, and D.J. Williams: A diepoxide cyclisation cascade initiated through the action of pig liver esterase. J. Chem. Soc., Chem. Comm. 351 (1987).

60. Hoye, T.R., and J.C. Suhadolnik: Symmetry-Assisted Synthesis of Triepoxide Stereoisomers of (E,Z,E)-dodeca-2,6,10-triene-1,12-diol and their cascade reactions to 2,5-linked bistetrahydrofurans. J. Amer. Chem. Soc. **107**, 5312 (1985).

61. Paterson, I., I. Boddy, and I. Mason: Studies in polyether synthesis using polyepoxide cyclisations. Tetrahedron Lett. **28**, 5205 (1987).

62. Paterson, I., and P.A. Craw: Studies in polyether synthesis: Controlled bisepoxide cyclisation using a β-diketone group. Tetrahedron Lett. **30**, 5799 (1989).

63. David, L., and H. Veschambre: Preparation d'oxydes de linalol par bioconversion. Tetrahedron Lett. **25**, 543 (1984).

64. Holmes, D.S., D.M. Ashworth, and J.A. Robinson: The bioconversion of (3RS, E)- and (3RS, Z)-nerolidol into oxygenated products by *S. cinnamonensis*. Possible

implications for the biosynthesis of the polyether antibiotic monensin A. Helv. Chim. Acta **73**, 260 (1990).

65. KLEIN, E., and W. ROJAHN: Die Permanganatoxydation von 1,5-Dienverbindungen. Tetrahedron **21**, 2353 (1965).

66. WALBA, D.M., M.D. WAND, and M.C. WILKES: Stereochemistry of the permanganate oxidation of 1,5-dienes. J. Amer. Chem. Soc. **101**, 4396 (1979).

67. BALDWIN, J.E., M.J. CROSSLEY, and E.-M.M. LEHTONEN: Stereospecificity of oxidative cycloaddition reactions of 1,5-dienes. J. Chem. Soc., Chem. Comm. 918 (1979).

68. WALBA, D.M., and P.D. EDWARDS: Total synthesis of ionophores. The monensin BC-rings via permanganate promoted stereospecific oxidative cyclisation. Tetrahedron Lett. **21**, 3531 (1980).

69. SPINO, C., and L. WEILER: A stereoselective synthesis of the tetrahydrofuran unit in ionomycin. Tetrahedron Lett. **28**, 731 (1987).

70. WALBA, D.M., C.A. PRZYBYLA, and C.B. WALKER: Total synthesis of ionophores 6. Asymmetric induction in the permanganate-promoted oxidative cyclisation of 1,5-dienes. J. Amer. Chem. Soc. **112**, 5624 (1990).

71. ROBINSON, J.A.: Enzymes of secondary metabolism in microorganisms. Chem. Soc. Revs. **17**, 383 (1988).

72. YUE, S., J.S. DUNCAN, Y. YAMAMOTO, and C.R. HUTCHINSON: Macrolide biosynthesis. Tylactone formation involves the processive addition of three carbon units. J. Amer. Chem. Soc. **109**, 1253 (1987).

73. CANE, D.E., and C.-C. YANG: Macrolide biosynthesis. 4. Intact incorporation of a chain-elongation intermediate into erythromycin. J. Amer. Chem. Soc. **109**, 1255 (1987).

74. CANE, D.E., and W.R. OTT: Macrolide biosynthesis. 5. Intact incorporation of a chain elongation intermediate into nargenicin. J. Amer. Chem. Soc. **110**, 4840 (1988).

75. KINOSHITA, K., S. TAKENAKA, and M. HAYASHI: Isolation of proposed intermediates in the biosynthesis of mycinamicins. J. Chem. Soc., Chem. Comm. 943 (1988).

76. TAKANO, S., Y. SEKIGUCHI, Y. SHIMAZAKI, and K. OGASAWARA: Stereochemistry of the proposed intermediates in the biosynthesis of mycinamicins. Tetrahedron Lett. **30**, 4001 (1989).

77. JONES, N.D., M.O. CHANEY, H.A. KIRST, G.M. WILD, R.H. BALTZ, R.L. HAMILL, and J.W. PASCHAL: Novel fermentation products from S. fradiae: X-ray crystal structure of 5-O-mycarosyltylactone and proof of the absolute configuration of tylosin. J. Antibiot. **35**, 420 (1982).

78. VANMIDDLESWORTH, F., D.V. PATEL, J. DONAUBAUER, F. GANNETT, and C.J. SIH: Synthesis of the putative biosynthetic triene precursor of monensin A. J. Amer. Chem. Soc. **107**, 2996 (1985).

79. PATEL, D.V., F. VANMIDDLESWORTH, J. DONAUBAUER, P. GANNETT, and C.J. SIH: Synthesis of the proposed penultimate biosynthetic triene intermediate of monensin A. J. Amer. Chem. Soc. **108**, 4603 (1986).

80. EVANS, D.A., and M. DIMARE: Asymmetric synthesis of premonensin, a potential intermediate in the biosynthesis of monensin. J. Amer. Chem. Soc. **108**, 2476 (1986).

81. HOLMES, D.S., U.C. DYER, S. RUSSELL, J.A. SHERRINGHAM, and J.A. ROBINSON: Total synthesis of a putative triene intermediate in monensin biosynthesis, activated as a caprylcysteamine thiol ester. Tetrahedron Lett. **29**, 6357 (1988).

82. HOLMES, D.S., J.A. SHERRINGHAM, U.C. DYER, S.T. RUSSELL, and J.A. ROBINSON: Synthesis of putative intermediates on the monensin biosynthetic pathway and incorporation experiments with the monensin producing organism. Helv. Chim. Acta **73**, 239 (1990).

83. BLOCK, M.H., and D.E. CANE: Synthesis of proposed chain-elongation intermediates of the monensin biosynthetic pathway. J. Org. Chem. **53**, 4923 (1988).

84. YOSHIZAWA, Y., Z. LI, P.B. REESE, and J.C. VEDERAS: Intact incorporation of acetate-derived di- and tetraketides during biosynthesis of dehydrocurvularin, a macrolide phytotoxin from *Aternaria cinerariae*. J. Amer. Chem. Soc. **112**, 3212 (1990).

85. WESTLEY, J.: Polyether ionophores: Versatile carboxylic acid ionophores produced by Streptomyces. Adv. Appl. Microbiol. **22**, 177 (1977).

86. WESTLEY, J.: Antibiotic structure and biosynthesis. J. Nat. Prod. **49**, 35 (1986).

87. CELMER, W.D.: Macrolide stereochemistry III. A configurational model for macrolide antibiotics. J. Amer. Chem. Soc. **87**, 1801 (1965).

88. CELMER, W.D.: Stereochemical problems in macrolide antibiotics. Pure Appl. Chem. **28**, 413 (1971).

89. WAKIL, S.J.: Fatty acid synthase, a proficient multifunctional enzyme. Biochemistry, **28**, 4523 (1989).

90. HUTCHINSON, C.R.: Biosynthetic studies of macrolide and polyether antibiotics. Acc. Chem. Res. **16**, 7 (1983).

91. RETEY, J., and J.A. ROBINSON: Stereospecificity in organic chemistry and enzymology. Vol. 13 in Monographs in Modern Chemistry. Weinheim: Verlag Chemie (1982).

92. HUTCHINSON, C.R., M.M. SHERMAN, A.G. MCINNES, J.A. WALTER, and J.C. VEDERAS: Biosynthesis of macrolides. 6. Mechanism of stereocontrol during the formation of lasalocid A. J. Amer. Chem. Soc. **103**, 5956 (1981).

93. SHERMANN, M.M., and C.R. HUTCHINSON: Biosynthesis of lasalocid A: Biochemical Mechanism for Assembly of the carbon framework. Biochemistry **26**, 438 (1987).

94. CANE, D.E., T.C. LIANG, P.B. TAYLOR, C. CHANG, and C.C. YANG: Macrolide Bio-synthesis. 3. Stereochemistry of the chain-elongation steps of erythromycin biosynthesis. J. Amer. Chem. Soc. **108**, 4957 (1986).

95. SOOD, G.R., J.A. ROBINSON, and A.A. AJAZ: Biosynthesis of the polyether antibiotic monensin A. Incorporation of $[2-^2H_2]-(R)-[2-^2H_1]-$ and $(S)-[2-^2H_1]$-propionate. J. Chem. Soc. Chem. Comm. 1421 (1984).

96. SOOD, G.R., D.M. ASHWORTH, A.A. AJAZ, and J.A. ROBINSON: Biosynthesis of the polyether antibiotic monensin A. Results from the incorporation of labelled acetate and propionate as a probe of the carbon chain assembly processes. J. Chem. Soc. Perkin I 3183 (1988).

97. HUNAITI, A.R., and P. E. KOLUTTUKUDY: Isolation and characterisation of an acyl-coenzyme A carboxylase from an erythromycin producing *Streptomyces erythreus*. Arch. Biochim. Biophys. **216**, 362 (1982).

98. VANCE, D., I. GOLGBERG, O. MITSUHASHI, K. BLOCH, S. OMURA, and S. NOMURA: Inhibition of fatty acid synthetases by the antibiotic cerulenin. Biochem. Biophys. Res. Comm. **48**, 649 (1972).

99. AGNOLO, G.D., I.S. ROSENFELD, J. AWAYA, S. OMURA, and P.R. VAGELOS: Inhibition .of fatty acid synthesis by the antibiotic cerulenin. Specific inactivation of β-ketoacyl-acyl carrier protein synthetase. Biochim. Biophys. Acta **326**, 155 (1973).

100. KAWAGUCHI, A., H. TOMODA, S. NOZOE, S. OMURA, and S. OKUDA: Mechanism of action of cerulenin on fatty acid synthase. Effect of cerulenin on iodoacetamide-induced malonyl-CoA decarboxylase activity. J. Biochemistry (Japan) **92**, 7 (1982).

101. FUNABASHI, H., A. KAWAGUCHI, H. TOMADA, S. OMURA, S. OKUDA, and S. IWASAKI: Binding site of cerulenin in fatty acid synthetase. J. Biochem. (Tokyo) **105**, 751 (1989).

102. OMURA, S., N. SADAKANE, Y. YANAKA, and H. MATSUBARA: Chimeramycins: new macrolide antibiotic produced by hybrid biosynthesis. J. Antibiot. **36**, 927 (1983).

103. SADAKANE, N., Y. TANAKA, and S. OMURA: Hybrid biosynthesis of derivative of

protylonolide and M-4365 by macrolide producing microorganisms. J. Antibiot. **35**, 680 (1982).

*104.* SADAKANE, N., Y. TANAKA, and S. OMURA: Hybrid biosynthesis of a new macrolide antibiotic by a daunomycin-producing microorganism. J. Antibiot. **36**, 921 (1983).

*105.* OMURA, S., N. SADAKANE, and H. MATSUBARA: Bioconversion and biosynthesis of 16-membered macrolide antibiotics. XXII. Biosynthesis of tylosin after protylonolide formation. Chem. Pharm. Bull. **30**, 223 (1982).

*106.* ROBERTS, G., and P.F. LEADLEY: [$^3$H]Tetrahydrocerulenin, a specific reagent for radio-labelling fatty acid synthases and related enzymes. FEBS Lett. **159**, 13 (1983).

*107.* ROBERTS, G., and P.F. LEADLEY: Use of [$^3$H]tetrahydrocerulenin to assay condensing enzyme activity in *Streptomyces erythraea*. Biochem. Soc. Trans. **12**, 642 (1984).

*108.* VOLPE, J.J., and P.R. VOGELOS: Saturated fatty acid biosynthesis and its regulation. Ann. Rev. Biochem. **42**, 21 (1973).

*109.* LYNEN, F: Biosynthesis of saturated fatty acids. Fed. Proc. **20**, 941 (1961).

*110.* FULCO, A.J.: Fatty acid metabolism in bacteria. Prog. Lipid Res. **22**, 133 (1983).

*111.* HOLAK, T.A., M. NILGES, J.H. PRESTEGARD, A.M. GRONENBORN, and G.M. CLORE: 3D Structure of acyl carrier protein in solution determined by NMR and the combined use of dynamical simulated annealing and distance geometry. Eur. J. Biochem. **175**, 9 (1988).

*112.* KIM, Y., and J.H. PRESTEGARD: A dynamic model for the structure of acyl carrier protein in solution. Biochemistry **28**, 8792 (1989).

*113.* MAYO, K.H., P.M. TYRELL, and J.H. PRESTEGARD: Acyl carrier protein from E. coli. I. Aspects of the solution structure as evidenced by proton nuclear overhauser experiments at 500 MHz. Biochemistry **22**, 4485 (1983).

*114.* MAYO, K.H., and J.H. PRESTEGARD: Acyl carrier protein from E. coli. Structural characterisation of short chain acylated acyl carrier proteins by NMR. Biochemistry **24**, 7834 (1985).

*115.* HOLAK, T.A., S.K. KEARSLEY, Y. KIM, and J.H. PRESTEGARD: 3D Structure of acyl carrier protein determined by NMR pseudoenergy and distance geometry calculations. Biochemistry **27**, 6135 (1988).

*116.* KIM, Y., and J.H. PRESTEGARD: Demonstration of a conformational equilibrium in acyl carrier protein from spinach using rotating frame NMR spectroscopy. J. Amer. Chem. Soc. **112**, 3707 (1990).

*117.* GARWIN, J.L., A.L. KLAGES, and J.E. CRONAN: Structural, enzymatic, and genetic studies of β-ketoacyl-acyl carrier protein synthases I and II of E. coli. J. Biol. Chem. **255**, 11949 (1980).

*118.* JACKOWSKI, S. and C.O. ROCK: Acetoacetyl-acyl carrier protein synthase, a potential regulator of fatty acid biosynthesis in bacteria. J. Biol. Chem. **262**, 7972 (1987).

*119.* JACKOWSKI, S., C.M. MURPHY, J.E. CRONAN, and C.O. ROCK: Acetoacetyl-acyl carrier protein synthase: A target for the antibiotic thiolactomycin. J. Biol. Chem. **264**, 7624 (1989).

*120.* KAUPPINEN, S., M. SIGGAARD-ANDERSEN, and P. VON WETTSTEIN-KNOWLES: β-Ketoacyl-ACP synthase I of E. coli: Nucleotide sequence of the *fabB* gene and identification of the cerulenin binding residue. Carlsberg Res. Comm. **53**, 357 (1988).

*121.* MATTICK, J.S. Y. TSUKAMOTO, J. NICKLESS, and S.J. WAKIL: The architecture of the animal fatty acid synthetase complex I. Proteolytic dissection and peptide mapping. J. Biol. Chem. **258**, 15291 (1983).

*122.* MATTICK, J.S., J.NICKLESS, M. MITZUGAKI, C.-Y. YOUNG, S. UCHIYAMA, and S.J. WAKIL: The architecture of the animal fatty acid synthetase complex II. Separation of the core and thioesterase functions and determination of the N–C orientation of the subunit. J. Biol. Chem. **258**, 15300 (1983).

123. Wong, H., J.S. Mattick, and S.J. Wakil: The architecture of the animal fatty acid synthetase complex III. Isolation and characterisation of β-ketoacyl reductase. J. Biol. Chem. **258**, 15305 (1983).

124. Tsukamoto, Y., H. Wong, J.S. Mattick, and S.J. Wakil: The architecture of the animal fatty acid synthetase complex IV. Mapping of active centres and model for the mechanism of action. J. Biol. Chem. **258**, 15312 (1983).

125. Holzer, K.P., W. Liu, and G.G. Hammes: Molecular cloning and sequencing of chicken liver fatty acid synthase cDNA. Proc. Nat. Acad. Sci. U.S.A. **86**, 4387 (1989).

126. Yuan, Z., W. Liu, and G.G. Hammes: Molecular cloning and sequencing of DNA complementary to chicken liver fatty acid synthase mRNA. Proc. Nat. Acad. Sci. U.S.A. **85**, 6328 (1988).

127. Amy, C.M., A. Witkowski, J. Naggert, B. Williams, Z. Randhawa, and S. Smith: Molecular cloning and sequencing of cDNAs encoding the entire rat fatty acid synthase. Proc. Nat. Acad. Sci. U.S.A. **86**, 3114 (1989).

128. Chirala, S.S., R. Kasturi, M. Pazirandeh, D.T. Stolow, W.-Y. Huang, and S.J. Wakil: A novel cDNA extension procedure. The isolation of chicken fatty acid synthase cDNA clones. J. Biol. Chem. **264**, 3750 (1989).

129. Chang, S.-Ik, and G. G. Hammes: Homology analysis of the protein sequences of fatty acid synthases from chicken liver, rat mammary gland, and yeast. Proc. Nat. Acad. Sci. U.S.A. **86**, 8373 (1989).

130. Pazirandeh, M., S.S. Chirala, W.-Y. Huang, and S.J. Wakil: Characterisation of recombinant thioesterase and acyl carrier protein domains of chicken fatty acid synthase expressed in E. coli. J. Biol. Chem. **264**, 18195 (1989).

131. Hopwood, D.A., and D.H. Sherman: Molecular genetics of polyketides and its comparison to fatty acid biosynthesis. Annu. Rev. Genet. **24**, 37 (1990).

132. Chirala, S.S., M.A. Kuziora, D.M. Spector, and S.G. Wakil: Complementation of mutations and nucleotide sequence of FAS1 gene encoding ß subunit of yeast fatty acid synthase. J. Biol. Chem. **262**, 4231 (1987).

133. Schweizer, M., L.M. Roberts, H.-J. Hoeltke, K. Takabayashi, E. Hoel-lerer, B. Hoffman, G. Muller, H. Kottig, and Schweizer: The pentafunctional FAS1 gene of yeasts, its nucleotide sequence and order of the catalytic domains. Mol. Gen. Genet. **203**, 479 (1986).

134. Mohamed, A.H., S.S. Chirala, N.H. Mody, W.Y. Huang, and S.J. Wakil: Primary structure of the multifunctional α subunit protein of yeast fatty acid synthase derived from FAS2 gene sequence. J. Biol. Chem. **262**, 12315 (1988).

135. Dimroth, P., H. Walter, and F. Lynen: Biosynthese von 6-Methylsalicylsäure. Eur. J. Biochem. **13**, 98 (1970).

136. Dimroth, P., E. Ringelman, and F. Lynen: 6-Methylsalicylic acid synthase from Penicillium patulum. Some catalytic properties of the enzyme and its relation to fatty acid synthase. Eur. J. Biochem. **68**, 591 (1976).

137. Beck, J., S. Ripka, A. Siegner, E. Schilltz, and E. Schweizer: The multifunctional 6-methylsalicyclic acid synthase gene of Penicillium patulum. Its gene structure relative to that of other polyketide synthases. Eur. J. Biochem. **192**, 487 (1990).

138. Jordan, P.M., and J.B. Spencer: Stereospecific manipulation of hydrogen atoms with opposite absolute orientations during the biosynthesis of the polyketide 6-methyl-salicylic acid from chiral malonates in Penicillium patulum. J. Chem. Soc., Chem. Comm. 238 (1990).

139. Spencer, J.B., and P.M. Jordan: Use of chiral malonates to determine the absolute configuration of the hydrogen atoms eliminated during the formation of 6-methyl-salicylic acid by 6-methylsalicylic acid synthase from Penicillium patulum. J. Chem. Soc., Chem. Comm. 1704 (1990).

*140.* WIESNER P., J. BECK, K.-F. BECK, S. RIPKA, and G. MUELLER: Isolation and sequence analysis of the fatty acid synthase FAS2 gene from *Penicillium patulum*. Eur. J. Biochem. **177**, 69 (1988).

*141.* MALPARTIDA, F., and D.A. HOPWOOD: Physical and genetic characterisation of the gene cluster for the antibiotic actinorhodin in *Streptomyces coelicolor* A3(2). Mol. Gen. Genet. **205**, 66 (1986).

*142.* MALPARTIDA, F., S.E. HALLAM, H.M. KIESER, H.MOTAMEDI, C.R. HUTCHINSON, M.J. BUTLER, D.A. SUGDEN, M. WARREN, C. MCKILLOP, C.R. BAILEY, G.O. HUMPREYS, and D.A. HOPWOOD: Homology between *Streptomyces* genes coding for synthesis of different polyketides used to clone antibiotic biosynthetic genes. Nature **325**, 818 (1987).

*143.* BUTLER, M.J., E.J. FRIEND, I. HUNTER, S. KACZMAREK, F.S. SUDGEN, and M.M. WARREN: Molecular cloning of resistance genes and architecture of a linked gene cluster involved in biosynthesis of oxytetracycline by *Streptomyces rimosus*. Mol. Gen. Genet. **215**, 231 (1989).

*144.* MOTAMEDI, H., and C.R. HUTCHINSON. Cloning and heterologous expression of a gene cluster for the biosynthesis of tetracenomycin C, anthracycline antitumor antibiotic of *Streptomyces glaucescens*. Proc. Nat. Acad. Sci. U.S.A. **84**, 4445 (1987).

*145.* STUTZMAN-ENGWALL, K.J., and C.R. HUTCHINSON: Multigene families for anthracycline antibiotic production in *Streptomyces peucetius*. Proc. Nat. Acad. Sci. U.S.A. **86**, 3135 (1989)

*146.* DONADIO, S. ET AL.: Genetic studies on erythromycin biosynthesis in *Saccharopolyspore erythraea*. In: Genetics and Molecular Biology of Industrial Microorganisms, pp. 53–59. Washington D.C.: Am. Soc. Microbiol. (1989).

*147.* STANZAK, R., P. MATSUSHIMA, R.H. BALTZ, and R.N. RAO.: Cloning and expression in *Streptomyces lividans* of clustered erythromycin biosynthesis genes from Streptomyces erythreus. Bio/Technology **4**, 229 (1986).

*148.* VARA, J., M. LEWANDOWSKI-SKARBEK, Y.-G. WANG, S. DONADIO, and C.R. HUTCHINSON: Cloning of genes governing the deoxysugar portion of the erythromycin pathway in *Saccharopolyspora erythraea*. J. Bacteriol. **171**, 5872 (1989).

*149.* DHILLON, N., R.S. HALE, J. CORTEZ, and P.F. LEADLEY: Molecular characterisation of a gene from *Saccharpolyspora erythraea* which is involved in erythromycin biosynthesis. Molec. Microbiol. **3**, 1405 (1989).

*150.* WEBER, J.M., and R. LOSICK: The use of a chromosome integration vector to map erythromycin resistance and production genes in *Saccharopolyspora erythraea*. Gene, **68**, 173 (1988).

*151.* STONESIFER, J., P. MATSUSHIMA, and R.H. BALTZ: High frequency conjugal transfer of tylosin genes and amplifiable DNA in *Streptomyces fradiae*. Mol. Gen. Genet. **202**, 348 (1986).

*152.* BECKMAN, R.J., K. COX, and E.T. SENO: A cluster of tylosin biosynthetic genes is interrupted by a structurally unstable segment containing four repeated sequences. See ref *146*, pp. 176–186.

*153.* RICHARDSON, M.A., S. KUHSTOSS, M. HUBER, L. FORD, O. GODFREY, J.R. TURNER, and R.A. RAO: Cloning of genes involved in spiramycin biosynthesis from *Streptomyces ambofaciens*. See ref. *146*, pp. 40–43.

*154.* EPP, J.K., M.L. HUBER, J.R. TURNER, and B.E. SCHONER: Molecular cloning and expression of carbomycin biosynthetic and resistance genes from *Streptomyces thermotolerans*. See ref. *146*, pp. 35–39.

*155.* STREICHER, S.L., ET AL.: Cloning genes for avermectin biosynthesis in *Streptomyces avermitilis*. See ref. *146*, pp. 44–52.

156. GIL. J.A., and D.A. HOPWOOD: Cloning and expression of a p-aminobenzoic acid synthetase gene of the candicidin-producing *Streptomyces griseus*. Gene **25**, 119 (1983).

157. HALLAM, S.E., F. MALPARTIDA, and D.A. HOPWOOD: Nucleotide sequence, transcription and deduced function of a gene involved in polyketide antibiotic biosynthesis in *Streptomyces coelicolor*. Gene **74**, 305 (1988).

158. SHERMAN, D.H., F. MALPARTIDA, M.J. BIBB, H.M. KIESER, M.J. BIBB, and D.A. HOPWOOD: Structure and deduced function of the granaticin-producing polyketide synthase gene cluster of *Streptomyces violaceoruber* TÜ22. EMBO J. **8**, 2717 (1989).

159. BIBB, M.J., S. BIRO, H. MOTAMEDI, J.F. COLLINS, and C.R. HUTCHINSON: Analysis of the nucleotide sequence of the *Streptomyces glaucescens tcmI* genes provides key information about the enzymology of polyketide antibiotic biosynthesis. EMBO J. **8**, 2727 (1989).

160. DAVIS, N.K., and K.F. CHATER: Spore colour in Streptomyces coelicolor A3(2) involves the developmentally regulated synthesis of a compound biosynthetically related to polyketide antibiotics. Molecul. Microbiol. **4**, 1679 (1990).

161. TUAN, J.S., J.M. WEBER, M.J. STAVER, J.O. LEUNG, S. DONADIO, and L. KATZ: Cloning of genes involved in erythromycin biosynthesis from *Saccharopolyspore erythraea* using a novel actinomycete-*E. coli* cosmid. Gene **90**, 21 (1990).

162. CORTES, J., S.F. HAYDOCK, G.A. ROBERTS, D.J. BEVITT, and P.F. LEADLAY: An unusually large multifunctional polypeptide in the erythromycin-producing polyketide synthase of *Saccharopolyspora erythraea*. Nature **348**, 176 (1990).

(*Received February 11, 1991*)

# Naturally Occurring Plant Coumarins

By R. D. H. Murray, Chemistry Department,
University of Glasgow, Glasgow, Scotland

## Contents

I. Scope of the Review . . . . . . . . . . . . . . . . . . . . . . 84

II. Progress in the Past Decade . . . . . . . . . . . . . . . . . . 84

III. Introduction to Tables . . . . . . . . . . . . . . . . . . . . 86

    Table 1. 7-Oxygenated Coumarins . . . . . . . . . . . . . . . . 88
        1.1. 6-Substituted-7-Oxygenated Coumarins. . . . . . . . . 101
        1.2. 8-Substituted-7-Oxygenated Coumarins. . . . . . . . . 114
        1.3. 5,6-Disubstituted-7-Oxygenated Coumarins . . . . . . . 131
        1.4. 6,8-Disubstituted-7-Oxygenated Coumarins . . . . . . . 132
    Table 2. 5,7-Dioxygenated Coumarins . . . . . . . . . . . . . . 133
    Table 3. 6,7-Dioxygenated Coumarins . . . . . . . . . . . . . . 145
    Table 4. 7,8-Dioxygenated Coumarins . . . . . . . . . . . . . . 151
    Table 5. 5,6,7-Trioxygenated Coumarins . . . . . . . . . . . . . 157
    Table 6. 5,7,8-Trioxygenated Coumarins . . . . . . . . . . . . . 159
    Table 7. 6,7,8-Trioxygenated Coumarins . . . . . . . . . . . . . 165
    Table 8. 5,6,7,8-Tetraoxygenated Coumarins . . . . . . . . . . . 174
    Table 9. 3-Substituted Coumarins . . . . . . . . . . . . . . . . 174
        9.1. 3-Aryl-Substituted Coumarins . . . . . . . . . . . . 178
    Table 10. 4-Substituted Coumarins . . . . . . . . . . . . . . . 180
        10.1. 4-Aryl-Substituted Coumarins . . . . . . . . . . . 182
    Table 11. Miscellaneous Coumarins. . . . . . . . . . . . . . . 195
        11.1. 3-Aryl Oxygenated Coumarins . . . . . . . . . . . 221
        11.2. Coumestans. . . . . . . . . . . . . . . . . . . . 223
    Table 12. Biscoumarins . . . . . . . . . . . . . . . . . . . . 230
    Table 13. Triscoumarins . . . . . . . . . . . . . . . . . . . . 236

Amendments/Additions to Entries in Reference *448*

    Table 1. 7-Oxygenated Coumarins . . . . . . . . . . . . . . . . 237
        1.1. 6-Substituted-7-Oxygenated Coumarins. . . . . . . . . 247
        1.2. 8-Substituted-7-Oxygenated Coumarins. . . . . . . . . 250
    Table 2. 5,7-Disubstituted Coumarins . . . . . . . . . . . . . . 254
    Table 3. 6,7-Disubstituted Coumarins . . . . . . . . . . . . . . 257

Table 4. 7,8-Disubstituted Coumarins . . . . . . . . . . . . . . . . .   259
Table 5. 5,6,7-Trisubstituted Coumarins . . . . . . . . . . . . . . .   260
Table 6. 5,7,8-Trisubstituted Coumarins . . . . . . . . . . . . . . .   261
Table 7. 6,7,8-Trisubstituted Coumarins . . . . . . . . . . . . . . .   262
Table 8. 5,6,7,8-Tetrasubstituted Coumarins . . . . . . . . . . . .   261
Table 9. 3-Substituted Coumarins . . . . . . . . . . . . . . . . .   263
Table 10. 4-Substituted Coumarins . . . . . . . . . . . . . . . . .   264
Table 11. Miscellaneous Coumarins . . . . . . . . . . . . . . . . .   270

Formula Index . . . . . . . . . . . . . . . . . . . . . . . . . . . .   271

Trivial Name Index . . . . . . . . . . . . . . . . . . . . . . . . .   275

References . . . . . . . . . . . . . . . . . . . . . . . . . . . . .   283

# I. Scope of the Review

This review of plant coumarins discovered between 1978 and 1989 has been compiled on the premise that the reader has access to the 1978 review in this journal by the author (*448*). For immediate access to every known plant coumarin the present review should be read side-by-side with the previous review.

In the earlier review, the 502 naturally occurring monomeric plant coumarins known in 1978 were tabulated principally according to the number and orientation of oxygen atoms on the benzenoid ring and then by the number of carbon atoms of a substituent and therein by the oxidation level of the substituent. For every entry, leading references to the isolation, structural elucidation, stereochemistry assignment where relevant, and synthesis where effected, of the coumarin were given. In setting out the information in the various tables herein, the author has presented the data in a similar format to that used earlier.

Whereas the previous review specifically excluded aryl-substituted and dimeric coumarins, these are now included. However, aflatoxins, benzocoumarins and ellagic acid derivatives have still not been included.

# II. Progress in the Past Decade

Comparison of the entries in each of Tables 1–9 in the present and the previous review (*448*) reveals the numbers of each type of coumarin discovered in the period 1978–1989 to be almost identical to those found during the 158 years since the isolation of coumarin itself in 1820.

An even more spectacular increase is revealed in Table 11. In 1978, the reviewer placed coumarins in which the pyrone ring was *O*- and/or *C*-

*References, pp. 283–316*

substituted in Tables 9, 10 and 11, commenting that these were growing rapidly in number. At that time six coumarins were known with an oxygen atom at C-4 and having 3- or 5-mono-, or 3,5-di-substituents. Quite remarkably the number of such coumarins, mostly from the Compositae, now exceeds 100.

The doubling in the number of known coumarins over the past decade is a reflection of improvements in isolation techniques and in the power and sophistication of spectroscopic techniques, especially high field nuclear magnetic resonance spectroscopy. Scrutiny of many of the publications cited bears witness to the elegance of such structural assignments on extremely small amounts of natural products. Apart from the many careful isolation procedures documented in the leading references, the reader's attention is directed towards some additional publications on high-performance liquid chromatography (*219, 220, 428, 627*) combined with ultrasensitive bioassay (*144*) and overpressure layer chromatography (*484*) which point the way to the isolation of even more coumarins in the future.

Many of the new coumarins isolated in the past decade could well have been anticipated as natural products being, for example, glycosides of a known coumarin aglycone with a different sugar, or a coumarin with a known side chain but at a higher or lower oxidation level. However, a number of new coumarins have been discovered the structures of which would have been less easy to predict. Amongst the latter the following deserve mention: the coumarin sulphates (**209**), (**234**) and (**358**) (*409*) from *Seseli libanotis*; ulismoncadin (**238**) (*200*) in which 7-hydroxycoumarin is prenylated at both C-5 and C-6; 7-acetoxy-4-methylcoumarin (**482**) (*94*), 7-phenylacetoxycoumarin (**69**) (*155*) and 7-acetoxycoumarin (**68**) (*634*), the first natural coumarin phenol esters; the first two natural 6-methyl-coumarins (**70** and **71**) (*569*); the non-oxygenated 3,4,7-trimethylcou-marin (**552**) (*371*) trivially named trigoforin; the first natural coumarins to carry isopropyl substituents (**665**) (*436*), (**483**) and (**673**) (*315*); and necatorin (**667**) (*607, 608*), a highly mutagenic azo compound from the wild edible mushroom, *Lactarias necator*.

Mention also should be made of the acrimarines (**112–114**) (*352*), the first naturally occurring acridone-coumarin dimers; naphthoherniarin (**111**) (*537, 538*), a link between coumarins and naphthoquinones and especially the structural (*45, 412, 413, 525*) and synthetic (*47, 411, 618, 619*) studies on the coumarinolignans (**443–448**) for example.

Many other papers on the synthesis of natural coumarins have been reported, a few of which are highlighted here: the many approaches to the antitumour agent geiparvarin (**792**) (*71, 160, 336, 343, 361, 547, 549, 630*), the intramolecular cycloaddition approach to eriobrucinol (**872**) (*172*), the syntheses of a vast array of *Mammea* coumarins (*170, 171*) including the insecticidal mammea E/BA (**939**), mammea E/BB (**937**) and surangin B (**938**) (*171*), the elegant development of multiple [3.3] sigmatropic rearrangements for constructing prenyl-substituted coumarins such as balsamiferone (**456**) (*134–136*) and a highly regioselective Fries rearrangement for geijerin (**838**) (*137*), the discovery of *exo*-dehydrochalepin (**459**) (*531*) in *Ruta chalepensis* by the specific synthesis of an authentic sample followed by a search for it in a root extract, and synthetic routes being established to natural coumarins such as the 1,1-dimethylallyl ether ponfolin (**299**) (*453*), and (**292**) (*452*) prior to their isolation from natural sources (*238, 332*).

The absolute stereochemistry of samarcandin (**825**) (*477*) and of many other bicyclofarnesyl ethers of 7-hydroxycoumarin has finally been resolved and the absolute configurations of the bicyclofarnesyl isofraxidin ethers (**423–442**) (*313*) have also been secured. The stereochemistry of the (*S*)-2-methylbutyryl moiety in mammea B/BB (**918**) (*83*) has been established synthetically. The structure of angelol, now called angelol A (**842**), has been revised (*54*) and its absolute stereochemistry and those of its congeners assigned (*53*).

The structure 7-methoxy-8-(15-hydroxypentadecyl)coumarin assigned to a constituent of *Erythrina stricta* bark (*589*), has been revised to octacosanyl 3'-hydroxy-4'-methoxycinnamate (*232*). Since the reported (*633*) ultraviolet maxima of rutalpinin (**489**), at 293 and 235 nm, differ significantly from those of 7,8-methylenedioxycoumarin, at 318, 262 and 254 nm (*448*), the reviewer believes the structural assignment must be in doubt. The reviewer is also of the opinion that libanotin A from *Libanotis buchtormensis* is not a 5-oxygenated coumarin as suggested (*648*) but is probably lomatin (**866**) (*448*), a well known *L. buchtormensis* constituent (*458*). Floribin, present in trace amounts in the bark of *Fraxinus floribunda*, has been assigned the structure 5-hydroxy-6-methoxycoumarin (*471*). If correct, this would represent the only 5,6-dioxygenated coumarin among some 1250 known monomeric plant coumarins. However, 6,7-dioxygenated coumarins are commonly found in *Fraxinus* species and, from the data presented for floribin and its intense blue fluorescence in ultraviolet light (*458*), the reviewer believes that it is impure scopoletin (*448*).

## III. Introduction to Tables

The arbitrary but biogenetically related classification of coumarins adopted in the earlier review (*448*) has again been used. It is based first on the number of nuclear oxygen atoms. Thereafter, within each Table, the entries are presented in the following order: (i) coumarins with acyclic substituents, (ii) dihydrofuranocoumarins, (iii) furanocoumarins, (iv) dihydropyranocoumarins, (v) pyranocoumarins. The coumarins of each subclass are listed in order of increasing number of carbon atom in the substituent and in increasing oxidation level within that group. Phenols are considered before ethers and glycosides and alcohols precede glycosides and esters.

In each case the plant source from which the coumarin was first isolated is given together with the year of isolation, trivial name and molecular formula. If one or more syntheses of a coumarin have been reported, these are included in the leading references. The $[\alpha]_\lambda^t$ and solvent columns refer to the specific rotation at t°C in the given solvent at a given wavelength, $\lambda$(nm). Where no wavelength is quoted the rotation has been measured at 589 nm.

Although many of the coumarins listed in the Tables have been isolated only from the plant source cited, some have been isolated from more than one plant. Such other sources are only given when another trivial name and/or different physical constants are given in the second publication.

The natural aryl-substituted coumarins which were not discussed previously (448) can be found in Tables 9.1, 10.1 and 11.1 with the 37 known coumestans in Table 11.2. The 34 biscoumarins isolated to date and recently reviewed (76) are recorded in Table 12.

The compound numbers given in parenthesis under (**786–943**) inclusive are the compound numbers which appear in the earlier review (448).

An asterisk (*) in the Structure column indicates that some aspect of the stereochemistry remains to be defined. In cases where the relative stereochemistry has been assigned the asterisk implies that the absolute sterochemistry has not yet been determined.

Table 1. 7-Oxygenated Coumarins

| Trivial name(s) | Year isolated | Structure | Formula | M.p. | $[\alpha]_\lambda^t$ | Solvent | Plant sources | Leading references |
|---|---|---|---|---|---|---|---|---|
| 1 | 1981 | | $C_{14}H_{14}O_4$ | 113–114 | $9^{21}$ | $CHCl_3$ | Coleonema album | (278) |
| 2 | 1981 | | $C_{14}H_{16}O_5$ | 94–95 | $31.1^{21}$ | $CHCl_3$ | Coleonema album | (278) |
| 3 | 1987 | | * $C_{14}H_{14}O_4$ | oil | | | Coleonema calycinum | (279) |
| 4 | 1987 | | $C_{15}H_{14}O_5$ | 170.5–172 | | | Coleonema calycinum | (279) |
| 5 | 1986 | | $C_{19}H_{22}O_4$ | 86 | | | Baccharis darwinii | (676) |
| 6 Prealtin A | 1989 | | $C_{21}H_{24}O_5$ | 163.3 | | | Aster prealtus | (653) |

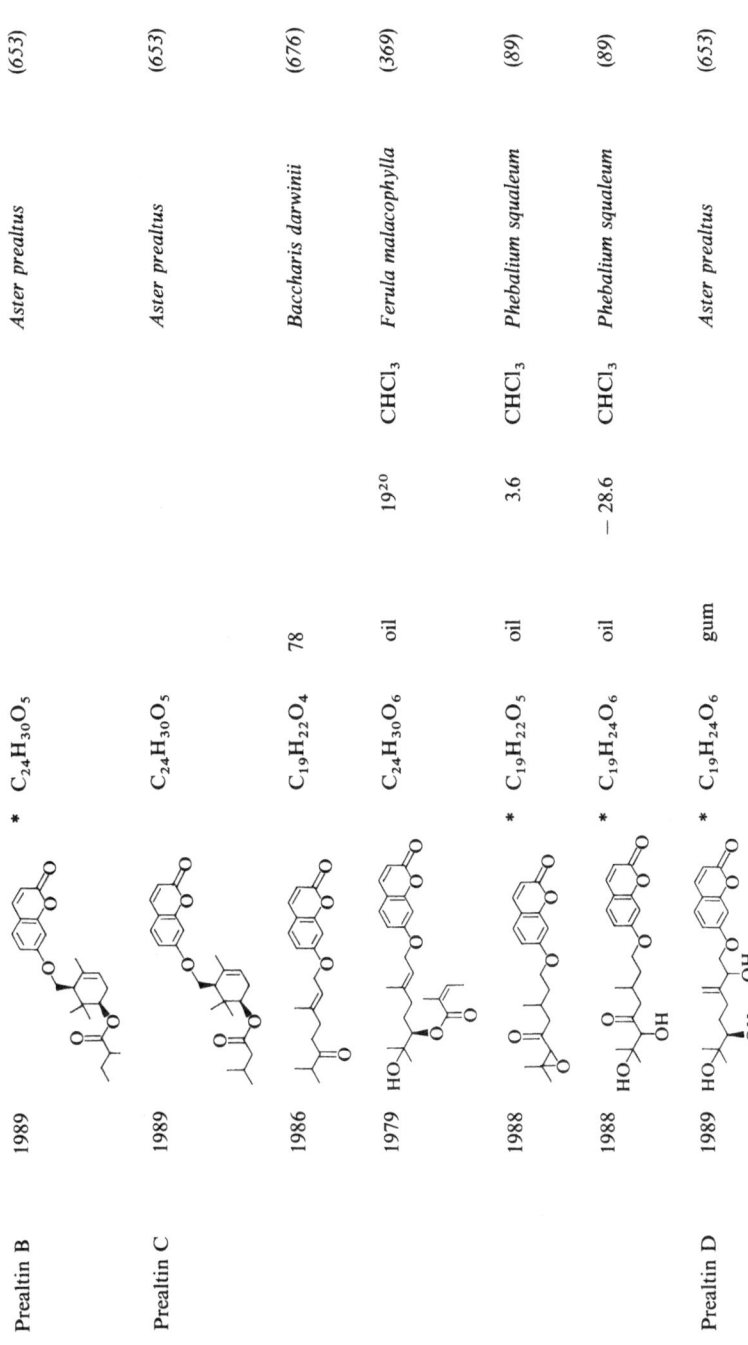

| | Name | Year | Formula | State | $[\alpha]_D$ | Solvent | Source | Ref. |
|---|---|---|---|---|---|---|---|---|
| 7 | Prealtin B | 1989 | * $C_{24}H_{30}O_5$ | | | | *Aster prealtus* | (653) |
| 8 | Prealtin C | 1989 | $C_{24}H_{30}O_5$ | | | | *Aster prealtus* | (653) |
| 9 | | 1986 | $C_{19}H_{22}O_4$ | 78 | | | *Baccharis darwinii* | (676) |
| 10 | | 1979 | $C_{24}H_{30}O_6$ | oil | $19^{20}$ | $CHCl_3$ | *Ferula malacophylla* | (369) |
| 11 | | 1988 | * $C_{19}H_{22}O_5$ | oil | 3.6 | $CHCl_3$ | *Phebalium squaleum* | (89) |
| 12 | | 1988 | * $C_{19}H_{24}O_6$ | oil | $-$ 28.6 | $CHCl_3$ | *Phebalium squaleum* | (89) |
| 13 | Prealtin D | 1989 | * $C_{19}H_{24}O_6$ | gum | | | *Aster prealtus* | (653) |

90

Table 1. (Continued)

| | Trivial name(s) | Year isolated | Structure | Formula | M.p. | $[\alpha]_\lambda^t$ | Solvent | Plant sources | Leading references |
|---|---|---|---|---|---|---|---|---|---|
| 14 | Tadshiferin | 1976 | | $C_{24}H_{30}O_4$ | 68–70 | $8^{23}$ | $CHCl_3$ | *Ferula tadshikorum* | (505) |
| 15 | Deacetyltadshikorin | 1984 | | $C_{24}H_{30}O_5$ | 64–66 | | | *Ferula tadshikorum* | (86) |
| 16 | Tadshikorin | 1976 | | $C_{26}H_{32}O_6$ | oil | $15^{23}$ | | *Ferula tadshikorum* | (505) |
| 17 | Asacoumarin A | 1989 | | $C_{24}H_{30}O_5$ | oil | 7.0 | $CHCl_3$ | *Ferula assafoetida* | (359) |
| 18 | Karatavicin | 1983 | | $C_{26}H_{34}O_6$ | 60–62 | $-21^{25}$ | EtOH | *Ferula karatavica* | (469) |
| 19 | Latilobinol | 1979 | | $C_{24}H_{30}O_4$ | 121–122 | $-63^{18}$ | $CHCl_3$ | *Prangos latiloba* | (1, 4) |

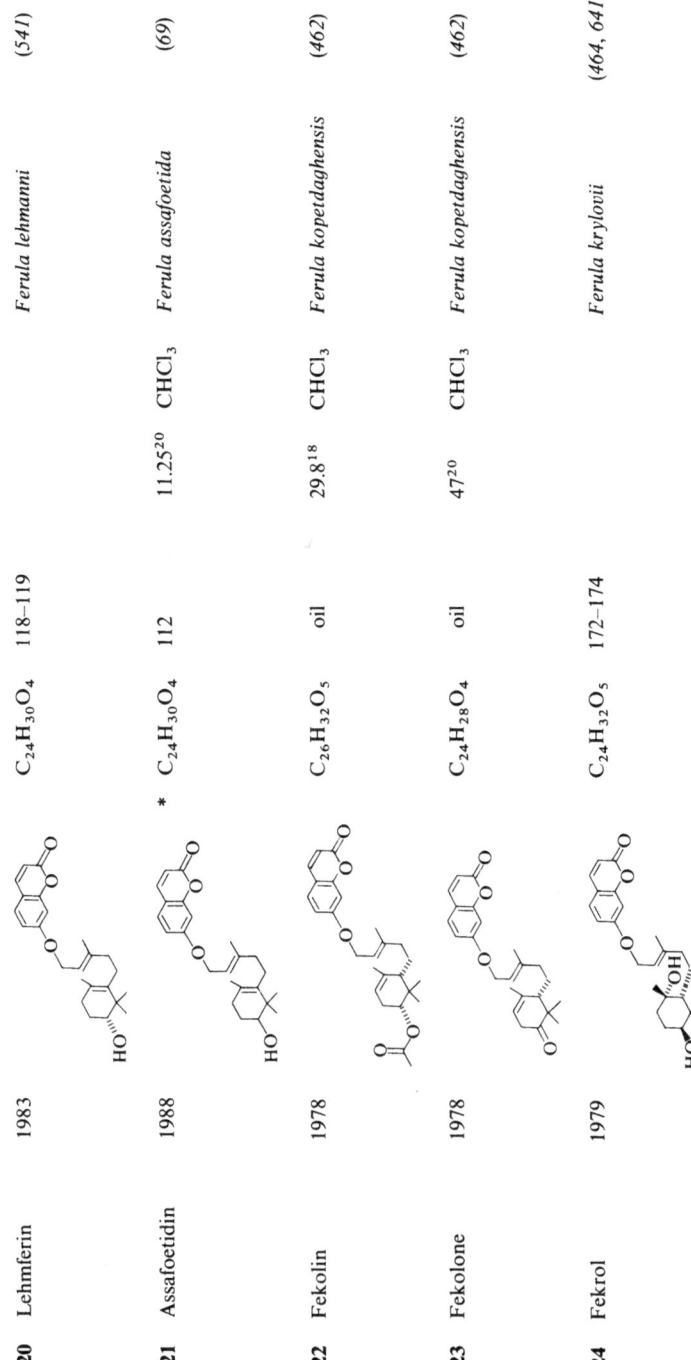

| | | | | | | | | |
|---|---|---|---|---|---|---|---|---|
| **20** | Lehmferin | 1983 | $C_{24}H_{30}O_4$ | 118–119 | | | *Ferula lehmanni* | (541) |
| **21** | Assafoetidin | 1988 | * $C_{24}H_{30}O_4$ | 112 | $11.25^{20}$ | CHCl$_3$ | *Ferula assafoetida* | (69) |
| **22** | Fekolin | 1978 | $C_{26}H_{32}O_5$ | oil | $29.8^{18}$ | CHCl$_3$ | *Ferula kopetdaghensis* | (462) |
| **23** | Fekolone | 1978 | $C_{24}H_{28}O_4$ | oil | $47^{20}$ | CHCl$_3$ | *Ferula kopetdaghensis* | (462) |
| **24** | Fekrol | 1979 | $C_{24}H_{32}O_5$ | 172–174 | | | *Ferula krylovii* | (464, 641) |

*Table 1.* (Continued)

| | Trivial name(s) | Year isolated | Structure | Formula | M.p. | $[\alpha]_\lambda^t$ | Solvent | Plant sources | Leading references |
|---|---|---|---|---|---|---|---|---|---|
| 25 | Kopeolone | 1982 | | $C_{24}H_{30}O_5$ | 125–126 | $70^{18}$ | EtOH | *Ferula kopetdaghensis* | (464) |
| 26 | Foliferin | 1978 | | * $C_{24}H_{34}O_6$ | 240–241 | $128^{21}$ | pyridine | *Ferula foliosa* | (356) |
| 27 | Asacoumarin B | 1989 | | * $C_{24}H_{30}O_5$ | | − 13.3 | CHCl$_3$ | *Ferula assafoetida* | (359) |
| 28 | Fekrynol | 1981 | | * $C_{24}H_{32}O_4$ | oil | $18^{16}$ | EtOH | *Ferula krylovii* | (644) |
| 29 | Fekrynol acetate | 1981 | | * $C_{26}H_{34}O_5$ | 80–82 | − $26.8^{22}$ | EtOH | *Ferula krylovii* | (644) |

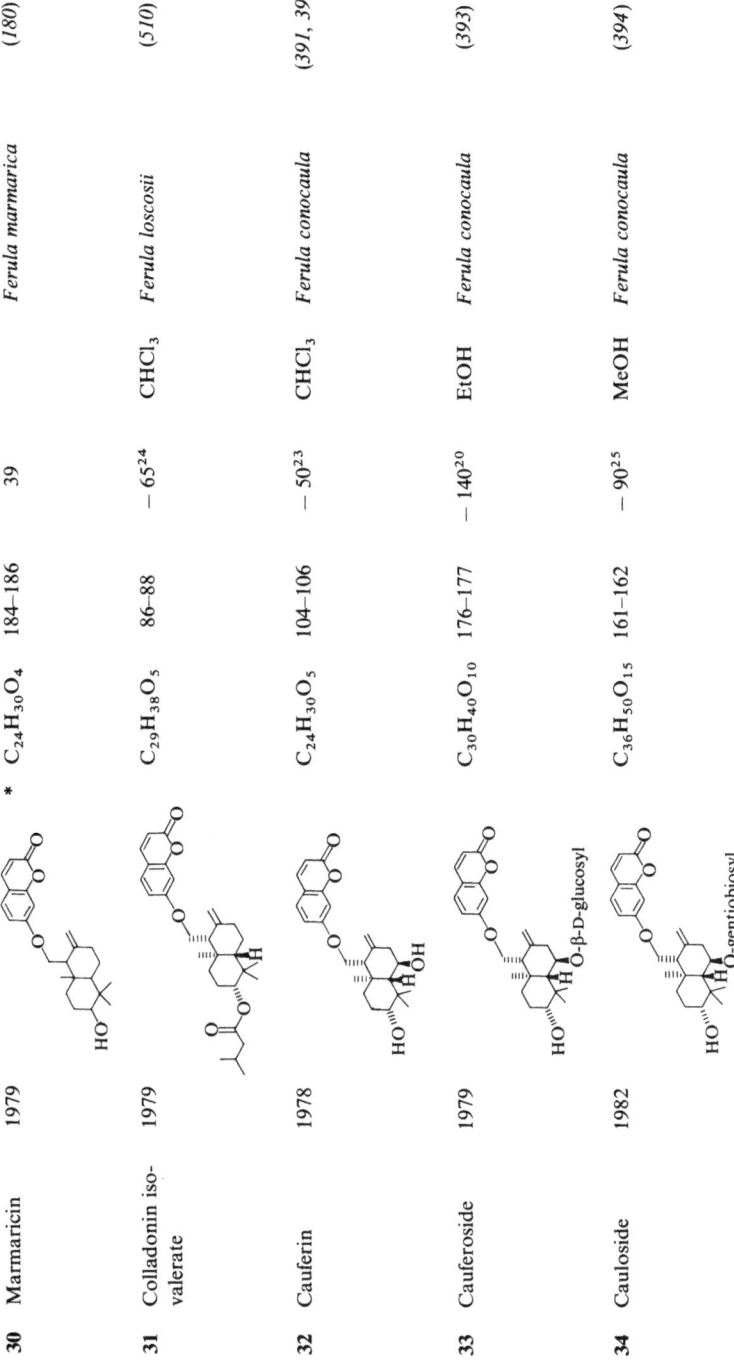

| No. | Name | Year | Structure | Formula | mp (°C) | $[\alpha]$ | Solvent | Source | Ref. |
|---|---|---|---|---|---|---|---|---|---|
| 30 | Marmaricin | 1979 | * | $C_{24}H_{30}O_4$ | 184–186 | 39 | | Ferula marmarica | (180) |
| 31 | Colladonin iso-valerate | 1979 | | $C_{29}H_{38}O_5$ | 86–88 | $-65^{24}$ | CHCl$_3$ | Ferula loscosii | (510) |
| 32 | Cauferin | 1978 | | $C_{24}H_{30}O_5$ | 104–106 | $-50^{23}$ | CHCl$_3$ | Ferula conocaula | (391, 392) |
| 33 | Cauferoside | 1979 | | $C_{30}H_{40}O_{10}$ | 176–177 | $-140^{20}$ | EtOH | Ferula conocaula | (393) |
| 34 | Cauloside | 1982 | | $C_{36}H_{50}O_{15}$ | 161–162 | $-90^{25}$ | MeOH | Ferula conocaula | (394) |

Table 1. (Continued)

| | Trivial name(s) | Year isolated | Structure | Formula | M.p. | $[\alpha]_\lambda^t$ | Solvent | Plant sources | Leading references |
|---|---|---|---|---|---|---|---|---|---|
| 35 | Feterin | 1978 | | $C_{26}H_{32}O_6$ | 155–158 | $-52.02^{20}$ | $CHCl_3$ | Ferula teterrima | (504) |
| 36 | Cauferidin | 1978 | | $C_{24}H_{28}O_4$ | 184–185.5 | $-60^{23}$ | $CHCl_3$ | Ferula conocaula | (391, 392, 541) |
| 37 | Lehmferidin Ferilin | 1983 1984 | | $C_{24}H_{28}O_4$ | 173–174 172–174 | $-66.9^{20}$ | $CHCl_3$ | Ferula lehmanni Ferula iliensis | (541) (640) |
| 38 | | 1986 | | $C_{29}H_{36}O_6$ | 91–94 | | | Ferula sinaica | (38) |

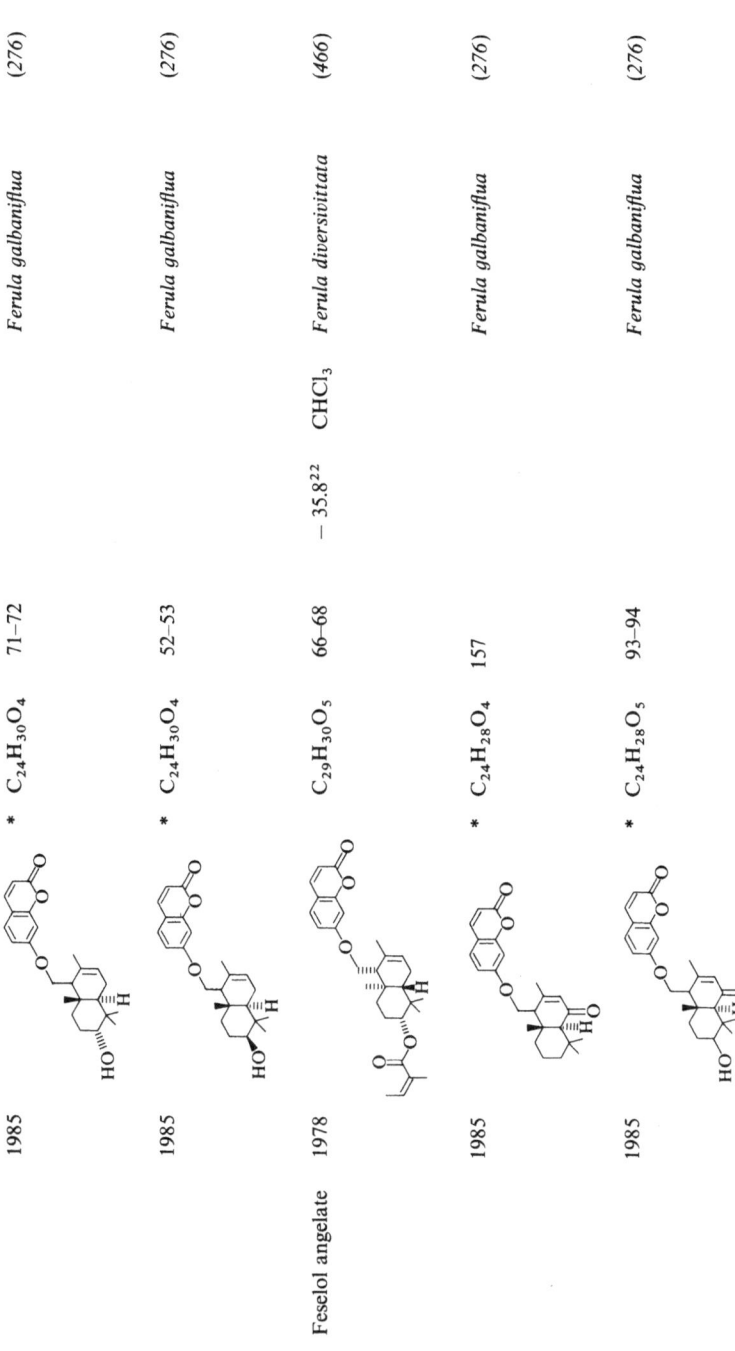

| 39 | | 1985 | | * | $C_{24}H_{30}O_4$ | 71–72 | | *Ferula galbaniflua* | (276) |
| 40 | | 1985 | | * | $C_{24}H_{30}O_4$ | 52–53 | | *Ferula galbaniflua* | (276) |
| 41 | Feselol angelate | 1978 | | | $C_{29}H_{30}O_5$ | 66–68 | $-35.8^{22}$ CHCl$_3$ | *Ferula diversivittata* | (466) |
| 42 | | 1985 | | * | $C_{24}H_{28}O_4$ | 157 | | *Ferula galbaniflua* | (276) |
| 43 | | 1985 | | * | $C_{24}H_{28}O_5$ | 93–94 | | *Ferula galbaniflua* | (276) |

*Table 1.* (Continued)

| | Trivial name(s) | Year isolated | Structure | Formula | M.p. | $[\alpha]_\lambda^t$ | Solvent | Plant sources | Leading references |
|---|---|---|---|---|---|---|---|---|---|
| **44** | Ferocaulidin | 1978 | | $C_{24}H_{28}O_5$ | 75–77 | $-75^{20}$ | EtOH | *Ferula conocaula* | (390) |
| **45** | Ferocaulicin | 1978 | | $C_{26}H_{30}O_6$ | 161–162.5 | $-120^{20}$ | $CHCl_3$ | *Ferula conocaula* | (390) |
| **46** | Ferocaulin | 1978 | | $C_{24}H_{28}O_5$ | 120–121 | $-20^{20}$ | EtOH | *Ferula conocaula* | (390) |
| **47** | Ferocaulinin | 1978 | | $C_{24}H_{28}O_5$ | 84–86 | $-40^{20}$ | EtOH | *Ferula conocaula* | (390) |

| No. | Name | Year | | Formula | mp (°C) | $[\alpha]$ | Solvent | Source | Ref. |
|---|---|---|---|---|---|---|---|---|---|
| 48 | Conferoside | 1979 | | $C_{30}H_{38}O_{10}$ | 195–197 | $-110^{20}$ | EtOH | *Ferula conocaula* | (393) |
| 49 | Tavimolidin | 1979 | | $C_{29}H_{34}O_6$ | 144–146 | $-110^{20}$ | $CHCl_3$ | *Peucedanum mogolta-vicum* | (368) |
| 50 | Feshurin | 1979 | | $C_{24}H_{32}O_5$ | 212–214 | $-50^{21}$ | pyridine | *Ferula schtschurow-skiana* | (357, 477, 542) |
| 51 | Kokanidin | 1982 | | $C_{26}H_{34}O_6$ | 189–191 | $-30^{18}$ | EtOH | *Ferula kokanica* | (465) |
| 52 | | 1985 | | * $C_{24}H_{32}O_5$ | 175 | | | *Ferula galbaniflua* | (276) |

Table 1. (Continued)

| Trivial name(s) | Year isolated | Structure | Formula | M.p. | $[\alpha]_\lambda^t$ | Solvent | Plant sources | Leading references |
|---|---|---|---|---|---|---|---|---|
| **53** Nevskone | 1978 | | $C_{24}H_{30}O_5$ | 180–181 | | | *Ferula nevskii* | (56, 477) |
| **54** Cauferinin | 1979 | | $C_{24}H_{32}O_6$ | 204–206 | $37.5^{20}$ | EtOH | *Ferula conocaula* | (392) |
| **55** Fepaldin | 1980 | | $C_{24}H_{32}O_5$ | 219–221 | $-55.5^{22}$ | CHCl$_3$ | *Ferula pallida* | (543) |
| **56** Ferukrin | 1977 | | $C_{24}H_{32}O_5$ | 211–213 | $30^{22}$ | EtOH | *Ferula krylovii* | (463, 503, 642) |

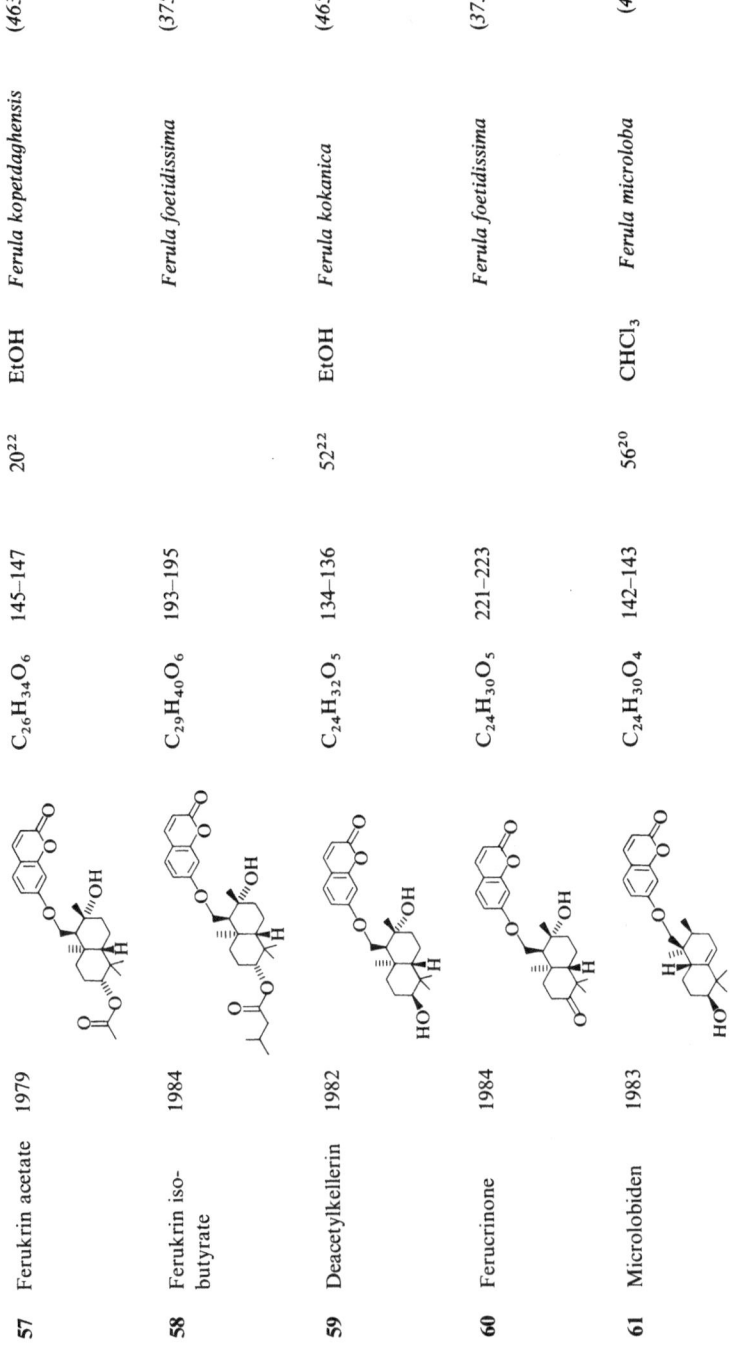

| | | | | | | |
|---|---|---|---|---|---|---|
| **57** | Ferukrin acetate | 1979 | $C_{26}H_{34}O_6$ | 145–147 | $20^{22}$ | EtOH | Ferula kopetdaghensis | (463) |
| **58** | Ferukrin iso-butyrate | 1984 | $C_{29}H_{40}O_6$ | 193–195 | | | Ferula foetidissima | (375) |
| **59** | Deacetylkellerin | 1982 | $C_{24}H_{32}O_5$ | 134–136 | $52^{22}$ | EtOH | Ferula kokanica | (465) |
| **60** | Ferucrinone | 1984 | $C_{24}H_{30}O_5$ | 221–223 | | | Ferula foetidissima | (375) |
| **61** | Microlobiden | 1983 | $C_{24}H_{30}O_4$ | 142–143 | $56^{20}$ | $CHCl_3$ | Ferula microloba | (468) |

Table 1. (Continued)

| Trivial name(s) | Year isolated | Structure | Formula | M.p. | $[\alpha]_\lambda^t$ | Solvent | Plant sources | Leading references |
|---|---|---|---|---|---|---|---|---|
| **62** Fecarpin | 1978 | | $C_{24}H_{32}O_4$ | 166–168 | $-20^{20}$ | $CHCl_3$ | *Ferula microcarpa* | (262) |
| **63** Microlobin | 1983 | | $C_{24}H_{30}O_5$ | 150–151 | $49^{20}$ | $CHCl_3$ | *Ferula microloba* | (467) |
| **64** Kamolonol | 1984 | | $C_{24}H_{30}O_5$ | | $17^{20}$ | $CHCl_3$ | *Ferula assafoetida* | (314) |
| **65** | 1985 | R = β-apiosyl-1 → 6-β-D-glucosyl | $C_{20}H_{24}O_{12}$ | 141–142 | $-38$ | | *Gmelina arborea* | (556) |
| Adicardin | 1986 | | | 138–139 | $-121.3$ | MeOH | *Adina cordifolia* | (49) |

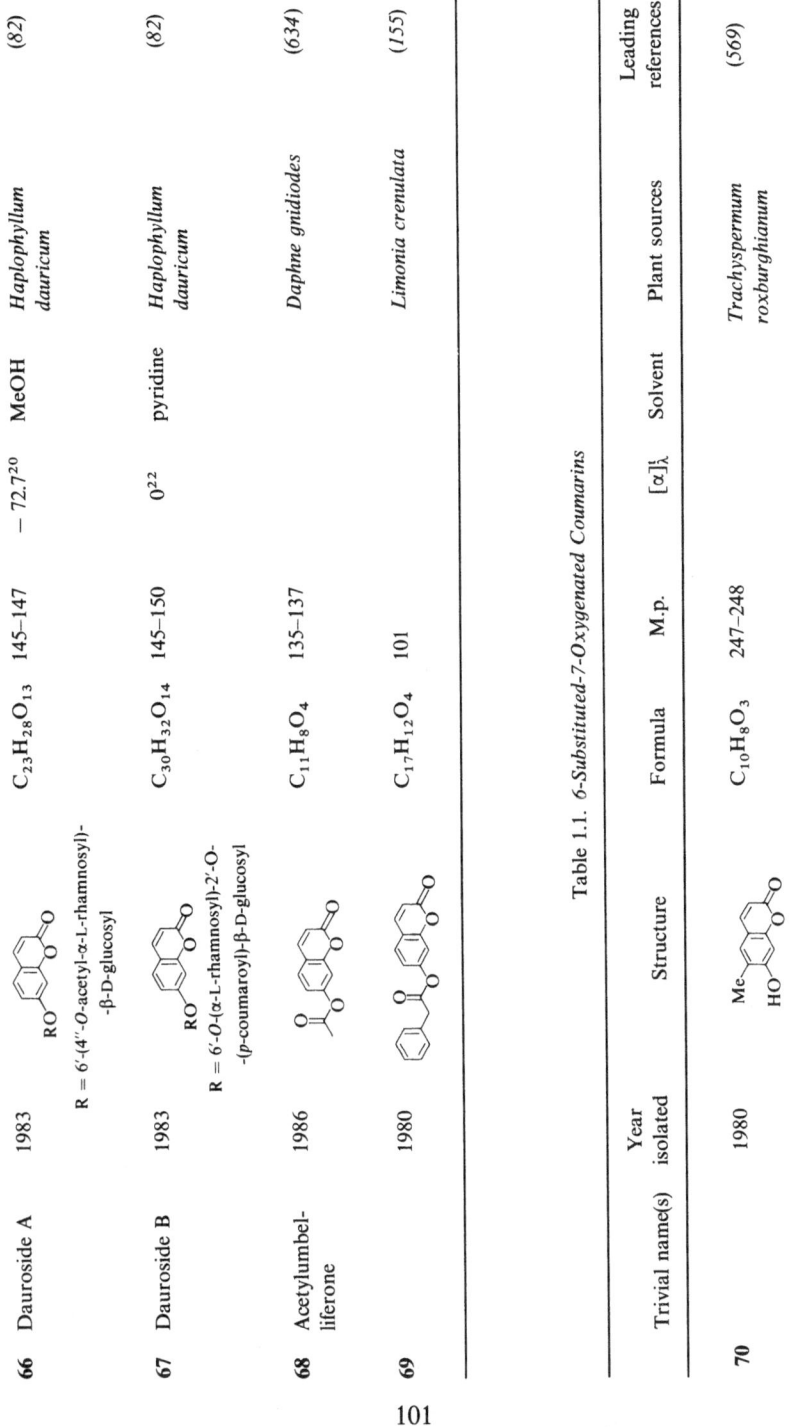

| | Trivial name(s) | Year isolated | Structure | Formula | M.p. | $[\alpha]_\lambda^t$ | Solvent | Plant sources | Leading references |
|---|---|---|---|---|---|---|---|---|---|
| 66 | Dauroside A | 1983 | RO... R = 6'-(4''-O-acetyl-α-L-rhamnosyl)-β-D-glucosyl | $C_{23}H_{28}O_{13}$ | 145–147 | $-72.7^{20}$ | MeOH | *Haplophyllum dauricum* | (82) |
| 67 | Dauroside B | 1983 | RO... R = 6'-O-(α-L-rhamnosyl)-2'-O-(p-coumaroyl)-β-D-glucosyl | $C_{30}H_{32}O_{14}$ | 145–150 | $0^{22}$ | pyridine | *Haplophyllum dauricum* | (82) |
| 68 | Acetylumbelliferone | 1986 | | $C_{11}H_8O_4$ | 135–137 | | | *Daphne gnidiodes* | (634) |
| 69 | | 1980 | | $C_{17}H_{12}O_4$ | 101 | | | *Limonia crenulata* | (155) |

Table 1.1. *6-Substituted-7-Oxygenated Coumarins*

| | Trivial name(s) | Year isolated | Structure | Formula | M.p. | $[\alpha]_\lambda^t$ | Solvent | Plant sources | Leading references |
|---|---|---|---|---|---|---|---|---|---|
| 70 | | 1980 | Me... HO | $C_{10}H_8O_3$ | 247–248 | | | *Trachyspermum roxburghianum* | (569) |

Table 1.1. (Continued)

| Trivial name(s) | Year isolated | Structure | Formula | M.p. | $[\alpha]_\lambda$ | Solvent | Plant sources | Leading references |
|---|---|---|---|---|---|---|---|---|
| 71 | 1980 | | $C_{11}H_{10}O_3$ | 135–137 | | | Trachyspermum roxburghianum | (423, 569) |
| 72 | 1988 | | $C_{11}H_{10}O_4$ | oil | | | Citrus funadoko | (351) |
| 73 | 1988 | | $C_{10}H_6O_4$ | oil | | | Citrus medica | (332) |
| 74 Buntansin | 1989 | | $C_{11}H_8O_5$ | 216–217 | | | Citrus grandis | (316) |
| 75 | 1981 | | $C_{11}H_8O_4$ | 179–181 | | | Apium petroselinum | (39) |
| 76 Tenuidin | 1980 | | $C_{14}H_{14}O_4$ | 74–75.5 | | | Haplophyllum tenue | (14) |
| 77 Funadonin | 1988 | | * $C_{14}H_{14}O_5$ | oil | 9.43 | CHCl$_3$ | Citrus funadoko | (351) |

| No. | Name | Year | Structure | Formula | mp (°C) | [α]D | Solvent | Source | Ref. |
|---|---|---|---|---|---|---|---|---|---|
| 78 | Swietenol | 1977 | | $C_{14}H_{16}O_4$ | | | | *Chloroxylon swietenia* | (442) |
| 79 | Dihydrosuberenol | 1982 | | $C_{15}H_{18}O_4$ | 105 | | | *Limonia acidissima* | (261) |
| 80 | | 1977 | | $C_{14}H_{14}O_4$ | 159–160 | | | *Boenning hausenia albiflora* | (578) |
| 81 | | 1977 | | $C_{14}H_{14}O_4$ | 139–140.5 | | | *Boenning hausenia albiflora* | (578) |
| 82 | (Z)-Suberenol | 1988 | | $C_{15}H_{16}O_4$ | oil | | | *Citrus funadoko* | (351) |
| 83 | | 1979 | | $C_{15}H_{16}O_4$ | 112–113 | $27.3^{25}$ | $CHCl_3$ | *Amyris balsamifera* | (132) |
| 84 | Tamarin | 1977 | * | $C_{15}H_{16}O_4$ | 112–113 | | | *Ruta pinnata* | (272) |
| 85 | (E)-Methyl-suberenol | 1988 | | $C_{16}H_{18}O_4$ | oil | | | *Citrus funadoko* | (351) |

103

*Table 1.1.* (Continued)

| Trivial name(s) | Year isolated | Structure | Formula | M.p. | $[\alpha]_\lambda$ | Solvent | Plant sources | Leading references |
|---|---|---|---|---|---|---|---|---|
| 86 (Z)-Methyl-suberenol | 1988 | (structure) | $C_{16}H_{18}O_4$ | oil | | | *Citrus funadoko* | (351) |
| 87 Ethylsuberenol | 1983 | (structure) | $C_{17}H_{20}O_4$ | 154–155 | | | *Citrus sinensis* | (660) |
| 88 | 1977 | (structure) | $C_{15}H_{16}O_4$ | 89–90 | | | *Ruta pinnata* | (272) |
| 89 | 1979 | (structure) | $C_{15}H_{18}O_5$ | 139–140 | 65.1 | MeOH | *Hippomarathrum cristatum* | (197) |
| 90 | 1978 | (structure) | $C_{20}H_{26}O_{10}$ | 205–205.5 | $-12.3^{24}$ | MeOH | *Seseli montanum* | (407) |
| 91 | 1978 1985 | (structure) | $C_{20}H_{26}O_{10}$ | 160–162 | $16.9^{23}$ | MeOH | *Seseli montanum Phlojodicarpus turczaninovii* | (407) (244) |
| 92 | 1978 | (structure) | $C_{20}H_{26}O_{10}$ | | $13.4^{25}$ | MeOH | *Seseli montanum* | (407) |

| No. | Name | Year | Structure | Formula | mp (°C) | $[\alpha]$ | Solvent | Source | Ref. |
|-----|------|------|-----------|---------|---------|------------|---------|--------|------|
| 93 | Grandivittinol | 1977 | * | $C_{19}H_{22}O_6$ | 143–145 | | | *Seseli grandivittatum* | (7) |
| 94 | Citrubuntin | 1988 | | $C_{15}H_{14}O_3$ | 115–117 | | | *Citrus grandis* | (659) |
| 95 | | 1983 | | $C_{15}H_{18}O_6$ | 139–141 | $-18.4^{18}$ | EtOH | *Angelica pubescens* | (385) |
| 96 | Angelol F | 1982 | mixture of 2 diastereoisomers | $C_{20}H_{26}O_7$ | oil | $-38.0^{22}$ | EtOH | *Angelica pubescens* | (53, 54) |
| 97 | Angelol C | 1982 | mixture of 2 diastereoisomers | $C_{20}H_{26}O_7$ | 113–114 | $-119.8^{18}$ | EtOH | *Angelica pubescens* | (53, 54) |
| 98 | Angelol E | 1982 | | $C_{20}H_{26}O_7$ | 83–84 | $-51.6^{22}$ | EtOH | *Angelica pubescens* | (53, 54) |

Table 1.1. (Continued)

| Trivial name(s) | Year isolated | Structure | Formula | M.p. | $[\alpha]_\lambda$ | Solvent | Plant sources | Leading references |
|---|---|---|---|---|---|---|---|---|
| 99 Angelol I | 1983 | | $C_{20}H_{26}O_7$ | 113–115 | $-127.6^{20}$ | EtOH | *Coelopleurum gmelinii* | (386) |
| 100 Angelol B | 1982 | | $C_{20}H_{24}O_7$ | 143–144 | $-229.1^{22}$ | EtOH | *Angelica pubescens* | (53, 54) |
| 101 Angeloside A | 1983 | R = β-D-glucosyl | $C_{26}H_{34}O_{12}$ | oil | $-91.7^{24}$ | EtOH | *Coelopleurum gmelinii* | (386) |
| 102 Angelol H | 1982 | | $C_{20}H_{26}O_7$ | oil | $-74.7^{22}$ | EtOH | *Angelica pubescens* | (53, 54) |

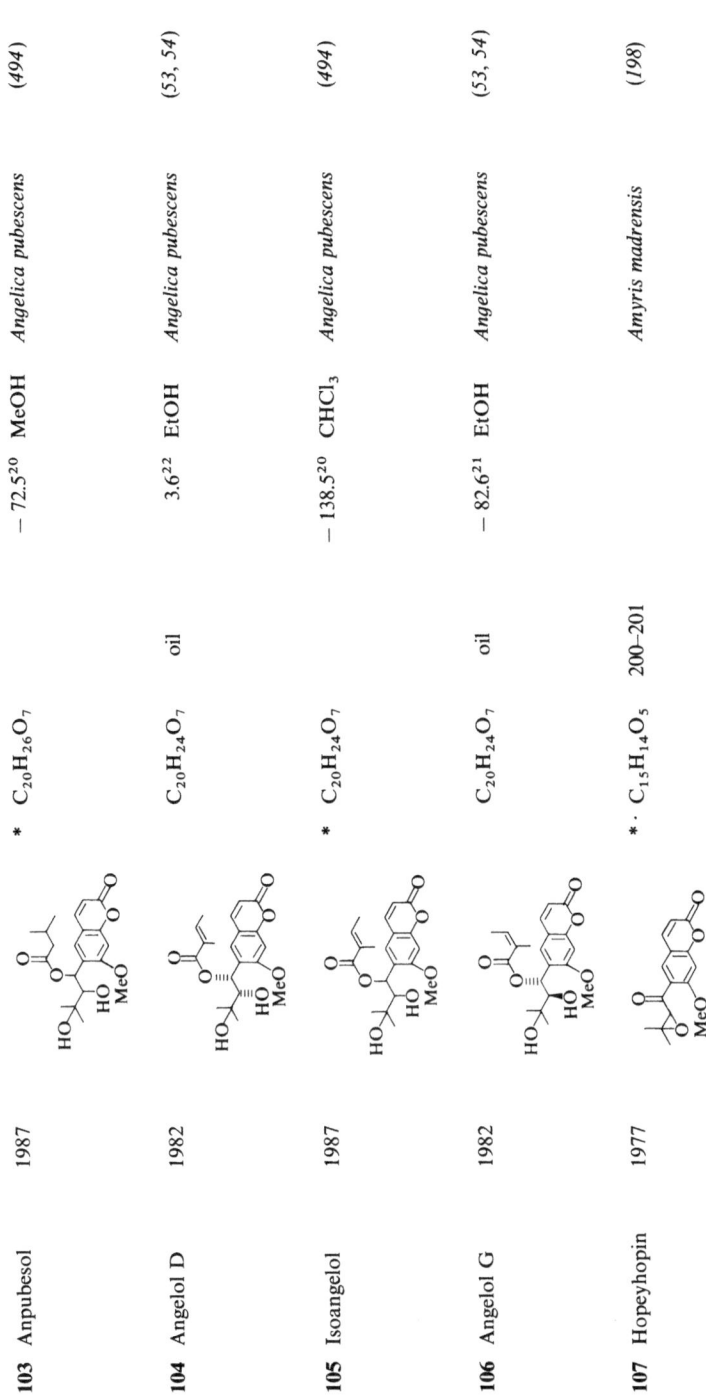

| | Name | Year | Formula | mp/oil | $[\alpha]$ | Solvent | Source | Ref. |
|---|---|---|---|---|---|---|---|---|
| 103 | Anpubesol | 1987 | * $C_{20}H_{26}O_7$ | | $-72.5^{20}$ | MeOH | *Angelica pubescens* | *(494)* |
| 104 | Angelol D | 1982 | $C_{20}H_{24}O_7$ | oil | $3.6^{22}$ | EtOH | *Angelica pubescens* | *(53, 54)* |
| 105 | Isoangelol | 1987 | * $C_{20}H_{24}O_7$ | | $-138.5^{20}$ | CHCl$_3$ | *Angelica pubescens* | *(494)* |
| 106 | Angelol G | 1982 | $C_{20}H_{24}O_7$ | oil | $-82.6^{21}$ | EtOH | *Angelica pubescens* | *(53, 54)* |
| 107 | Hopeyhopin | 1977 | *· $C_{15}H_{14}O_5$ | 200–201 | | | *Amyris madrensis* | *(198)* |

107

Table 1.1. (Continued)

| No. | Trivial name(s) | Year isolated | Structure | Formula | M.p. | $[\alpha]_\lambda^t$ | Solvent | Plant sources | Leading references |
|---|---|---|---|---|---|---|---|---|---|
| 108 | Dihydromelin A | 1984 | | * $C_{15}H_{14}O_6$ | gum | | | *Micromelum minutum* | (177) |
| 109 | Dihydromelin B | 1984 | | * $C_{15}H_{14}O_6$ | gum | | | *Micromelum minutum* | (177) |
| 110 | Acetyldihydro-melin A | 1984 | | * $C_{17}H_{16}O_7$ | 66–68 | | | *Micromelum minutum* | (177) |
| 111 | Naphtho-herniarin | 1980 | | $C_{22}H_{16}O_6$ | 242–244 | | | *Ruta graveolens* | (528, 537, 538) |

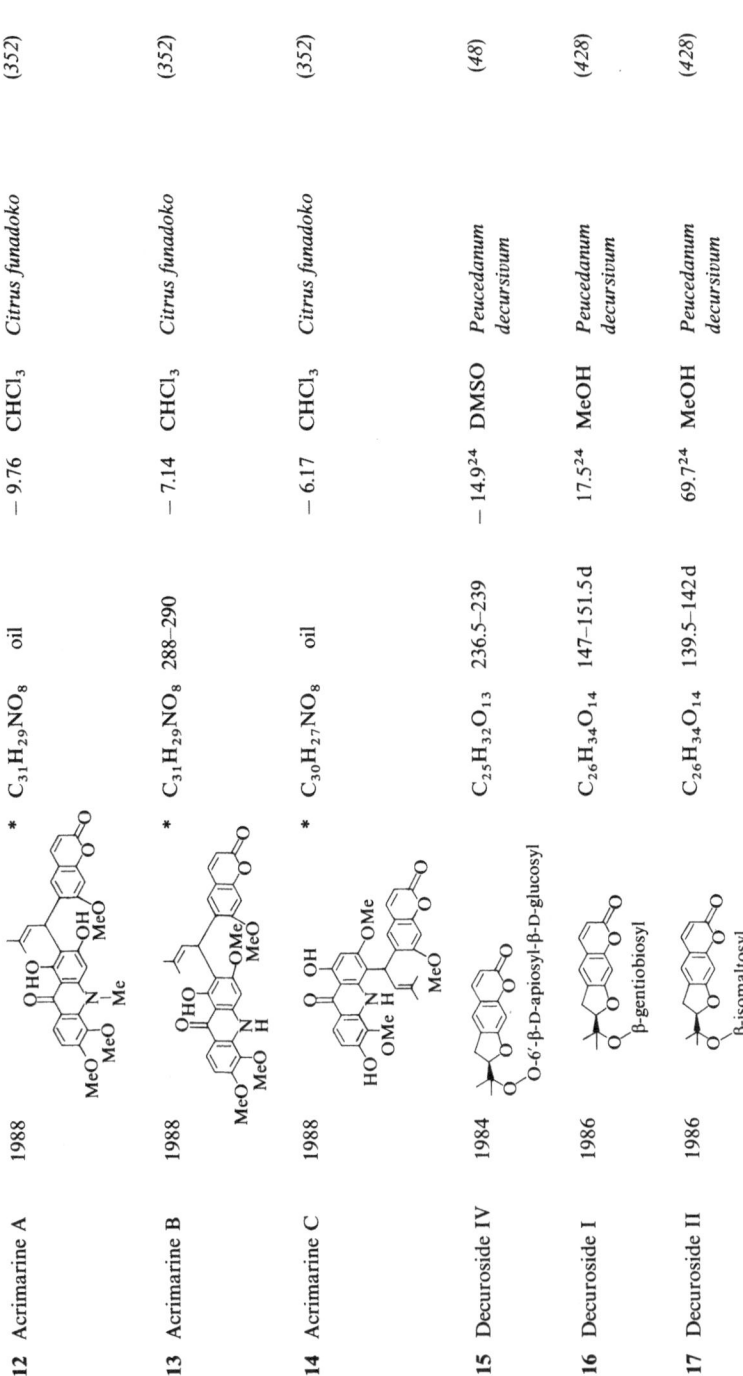

| No. | Name | Year | | Formula | mp | [α] | Solvent | Source | Ref. |
|---|---|---|---|---|---|---|---|---|---|
| 112 | Acrimarine A | 1988 | * | $C_{31}H_{29}NO_8$ | oil | − 9.76 | $CHCl_3$ | *Citrus funadoko* | (352) |
| 113 | Acrimarine B | 1988 | * | $C_{31}H_{29}NO_8$ | 288–290 | − 7.14 | $CHCl_3$ | *Citrus funadoko* | (352) |
| 114 | Acrimarine C | 1988 | * | $C_{30}H_{27}NO_8$ | oil | − 6.17 | $CHCl_3$ | *Citrus funadoko* | (352) |
| 115 | Decuroside IV | 1984 | | $C_{25}H_{32}O_{13}$ | 236.5–239 | − 14.9[24] | DMSO | *Peucedanum decursivum* | (48) |
| 116 | Decuroside I | 1986 | | $C_{26}H_{34}O_{14}$ | 147–151.5d | 17.5[24] | MeOH | *Peucedanum decursivum* | (428) |
| 117 | Decuroside II | 1986 | | $C_{26}H_{34}O_{14}$ | 139.5–142d | 69.7[24] | MeOH | *Peucedanum decursivum* | (428) |

110

Table 1.1. (Continued)

| Trivial name(s) | Year isolated | Structure | Formula | M.p. | $[\alpha]_\lambda^t$ | Solvent | Plant sources | Leading references |
|---|---|---|---|---|---|---|---|---|
| **118** Decuroside III | 1986 | β-maltosyl | $C_{26}H_{34}O_{14}$ | 143.5–146.5 d | $85.5^{24}$ | MeOH | *Peucedanum decursivum* | (428) |
| **119** Marmesin acetate | 1981 | | $C_{16}H_{16}O_5$ | 135–136 | $13.5^{25}$ | $CHCl_3$ | *Amyris elemifera* | (133) |
| **120** Marmesin iso-valerate | 1979 | | $C_{19}H_{22}O_5$ | 115–116 | $-10.5$ | $CHCl_3$ | *Ferulago granatensis* | (194) |
| **121** (−)-Sprenge-lianin | 1986 1985 | | $C_{19}H_{20}O_5$ | 111.5–113 | $-33^{23}$ | $CHCl_3$ | *Cachrys sicula* *Eryngium ilicifolium* | (277) (509) |
| **122** Tortuosinin | 1983 | racemic | $C_{19}H_{20}O_5$ | 109 | | | *Seseli tortuosum* | (5) |

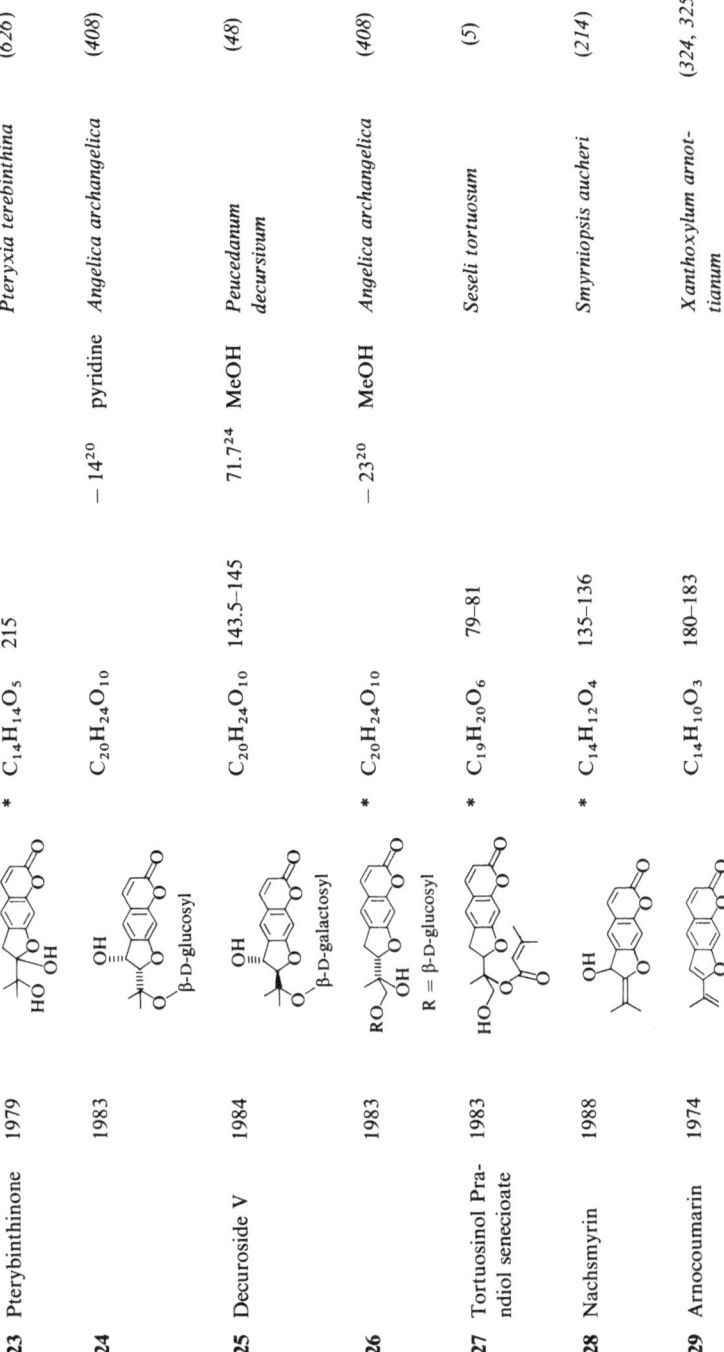

| No. | Name | Year | | Formula | mp | [α] | Solvent | Species | Ref. |
|---|---|---|---|---|---|---|---|---|---|
| 123 | Pterybinthinone | 1979 | * | $C_{14}H_{14}O_5$ | 215 | | | Pteryxia terebinthina | (626) |
| 124 | | 1983 | | $C_{20}H_{24}O_{10}$ | | $-14^{20}$ | pyridine | Angelica archangelica | (408) |
| 125 | Decuroside V | 1984 | | $C_{20}H_{24}O_{10}$ | 143.5–145 | $71.7^{24}$ | MeOH | Peucedanum decursivum | (48) |
| 126 | | 1983 | * | $C_{20}H_{24}O_{10}$ | | $-23^{20}$ | MeOH | Angelica archangelica | (408) |
| 127 | Tortuosinol Pradiol senecioate | 1983 | * | $C_{19}H_{20}O_6$ | 79–81 | | | Seseli tortuosum | (5) |
| 128 | Nachsmyrin | 1988 | * | $C_{14}H_{12}O_4$ | 135–136 | | | Smyrniopsis aucheri | (214) |
| 129 | Arnocoumarin | 1974 | | $C_{14}H_{10}O_3$ | 180–183 | | | Xanthoxylum arnotianum | (324, 325) |

Table 1.1. (Continued)

| Trivial name(s) | Year isolated | Structure | Formula | M.p. | $[\alpha]_\lambda^i$ | Solvent | Plant sources | Leading references |
|---|---|---|---|---|---|---|---|---|
| **130** Dihydroxanthyletin | 1982 | | $C_{14}H_{14}O_3$ | 122–124 | | | *Seseli tortuosum* | (269) |
| **131** (−)-3'R-Decursinol | 1977 | | $C_{14}H_{14}O_4$ | 180–181.5 | $-8.98^{22}$ | $CHCl_3$ | *Seseli grandivittatum* | (7) |
| Aegelinol | 1977 1978 | | | 179–180 175–177 | $-96^{23}$ $-12^{20}$ | pyridine $CHCl_3$ | *Cachrys libanotis* *Aegle marmelos* | (508) (156) |
| **132** Grandivittin | 1977 | | $C_{19}H_{20}O_5$ | oil | $-83.2^{22}$ | $CHCl_3$ | *Seseli grandivittatum* | (7) |
| **133** Aegelinol benzoate | 1985 | | $C_{21}H_{18}O_5$ | | $-95.6^{20}$ | $CHCl_3$ | *Eryngium campestre* | (221) |
| **134** (−)-Methyldecursidinol | 1978 | | $C_{15}H_{16}O_5$ | 93.5–94.5 | $-92.14^{27}$ | MeOH | *Peucedanum arenarium* | (402) |
| **135** Pd-C-I | 1982 | | $C_{19}H_{20}O_6$ | 194–196.5 | $138.5^{25}$ | dioxan | *Peucedanum decursivum* | (550) |

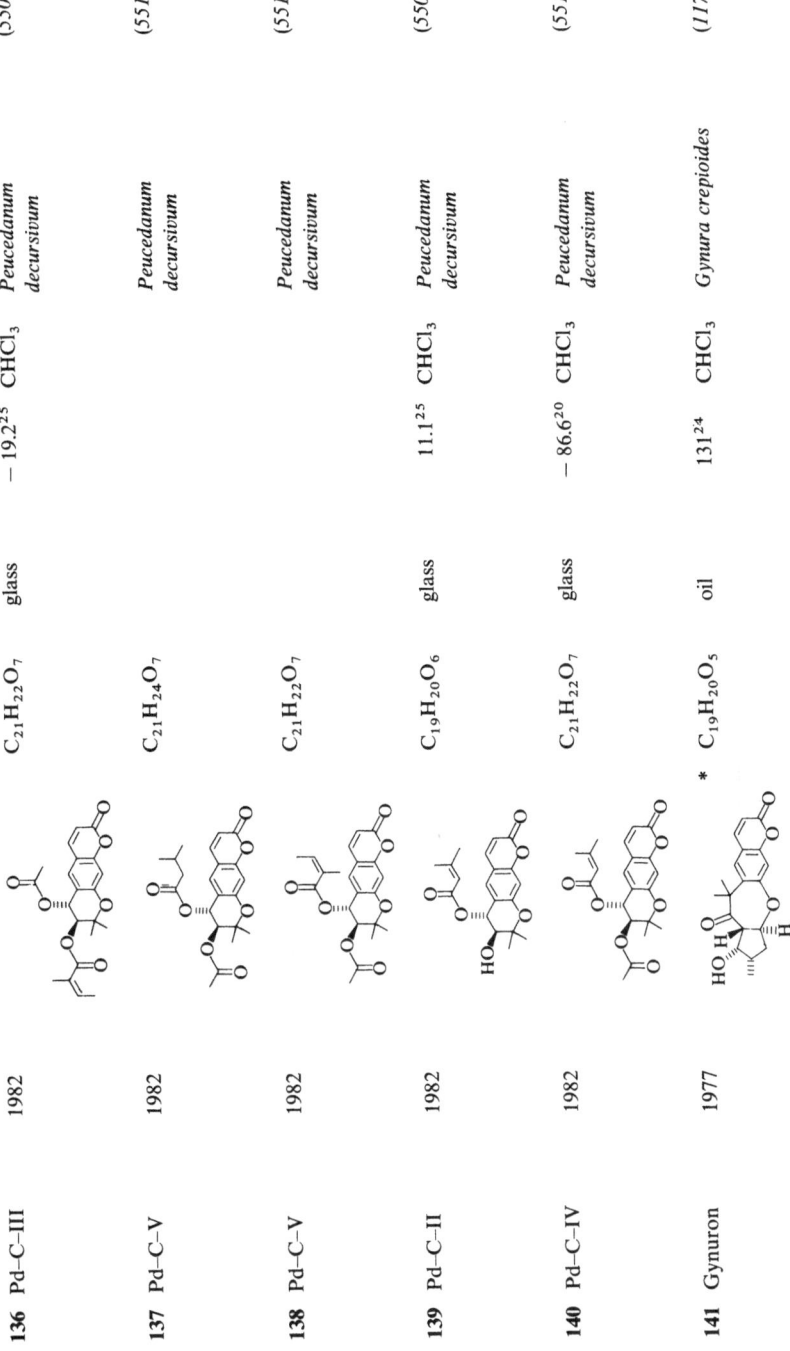

| No. | Name | Year | Formula | State | $[\alpha]_D$ | Solvent | Source | Ref. |
|---|---|---|---|---|---|---|---|---|
| 136 | Pd-C-III | 1982 | $C_{21}H_{22}O_7$ | glass | $-19.2^{25}$ | $CHCl_3$ | *Peucedanum decursivum* | (550) |
| 137 | Pd-C-V | 1982 | $C_{21}H_{24}O_7$ | | | | *Peucedanum decursivum* | (551) |
| 138 | Pd-C-V | 1982 | $C_{21}H_{22}O_7$ | | | | *Peucedanum decursivum* | (551) |
| 139 | Pd-C-II | 1982 | $C_{19}H_{20}O_6$ | glass | $11.1^{25}$ | $CHCl_3$ | *Peucedanum decursivum* | (550) |
| 140 | Pd-C-IV | 1982 | $C_{21}H_{22}O_7$ | glass | $-86.6^{20}$ | $CHCl_3$ | *Peucedanum decursivum* | (551) |
| 141 | Gynuron | 1977 | * $C_{19}H_{20}O_5$ | oil | $131^{24}$ | $CHCl_3$ | *Gynura crepioides* | (117) |

Table 1.2. 8-Substituted-7-Oxygenated Coumarins

| | Trivial name(s) | Year isolated | Structure | Formula | M.p. | $[\alpha]_\lambda^t$ | Solvent | Plant sources | Leading references |
|---|---|---|---|---|---|---|---|---|---|
| 142 | Murrayacarpin A | 1989 | | $C_{11}H_{10}O_4$ | 163–166 d | | | Murraya paniculata | (663) |
| 143 | | 1977 | | $C_{11}H_8O_4$ | 110–112 | | | Peucedanum hispanicum | 263 |
| | Paniculal | 1987 | | | 214–216 | | | Murraya paniculata | (322) |
| 144 | Kiyomal | 1989 | | $C_{12}H_{10}O_4$ | | | | Citrus unshiu | (333) |
| 145 | Hassanon | 1988 | | $C_{14}H_{14}O_4$ | 132–134 | | | Citrus medica | (332) |
| 146 | Osthenon | 1987 | | $C_{14}H_{12}O_4$ | 134–136 | | | Murraya exotica | (328) |
| | | 1989 | | | 141–142 | | | Citrus tamurana | (354) |
| 147 | cis-Osthenon | 1987 | | $C_{14}H_{12}O_4$ | oil | | | Murraya paniculata | (331) |

| No. | Compound | Year | Structure | Formula | mp/oil | [α] | Solvent | Source | Ref. |
|---|---|---|---|---|---|---|---|---|---|
| 148 | Isomurralonginol acetate | 1987 | | * $C_{17}H_{18}O_5$ | oil | 18.8 | $CHCl_3$ | *Murraya exotica* | (328) |
| 149 | Isomurralonginol nicotinate | 1987 | | * $C_{21}H_{19}NO_5$ | oil | 31.8 | $CHCl_3$ | *Murraya paniculata* | (331) |
| 150 | Microminutin | 1983 | | $C_{15}H_{12}O_5$ | 154–155 | | | *Micromelum minutum* | (622, 647) |
| 151 | Paniculin | 1987 | | $C_{15}H_{16}O_5$ | 236–238 | | | *Murraya paniculata* | (321, 322) |
| 152 | | 1980 | | $C_{26}H_{34}O_{13}$ | | | | *Glehnia littoralis* | (554) |
| 153 | Myrsellin | 1977 | | * $C_{19}H_{22}O_4$ | 99 | | | *Myrtopsis sellingii* | (305) |

115

## Table 1.2. (Continued)

| Trivial name(s) | Year isolated | Structure | Formula | M.p. | $[\alpha]_\lambda^t$ | Solvent | Plant sources | Leading references |
|---|---|---|---|---|---|---|---|---|
| 154 Myrsellinol | 1977 | | * $C_{19}H_{24}O_5$ | | 44 | $CHCl_3$ | *Myrtopsis sellingii* | (305) |
| 155 | 1985 | | $C_{24}H_{30}O_3$ | 44–45 | | | *Sargentia greggii* | (438) |
| 156 Anisocoumarin F | 1989 | | * $C_{19}H_{22}O_4$ | oil | $27.5^{25}$ | $CHCl_3$ | *Clausena anisata* | (478) |
| 157 Tortuosidin | 1983 | | * $C_{24}H_{30}O_5$ | | $45^{22}$ | EtOH | *Selseli tortuosum* | (5) |
| 158 | 1978 | | * $C_{14}H_{14}O_4$ | | | | *Seseli tortuosum* | (265) |

| No. | Name | Year | Structure | Formula | mp (°C) | [α] | Solvent | Source | Ref. |
|---|---|---|---|---|---|---|---|---|---|
| 159 | CM-c$_2$<br>Murraol | 1985<br>1987 | | $C_{15}H_{16}O_4$ | 138–141<br>105–107 | | | *Cnidium monnieri*<br>*Murraya exotica* | (460, 665)<br>(328, 330, 528) |
| 160 | Anisocoumarin E | 1989 | | $C_{19}H_{22}O_4$ | oil | | | *Clausena anisata* | (478) |
| 161 | Triphasiol | 1981 | * | $C_{19}H_{24}O_6$ | 85 | | | *Triphasia trifoliata* | (195) |
| 162 | Merillin | 1989 | | $C_{15}H_{18}O_5$ | 68 | $-15.6^{20}$ | $CHCl_3$ | *Merrillia caloxylon* | (674) |
| 163 | | 1978 | * | $C_{14}H_{16}O_5$ | 142–143 | 117 | pyridine | *Seseli tortuosum* | (265) |
| 164 | | 1986 | * | $C_{20}H_{26}O_{10}$ | 224–226 | | | *Phlojodicarpus sibiricus* | (245) |

Table 1.2. (Continued)

| Trivial name(s) | Year isolated | Structure | Formula | M.p. | $[\alpha]_\lambda^t$ | Solvent | Plant sources | Leading references |
|---|---|---|---|---|---|---|---|---|
| 165 Anisocou-marin G | 1989 | | * $C_{19}H_{24}O_5$ | oil | $32.5^{25}$ | $CHCl_3$ | Clausena anisata | (478) |
| 166 Tortuoside | 1989 | | $C_{20}H_{26}O_{10}$ | 212–216 | 18.9 | MeOH | Seseli tortuosum | (143) |
| 167 | 1987 | | * $C_{21}H_{28}O_{10}$ | | | | | Citrus flavedo | (434) |
| 168 Coumurrin | 1987 | | * $C_{16}H_{18}O_6$ | 128–130 | 14.1 | | Murraya paniculata | (321, 322) |
| 169 Murrayatin | 1983 | | $C_{20}H_{26}O_6$ | 108–110 | $104.7^{22}$ | $CHCl_3$ | Murraya exotica | (72) |

118

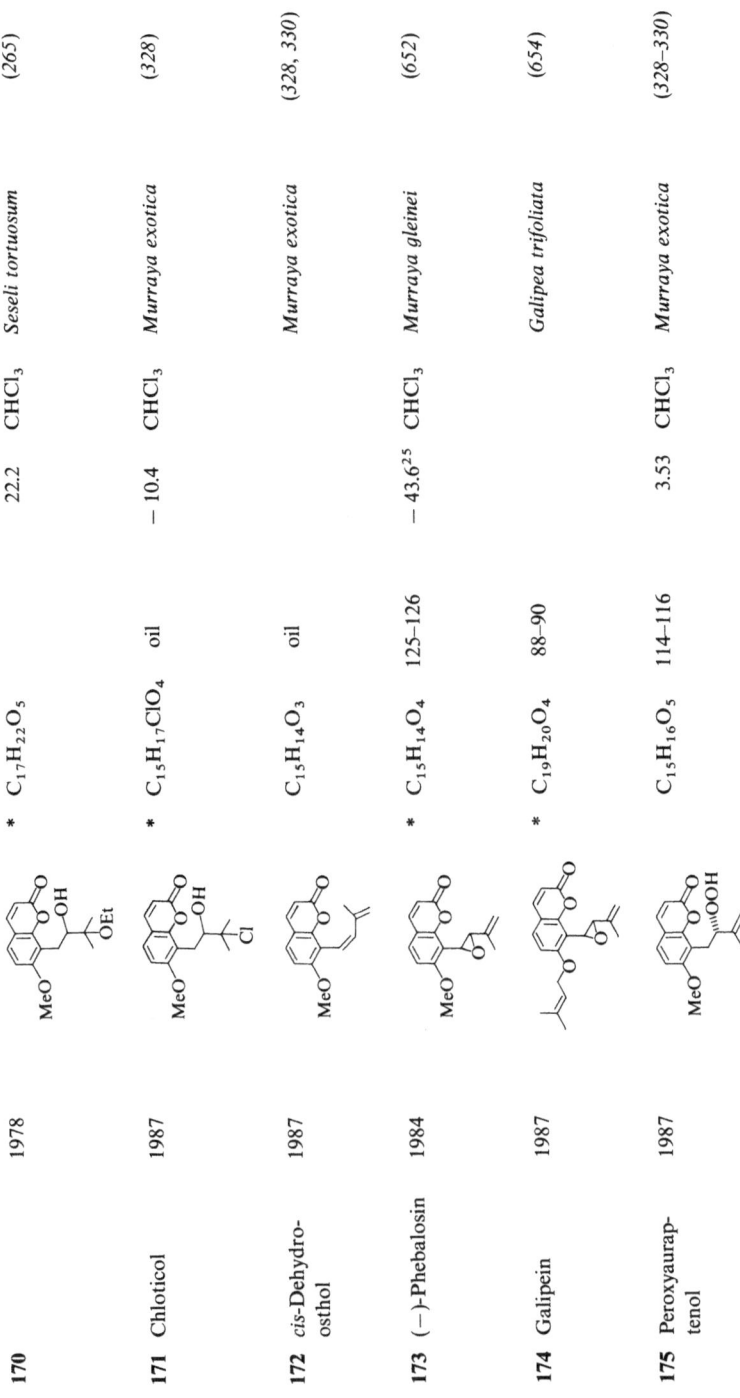

| No. | Name | Year | Structure | Formula | mp (°C) | $[\alpha]_D$ | Solvent | Source | Ref. |
|---|---|---|---|---|---|---|---|---|---|
| 170 | | 1978 | | * $C_{17}H_{22}O_5$ | | 22.2 | CHCl₃ | *Seseli tortuosum* | (265) |
| 171 | Chloticol | 1987 | | * $C_{15}H_{17}ClO_4$ | oil | −10.4 | CHCl₃ | *Murraya exotica* | (328) |
| 172 | *cis*-Dehydro-osthol | 1987 | | $C_{15}H_{14}O_3$ | oil | | | *Murraya exotica* | (328, 330) |
| 173 | (−)-Phebalosin | 1984 | | * $C_{15}H_{14}O_4$ | 125–126 | $-43.6^{25}$ | CHCl₃ | *Murraya gleinei* | (652) |
| 174 | Galipein | 1987 | | * $C_{19}H_{20}O_4$ | 88–90 | | | *Galipea trifoliata* | (654) |
| 175 | Peroxyaurap-tenol | 1987 | | $C_{15}H_{16}O_5$ | 114–116 | 3.53 | CHCl₃ | *Murraya exotica* | (328–330) |

119

120

Table 1.2. (Continued)

| Trivial name(s) | Year isolated | Structure | Formula | M.p. | $[\alpha]_\lambda^t$ | Solvent | Plant sources | Leading references |
|---|---|---|---|---|---|---|---|---|
| **176** Peroxymurraol | 1989 | | $C_{15}H_{16}O_5$ | oil | | | *Murraya exotica* | (329) |
| **177** Murpanidin Minumicrolin | 1983 1984 | | $C_{15}H_{16}O_5$ | 163–164 132–135 | $14.6^{20}$ 17.5 | $CHCl_3$ $CHCl_3$ | *Murraya paniculata* *Micromelum minutum* | (666) (177, 328) |
| **178** (−)-Murracarpin | 1989 | | * $C_{16}H_{18}O_5$ | 164–165 | − 15.6 | $CHCl_3$ | *Murraya paniculata* | (663) |
| **179** Murpanicin Murraxocin | 1983 | | * $C_{17}H_{20}O_5$ | 124–125 | | | *Murraya paniculata* *Murraya exotica* | (666) (74) |
| **180** | 1984 | | $C_{34}H_{32}O_{12}$ | 182–186 | | | *Micromelum minutum* | (177, 328) |

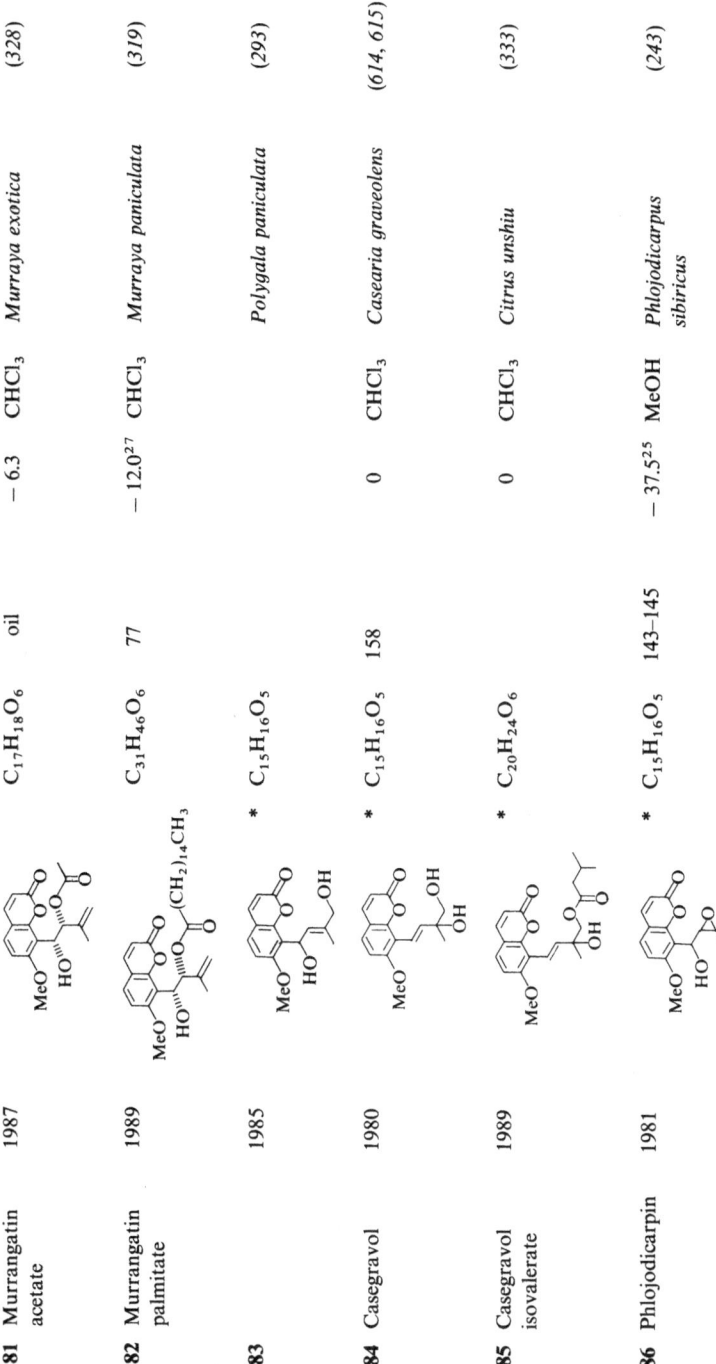

| No. | Name | Year | Structure | Formula | mp (°C) | $[\alpha]$ | Solvent | Source | Ref. |
|---|---|---|---|---|---|---|---|---|---|
| 181 | Murrangatin acetate | 1987 | | $C_{17}H_{18}O_6$ | oil | − 6.3 | $CHCl_3$ | *Murraya exotica* | *(328)* |
| 182 | Murrangatin palmitate | 1989 | | $C_{31}H_{46}O_6$ | 77 | − 12.0$^{27}$ | $CHCl_3$ | *Murraya paniculata* | *(319)* |
| 183 | | 1985 | | * $C_{15}H_{16}O_5$ | | | | *Polygala paniculata* | *(293)* |
| 184 | Casegravol | 1980 | | * $C_{15}H_{16}O_5$ | 158 | 0 | $CHCl_3$ | *Casearia graveolens* | *(614, 615)* |
| 185 | Casegravol isovalerate | 1989 | | * $C_{20}H_{24}O_6$ | | 0 | $CHCl_3$ | *Citrus unshiu* | *(333)* |
| 186 | Phlojodicarpin | 1981 | | * $C_{15}H_{16}O_5$ | 143–145 | − 37.5$^{25}$ | MeOH | *Phlojodicarpus sibiricus* | *(243)* |

Table 1.2. (Continued)

| | Trivial name(s) | Year isolated | Structure | Formula | M.p. | $[\alpha]_\lambda^t$ | Solvent | Plant sources | Leading references |
|---|---|---|---|---|---|---|---|---|---|
| 187 | Isophlojodi-carpin | 1981 | | * $C_{15}H_{16}O_5$ | 132–134 | $-102.5^{25}$ | MeOH | *Phlojodicarpus sibiricus* | (243) |
| 188 | Murranganon Murpanicol | 1987 1987 | | * $C_{15}H_{16}O_5$ | oil 102–104 | 105.8 0 | CHCl$_3$ | *Murraya exotica* *Murraya paniculata* | (328) (321, 322) |
| 189 | Hainanmurpanin | 1984 | | * $C_{17}H_{18}O_6$ | 98–101 | $7^{28}$ | CHCl$_3$ | *Murraya paniculata* | (667) |
| 190 | Isomurranganon senecioate | 1987 | | * $C_{20}H_{22}O_6$ | oil | 60.6 | CHCl$_3$ | *Murraya exotica* | (328) |
| 191 | Paniculonol isovalerate | 1989 | | $C_{20}H_{24}O_6$ | oil | | | *Murraya paniculata* | (329) |

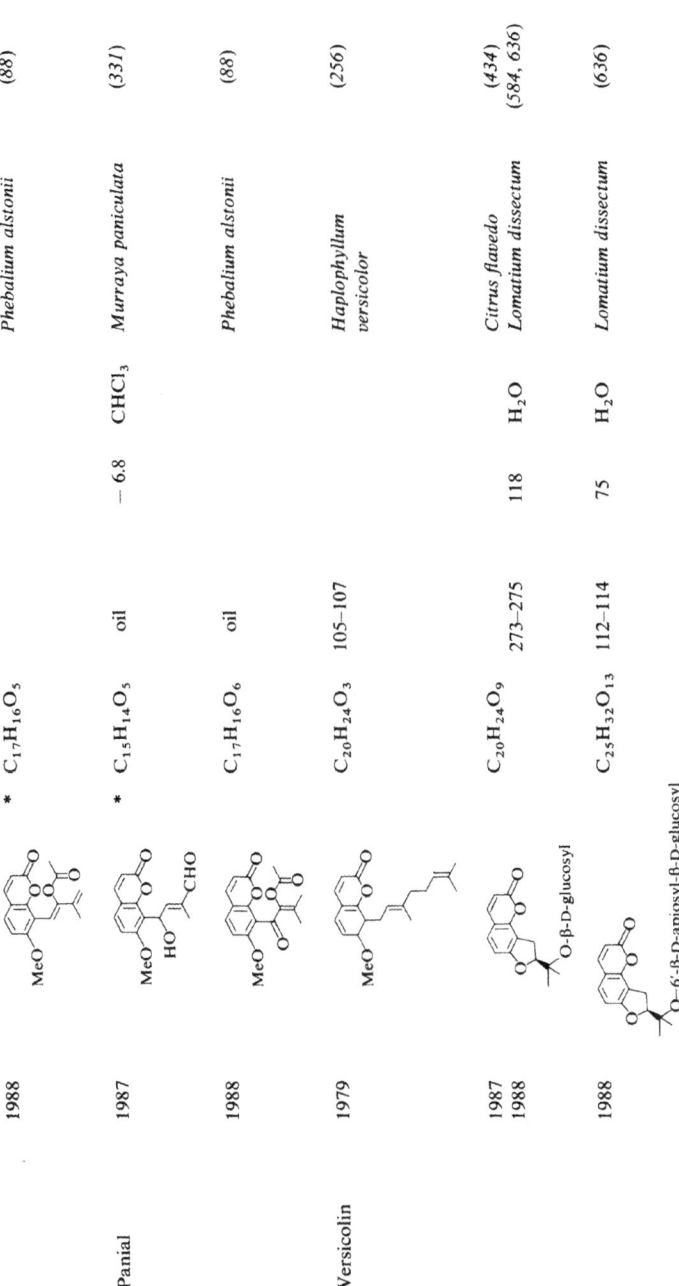

| No. / Name | Year | Structure | Formula | mp (°C) | [α] | Solvent | Source | Ref. |
|---|---|---|---|---|---|---|---|---|
| **192** | 1988 | | $*$ $C_{17}H_{16}O_5$ | | | | *Phebalium alstonii* | *(88)* |
| **193** Panial | 1987 | | $*$ $C_{15}H_{14}O_5$ | oil | − 6.8 | CHCl₃ | *Murraya paniculata* | *(331)* |
| **194** | 1988 | | $C_{17}H_{16}O_6$ | oil | | | *Phebalium alstonii* | *(88)* |
| **195** Versicolin | 1979 | | $C_{20}H_{24}O_3$ | 105–107 | | | *Haplophyllum versicolor* | *(256)* |
| **196** | 1987 1988 | | $C_{20}H_{24}O_9$ | 273–275 | 118 | H₂O | *Citrus flavedo* *Lomatium dissectum* | *(434)* *(584, 636)* |
| **197** | 1988 | | $C_{25}H_{32}O_{13}$ | 112–114 | 75 | H₂O | *Lomatium dissectum* | *(636)* |

O-β-D-glucosyl

O-6'-β-D-apiosyl-β-D-glucosyl

*Table 1.2.* (Continued)

| Trivial name(s) | Year isolated | Structure | Formula | M.p. | $[\alpha]_\lambda^t$ | Solvent | Plant sources | Leading references |
|---|---|---|---|---|---|---|---|---|
| **198** | 1988 | | $C_{26}H_{34}O_{14}$ | 114–118 | 210 | $H_2O$ | *Lomatium dissectum* | (636) |
| **199** | 1982 | | * $C_{16}H_{18}O_4$ | 145–148 | | | | *Ammi majus* | (224) |
| **200** Secrolin | 1979 | | $C_{18}H_{18}O_5S$ | 78–82 | $233.5^{20}$ | $CHCl_3$ | *Seseli mucronatum* | (209) |
| **201** | 1978 | | $C_{19}H_{20}O_6$ | 78–80 | $200^{27}$ | MeOH | *Angelica keiskei* | (387) |
| **202** Cniforin B | 1985 | | $C_{23}H_{26}O_7$ | 105–106 | $85.2^{16}$ | $CHCl_3$ | *Cnidium monnieri* | (52) |

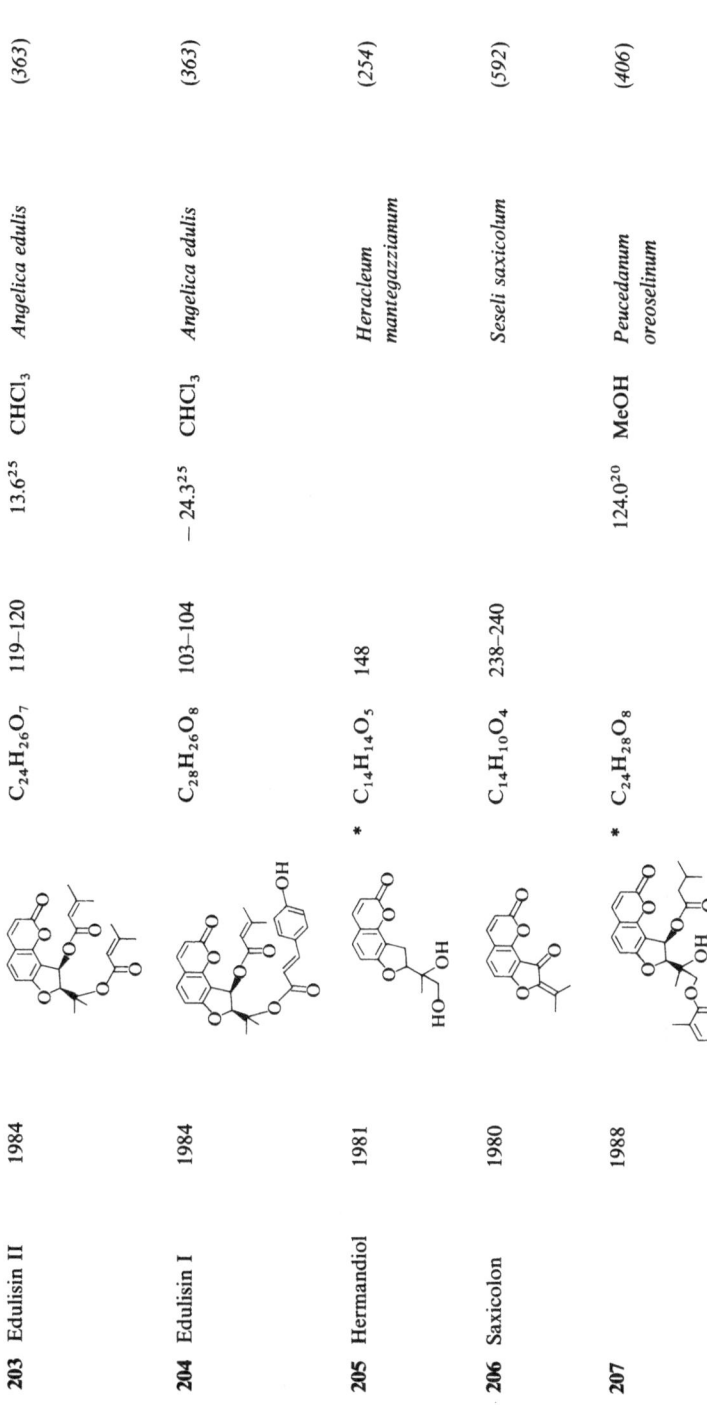

| | | | | | | | | |
|---|---|---|---|---|---|---|---|---|
| **203** Edulisin II | 1984 | | $C_{24}H_{26}O_7$ | 119–120 | $13.6^{25}$ | $CHCl_3$ | *Angelica edulis* | (363) |
| **204** Edulisin I | 1984 | | $C_{28}H_{26}O_8$ | 103–104 | $-24.3^{25}$ | $CHCl_3$ | *Angelica edulis* | (363) |
| **205** Hermandiol | 1981 | * | $C_{14}H_{14}O_5$ | 148 | | | *Heracleum mantegazzianum* | (254) |
| **206** Saxicolon | 1980 | | $C_{14}H_{10}O_4$ | 238–240 | | | *Seseli saxicolum* | (592) |
| **207** | 1988 | * | $C_{24}H_{28}O_8$ | | $124.0^{20}$ | MeOH | *Peucedanum oreoselinum* | (406) |

125

Table 1.2. (Continued)

| | Trivial name(s) | Year isolated | Structure | Formula | M.p. | $[\alpha]_\lambda^i$ | Solvent | Plant sources | Leading references |
|---|---|---|---|---|---|---|---|---|---|
| 208 | Praeroside IV | 1988 | O-β-D-glucosyl | $C_{20}H_{24}O_9$ | 115–116.5 | $4^{24}$ | MeOH | Peucedanum praeruptorum | (613) |
| 209 | | 1984 | OSO$_3$K | $C_{14}H_{13}KO_8S$ | | $27^{23}$ | MeOH | Seseli libanotis | (409) |
| 210 | Praeroside V | 1988 | * O-β-D-glucosyl | $C_{20}H_{24}O_9$ | 130–131.5 | $0^{24}$ | | Peucedanum praeruptorum | (613) |
| 211 | | 1986 | OMe, OH | $C_{15}H_{16}O_5$ | 161–162 | | | Phlojodicarpus sibiricus | (246) |
| 212 | | 1986 | OMe, OH | $C_{15}H_{16}O_5$ | 136–137 | | | Phlojodicarpus sibiricus | (246) |
| 213 | | 1985 | * OH | $C_{19}H_{22}O_6$ | | | | Musineon divaricatum | (609) |

126

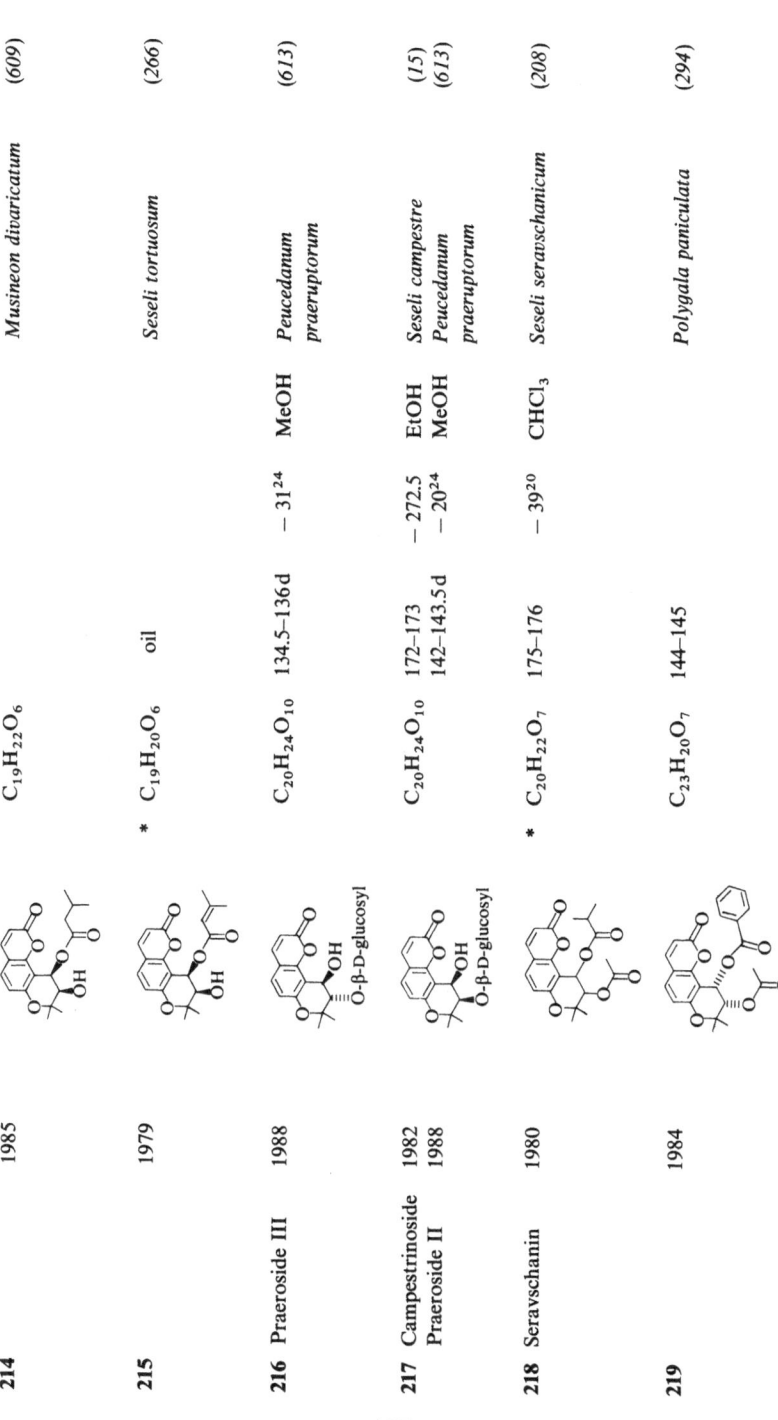

| | | | | | | | |
|---|---|---|---|---|---|---|---|
| **214** | | 1985 | | $C_{19}H_{22}O_6$ | | | *Musineon divaricatum* | (609) |
| **215** | | 1979 | * | $C_{19}H_{20}O_6$ | oil | | *Seseli tortuosum* | (266) |
| **216** Praeroside III | | 1988 | | $C_{20}H_{24}O_{10}$ | 134.5–136 d | $-31^{24}$ | MeOH | *Peucedanum praeruptorum* | (613) |
| **217** Campestrinoside Praeroside II | | 1982 1988 | | $C_{20}H_{24}O_{10}$ | 172–173 142–143.5 d | $-272.5$ $-20^{24}$ | EtOH MeOH | *Seseli campestre Peucedanum praeruptorum* | (15) (613) |
| **218** Seravschanin | | 1980 | * | $C_{20}H_{22}O_7$ | 175–176 | $-39^{20}$ | $CHCl_3$ | *Seseli seravschanicum* | (208) |
| **219** | | 1984 | | $C_{23}H_{20}O_7$ | 144–145 | | | *Polygala paniculata* | (294) |

Table 1.2. (Continued)

| Trivial name(s) | Year isolated | Structure | Formula | M.p. | $[\alpha]_\lambda^t$ | Solvent | Plant sources | Leading references |
|---|---|---|---|---|---|---|---|---|
| **220** | 1985 | | $C_{18}H_{20}O_6$ | | | | Musineon divaricatum | (609) |
| **221** Bocconin | 1986 | | * $C_{20}H_{22}O_7$ | 147–148 | 7.81[20] | EtOH | Seseli bocconi | (85) |
| **222** Junosmarin | 1986 | | $C_{19}H_{22}O_6$ | oil | − 16.4 | CHCl₃ | Citrus junos | (353) |
| **223** Campestrol | 1982 | | $C_{19}H_{22}O_6$ | oil | 32[20] | CHCl₃ | Seseli campestre | (15) |
| **224** Turgeniifolin B | 1981 | | * $C_{19}H_{22}O_6$ | 161–162 | | | Peucedanum turgeniifolium | (605) |

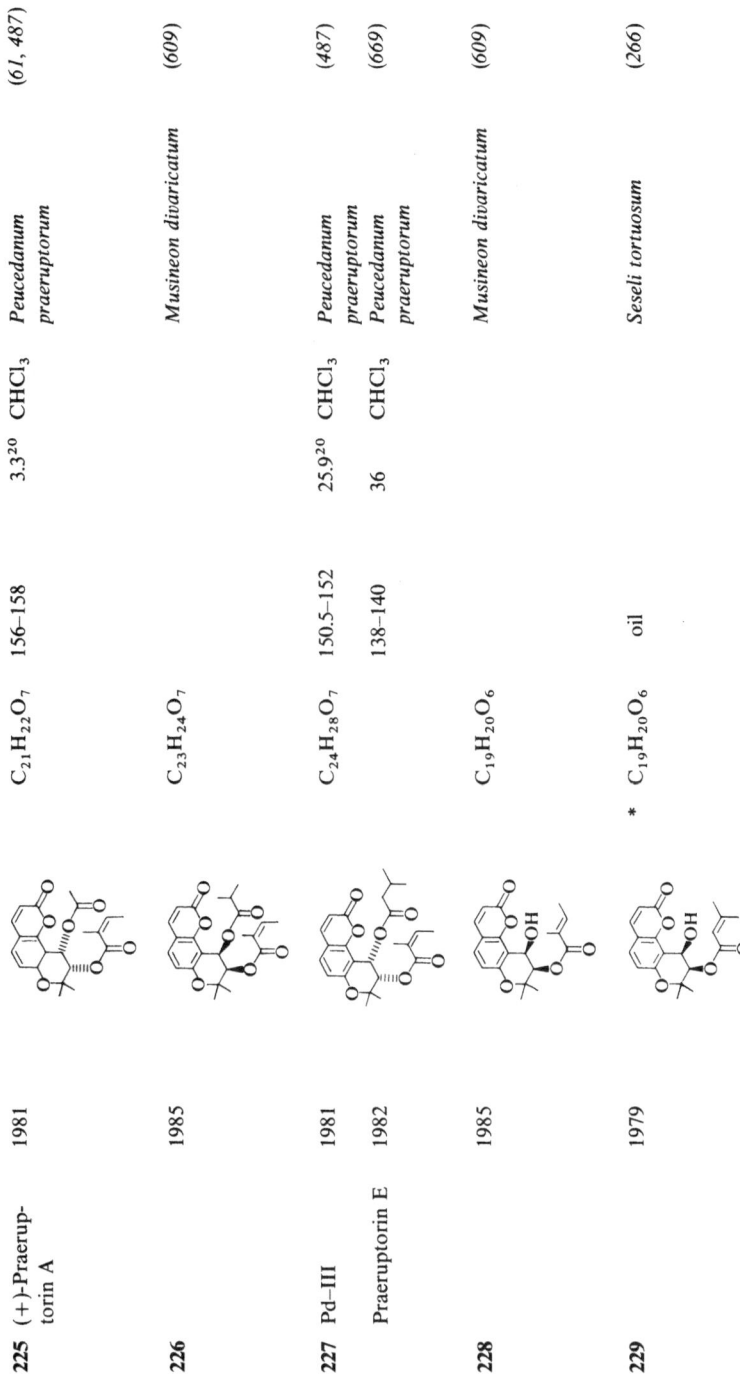

| | | | | | | | | |
|---|---|---|---|---|---|---|---|---|
| **225** (+)-Praeruptorin A | 1981 | | $C_{21}H_{22}O_7$ | 156–158 | $3.3^{20}$ | $CHCl_3$ | *Peucedanum praeruptorum* | (61, 487) |
| **226** | 1985 | | $C_{23}H_{24}O_7$ | | | | *Musineon divaricatum* | (609) |
| **227** Pd–III | 1981 | | $C_{24}H_{28}O_7$ | 150.5–152 | $25.9^{20}$ | $CHCl_3$ | *Peucedanum praeruptorum* | (487) |
| Praeruptorin E | 1982 | | | 138–140 | 36 | $CHCl_3$ | *Peucedanum praeruptorum* | (669) |
| **228** | 1985 | | $C_{19}H_{20}O_6$ | | | | *Musineon divaricatum* | (609) |
| **229** | 1979 | | * $C_{19}H_{20}O_6$ | oil | | | *Seseli tortuosum* | (266) |

129

Table 1.2. (Continued)

| Trivial name(s) | Year isolated | Structure | Formula | M.p. | $[\alpha]_\lambda^t$ | Solvent | Plant sources | Leading references |
|---|---|---|---|---|---|---|---|---|
| 230 Turgeniifolin C | 1981 | | * $C_{19}H_{20}O_6$ | 202–204 | | | Peucedanum turgeniifolium | (605) |
| 231 (+)-Samidin | 1982 | | $C_{21}H_{22}O_7$ | | $13.8^{30}$ | EtOH | Peucedanum japonicum | (579) |
| 232 | 1984 | | * $C_{24}H_{26}O_9$ | 163 | $-70^{21}$ | $CHCl_3$ | Peucedanum austriaca | (603) |
| 233 Campestrinol | 1982 | | $C_{24}H_{26}O_7$ | 116–118 | | | Seseli campestre | (15) |
| 234 | 1984 | | $C_{14}H_{13}KO_9S$ | | $-51^{23}$ | MeOH | Seseli libanotis | (409) |

| | Year isolated | Structure | Formula | M.p. | $[\alpha]_\lambda$ | Solvent | Plant sources | |
|---|---|---|---|---|---|---|---|---|
| **235** Pd–Ib | 1981 | | $C_{19}H_{18}O_6$ | 209.5–212.5 | 49.8[20] | dioxan | *Peucedanum praeruptorum* | (487) |
| **236** | 1987 | | $C_{19}H_{18}O_6$ | 198–201 | $-30^{25}$ | $CHCl_3$ | *Arracacia nelsonii* | (185) |
| **237** Turgeniifolin A | 1981 | | * $C_{19}H_{18}O_6$ | | | | *Peucedanum turgeniifolium* | (605) |

Table 1.3. *5,6-Disubstituted-7-Oxygenated Coumarins*

| Trivial name(s) | Year isolated | Structure | Formula | M.p. | $[\alpha]_\lambda$ | Solvent | Plant sources | Leading references |
|---|---|---|---|---|---|---|---|---|
| **238** Ulismoncadin | 1984 | | $C_{19}H_{22}O_3$ | 132–133 | | | *Thamnosma texana* | (200) |

131

Table 1.4. 6,8-Disubstituted-7-Oxygenated Coumarins

| Trivial name(s) | Year isolated | Structure | Formula | M.p. | [α]λ | Solvent | Plant sources | Leading references |
|---|---|---|---|---|---|---|---|---|
| **239** 8-Prenylnodakenetin | 1977 | | * $C_{19}H_{22}O_4$ | 111 | | | *Chloroxylon swietenia* | (100) |
| **240** Swietenocoumarin A | 1977 | | $C_{16}H_{14}O_3$ | 113 | | | *Chloroxylon swietenia* | (100) |
| **241** Swietenocoumarin C | 1977 | | * $C_{16}H_{14}O_4$ | 155 | | | *Chloroxylon swietenia* | (100) |
| **242** Swietenocoumarin H | 1980 | | $C_{16}H_{14}O_4$ | 157–158 | | | *Chloroxylon swietenia* | (520) |
| **243** Swietenocoumarin E | 1977 | | * $C_{16}H_{16}O_5$ | 164–166 | | | *Chloroxylon swietenia* | (100) |

Table 2. 5,7-Dioxygenated Coumarins

| Trivial name(s) | Year isolated | Structure | Formula | M.p. | $[\alpha]_\lambda^t$ | Solvent | Plant sources | Leading references |
|---|---|---|---|---|---|---|---|---|
| 244 | 1978 1982 | | $C_9H_6O_4$ | 285–286 283–285 | | | *Rumex conglomeratus* *Haplophyllum dauricum* | (248) (81) |
| 245 | 1978 | | $C_{10}H_8O_4$ | 226–227 | | | *Haplophyllum bungei* | (257) |
| 246 Anisocoumarin B | 1989 | | $C_{14}H_{14}O_4$ | 94–95 | | | *Clausena anisata* | (479) |
| 247 | 1983 | | $C_{12}H_{10}O_5$ | 194–195 | | | *Toddalia asiatica* | (326) |
| 248 | 1982 | | $C_{15}H_{14}O_5$ | 176–180 199–200 | | | *Toddalia aculeata* | (532) (326) |
| 249 Toddanol | 1981 | | * $C_{16}H_{18}O_5$ | 125 | $-93.3^{20}$ | CHCl$_3$ | *Toddalia asiatica* | (572, 574) |

133

134

*Table 2.* (Continued)

| Trivial name(s) | Year isolated | Structure | Formula | M.p. | $[\alpha]_\lambda^t$ | Solvent | Plant sources | Leading references |
|---|---|---|---|---|---|---|---|---|
| **250** Toddanone | 1981 1982 | | $C_{16}H_{18}O_5$ | 116 114 | | | *Toddalia asiatica* *Toddalia aculeata* | (574) (532, 572) |
| **251** | 1982 | | * $C_{17}H_{22}O_6$ | oil | | | *Toddalia aculeata* | (532) |
| **252** Dauroside D | 1982 | | $C_{15}H_{16}O_9$ | 214–215 | 108.6 | pyridine | *Haplophyllum dauricum* | (81, 637) |
| Mulberroside **B** | 1985 | | | 220–222 | $55^{18}$ | pyridine | *Morus lhou* | (307) |
| **253** Celereoin | 1986 | | * $C_{14}H_{14}O_5$ | 201 | $-38.0^{21}$ | MeOH | *Apium graveolens* | (339, 535, 536) |
| **254** Jumutinol | 1980 | | * $C_{24}H_{30}O_6$ | 148–150 | | | *Seseli jomuticum* | (2) |
| **255** Celereoside | 1980 | | * $C_{20}H_{24}O_{10}$ | 200–201 | $-37.27^{24}$ | MeOH | *Apium graveolens* | (253) |

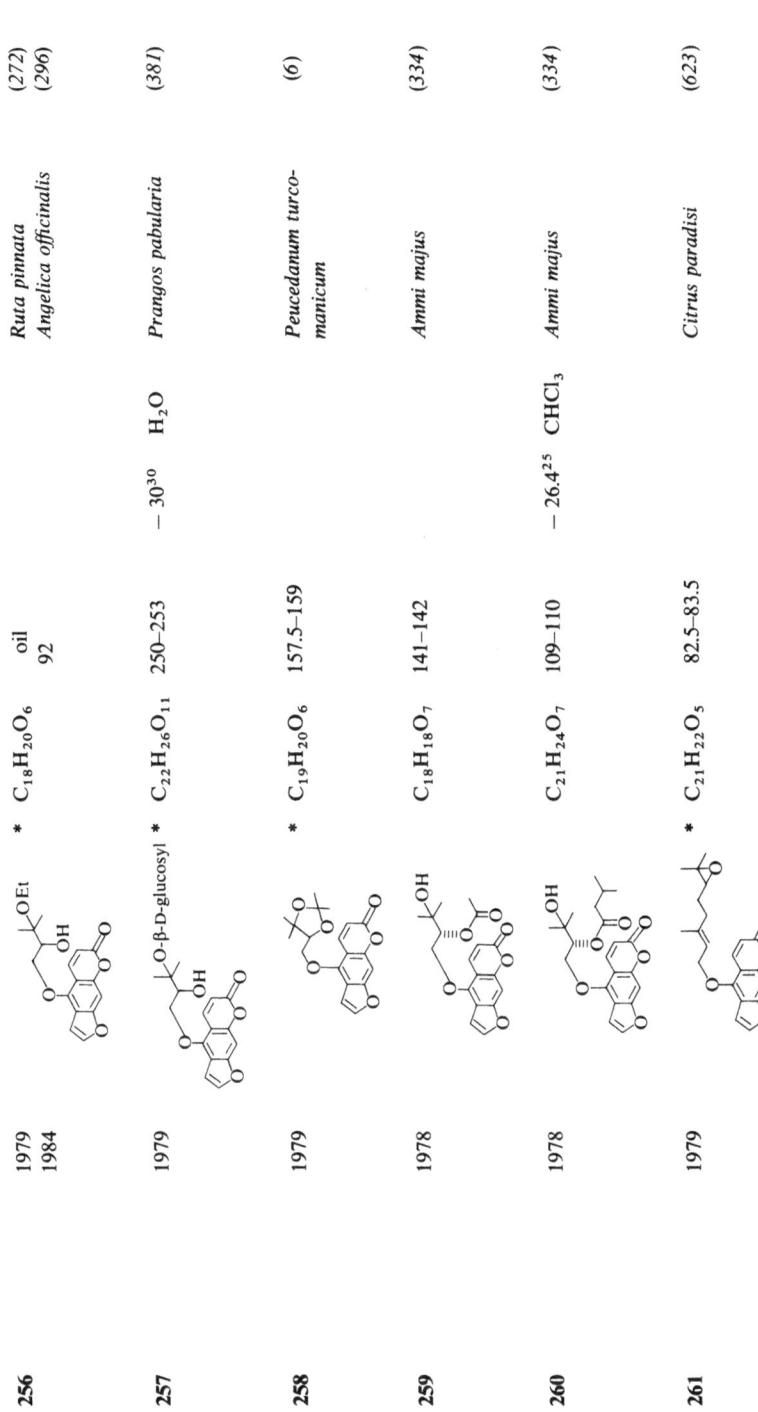

| No. | Year | | Formula | mp | [α] | Solvent | Source | Ref. |
|---|---|---|---|---|---|---|---|---|
| 256 | 1979 1984 | * | $C_{18}H_{20}O_6$ | oil 92 | | | *Ruta pinnata* *Angelica officinalis* | (272) (296) |
| 257 | 1979 | * | $C_{22}H_{26}O_{11}$ | 250–253 | $-30^{30}$ | $H_2O$ | *Prangos pabularia* | (381) |
| 258 | 1979 | * | $C_{19}H_{20}O_6$ | 157.5–159 | | | *Peucedanum turcomanicum* | (6) |
| 259 | 1978 | | $C_{18}H_{18}O_7$ | 141–142 | | | *Ammi majus* | (334) |
| 260 | 1978 | | $C_{21}H_{24}O_7$ | 109–110 | $-26.4^{25}$ | $CHCl_3$ | *Ammi majus* | (334) |
| 261 | 1979 | * | $C_{21}H_{22}O_5$ | 82.5–83.5 | | | *Citrus paradisi* | (623) |

*Table 2.* (Continued)

| No. | Trivial name(s) | Year isolated | Structure | | Formula | M.p. | $[\alpha]_\lambda^t$ | Solvent | Plant sources | Leading references |
|---|---|---|---|---|---|---|---|---|---|---|
| 262 | Tortuosin | 1983 | | * | $C_{21}H_{22}O_6$ | 156–157 | | | *Seseli tortuosum* | (5) |
| 263 | Notopterol | 1983 | | * | $C_{21}H_{22}O_5$ | 90–92 | 0 | | *Notopterygium incisum* | (385) |
| 264 | Notoptol | 1983 | | | $C_{21}H_{22}O_4$ | 73 | | | *Notopterygium incisum* | (385) |
| 265 | Anhydro-notoptol | 1983 | | | $C_{21}H_{20}O_4$ | oil | | | *Notopterygium incisum* | (385) |
| 266 | Anisolactone | 1984 | | * | $C_{21}H_{18}O_6$ | 147–148 | $19.8^{25}$ | $CHCl_3$ | *Clausena anisata* | (403) |

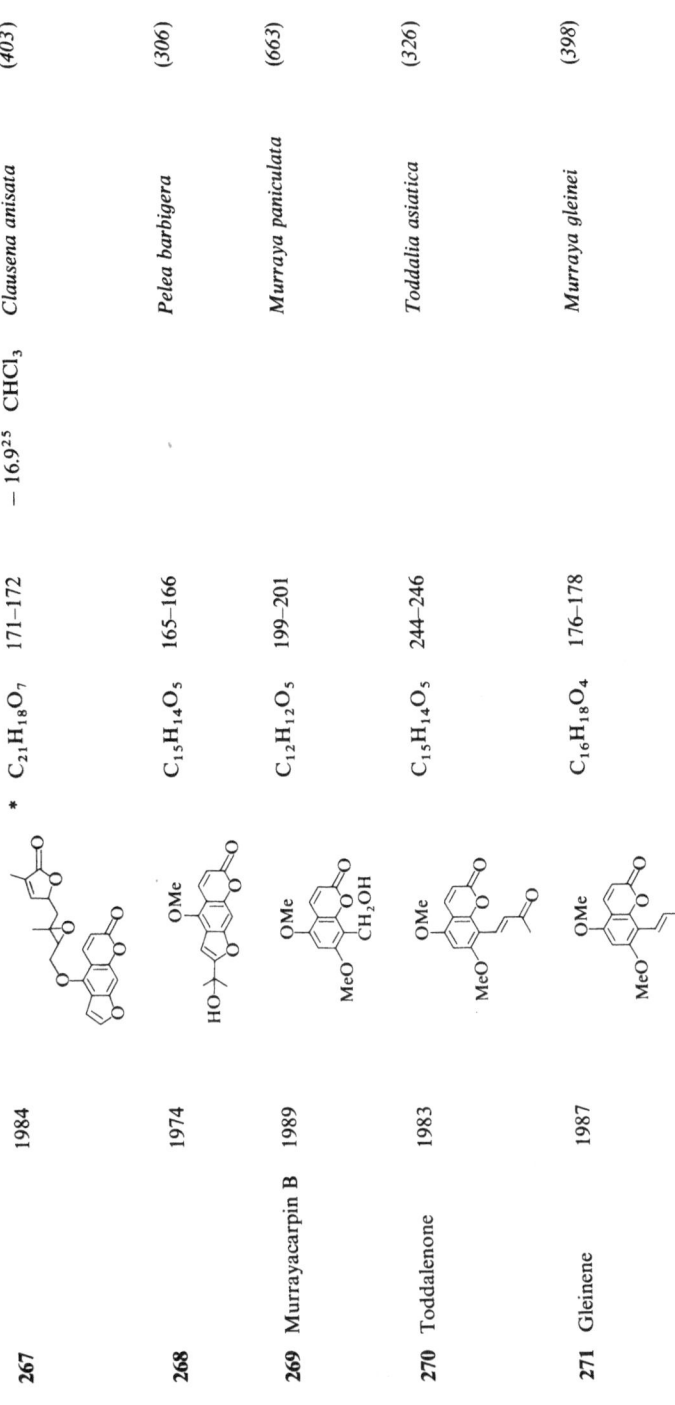

| No. | Name | Year | Formula | mp (°C) | [α] | Source | Ref. |
|---|---|---|---|---|---|---|---|
| 267 | | 1984 | * $C_{21}H_{18}O_7$ | 171–172 | $-16.9^{25}$ $CHCl_3$ | *Clausena anisata* | (403) |
| 268 | | 1974 | $C_{15}H_{14}O_5$ | 165–166 | | *Pelea barbigera* | (306) |
| 269 | Murrayacarpin B | 1989 | $C_{12}H_{12}O_5$ | 199–201 | | *Murraya paniculata* | (663) |
| 270 | Toddalenone | 1983 | $C_{15}H_{14}O_5$ | 244–246 | | *Toddalia asiatica* | (326) |
| 271 | Gleinene | 1987 | $C_{16}H_{18}O_4$ | 176–178 | | *Murraya gleinei* | (398) |

138

## Table 2. (Continued)

| Trivial name(s) | Year isolated | Structure | Formula | M.p. | $[\alpha]^t_\lambda$ | Solvent | Plant sources | Leading references |
|---|---|---|---|---|---|---|---|---|
| **272** Sibiricol | 1978 | | $C_{15}H_{16}O_4$ | 193–194 | | | *Seseli sibiricum* | (395, 449, 454) |
| **273** (−)-Sibiricin | 1984 | | * $C_{16}H_{18}O_5$ | 148–149 | $-59.7^{25}$ | $CHCl_3$ | *Murraya gleinei* | (397, 652) |
| **274** Sesebrin | 1978 | | * $C_{20}H_{24}O_5$ | 112–113 | 5 | EtOH | *Seseli sibiricum* | (395) |
| **275** | 1978 | | * $C_{15}H_{16}O_5$ | | | | *Seseli tortuosum* | (265) |

| No. | Name | Year | Formula | mp (°C) | $[\alpha]$ | Solvent | Source | Ref. |
|---|---|---|---|---|---|---|---|---|
| 276 | Sibirinol<br>Omphamurin | 1980<br>1981 | * $C_{16}H_{18}O_5$ | 121–122<br>131–132 | $-22^{20}$ | $CHCl_3$ | *Seseli sibiricum*<br>*Murraya omphalo-carpa* | (66)<br>(657) |
| 277 | Sesibiricol | 1980 | * $C_{20}H_{24}O_5$ | 110–111 | | | *Seseli sibiricum* | (67) |
| 278 | Seselinal | 1980 | $C_{16}H_{18}O_5$ | 174–175 | | | *Seseli sibiricum* | (67) |
| 279 | Murraculatin | 1988 | $C_{16}H_{18}O_6$ | 217–218 | | | *Murraya paniculata* | (658) |
| 280 | Gleinadiene | 1985 | $C_{16}H_{16}O_4$ | 120–121 | | | *Murraya gleinei* | (397, 398) |

139

Table 2. (Continued)

| Trivial name(s) | Year isolated | Structure | Formula | M.p. | $[\alpha]_\lambda$ | Solvent | Plant sources | Leading references |
|---|---|---|---|---|---|---|---|---|
| **281** Isomexoticin (−)-Mexoticin | 1983 1984 | | * $C_{16}H_{20}O_6$ | 194–195 191–192 | $-36^{20}$ $-31.1^{25}$ | $CHCl_3$ $CHCl_3$ | *Murraya paniculata* *Murraya gleinei* | (666) (397, 652) |
| **282** Sesebrinol | 1978 | | * $C_{20}H_{26}O_6$ | 114–115 | 56 | EtOH | *Seseli sibiricum* | (395) |
| **283** Omphalocarpin | 1989 | | * $C_{17}H_{22}O_6$ | 159–160 | −44.1 | $CHCl_3$ | *Murraya paniculata* | (663) |
| **284** Angelin | 1980 | | $C_{20}H_{22}O_5$ | 143–144 | | | *Angelica pubescens* | (383) |

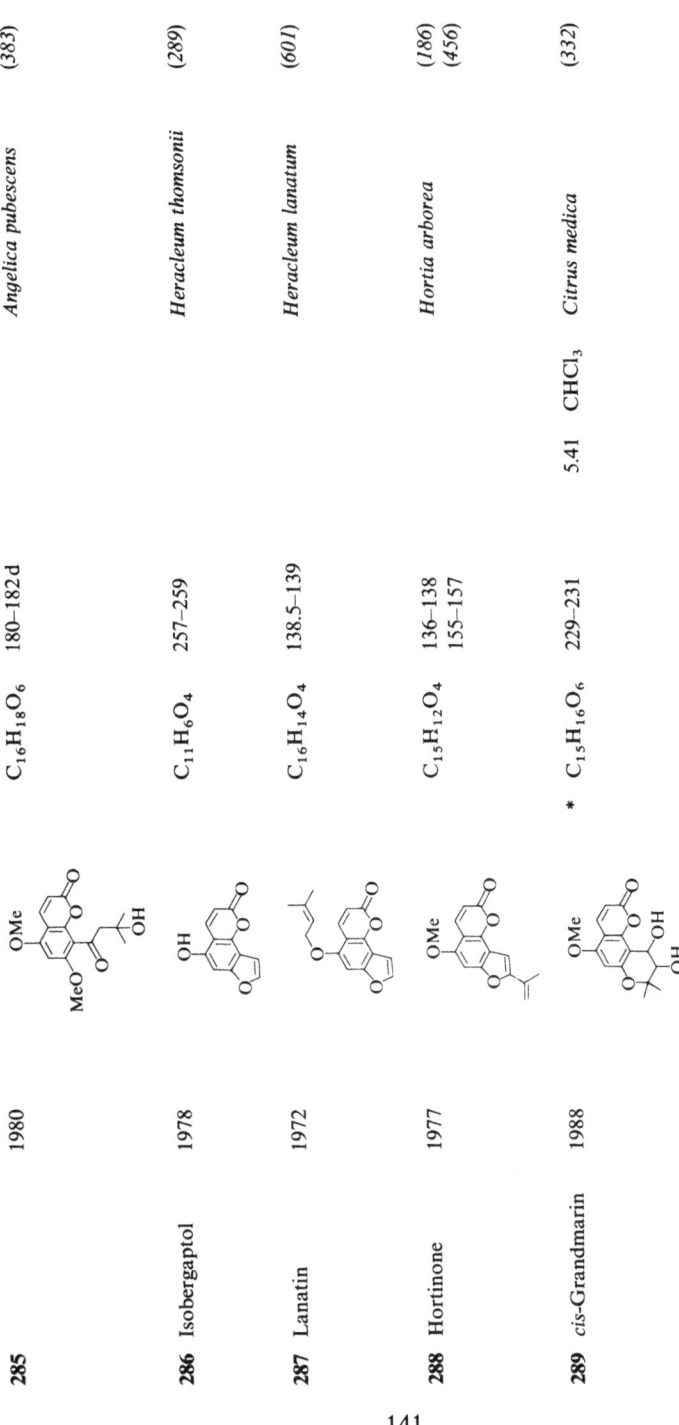

| | | | | | | |
|---|---|---|---|---|---|---|
| **285** | 1980 | | $C_{16}H_{18}O_6$ | 180–182d | *Angelica pubescens* | (383) |
| **286** Isobergaptol | 1978 | | $C_{11}H_6O_4$ | 257–259 | *Heracleum thomsonii* | (289) |
| **287** Lanatin | 1972 | | $C_{16}H_{14}O_4$ | 138.5–139 | *Heracleum lanatum* | (601) |
| **288** Hortinone | 1977 | | $C_{15}H_{12}O_4$ | 136–138 155–157 | *Hortia arborea* | (186) (456) |
| **289** *cis*-Grandmarin | 1988 | * | $C_{15}H_{16}O_6$ | 229–231 | 5.41 CHCl₃ *Citrus medica* | (332) |

*Table 2. (Continued)*

| | Trivial name(s) | Year isolated | Structure | Formula | M.p. | $[\alpha]_\lambda^i$ | Solvent | Plant sources | Leading references |
|---|---|---|---|---|---|---|---|---|---|
| **290** | *trans-O*-Methyl-grandmarin | 1989 | | * $C_{16}H_{18}O_6$ | | $-13.4$ | $CHCl_3$ | *Citrus unshiu* | *(333)* |
| **291** | *trans*-Grand-marin iso-valerate | 1988 | | * $C_{20}H_{24}O_7$ | oil | 24.1 | $CHCl_3$ | *Citrus medica* | *(332)* |
| **292** | | 1988 | | $C_{14}H_{12}O_4$ | 210–213 | | | *Citrus medica* | *(332, 452)* |
| **293** | | 1983 | | $C_{15}H_{14}O_4$ | 162–164 | | | *Citrus grandis* | *(452, 662)* |

| No. | Name | Year | Structure | Formula | mp | Source | (refs) |
|---|---|---|---|---|---|---|---|
| 294 | | 1980 | OH | $C_{16}H_{14}O_4$ | 235–236 | *Heracleum thomsonii* | *(66, 289)* |
| 295 | Swieteno-coumarin B | 1977 | OMe | $C_{17}H_{16}O_4$ | 145 | *Chloroxylon swietenia* | *(100)* |
| 296 | Swieteno-coumarin D | 1977 | OMe racemic | $C_{17}H_{16}O_5$ | 150 | *Chloroxylon swietenia* | *(100)* |
| 297 | Swieteno-coumarin G | 1980 | OMe OH | $C_{17}H_{16}O_5$ | 190 | *Chloroxylon swietenia* | *(460, 520)* |
| 298 | Swieteno-coumarin F | 1977 | OMe racemic OH OH | $C_{17}H_{18}O_6$ | 178–179 | *Chloroxylon swietenia* | *(100)* |

143

*Table 2.* (Continued)

| Trivial name(s) | Year isolated | Structure | Formula | M.p. | $[\alpha]_\lambda^t$ | Solvent | Plant sources | Leading references |
|---|---|---|---|---|---|---|---|---|
| **299** Ponfolin | 1986 | | $C_{24}H_{28}O_4$ | oil | | | *Poncirus trifoliata* | *(238, 453)* |
| **300** Kinocoumarin | 1988 | | $C_{24}H_{28}O_4$ | oil | | | *Citrus medica* | *(332)* |
| **301** | 1978 | | $C_{16}H_{14}O_4$ | 145–146 | | | *Heracleum thomsonii* | *(66, 289)* |
| **302** Honydusin | 1988 | | $C_{19}H_{20}O_4$ | 179–180 | | | *Citrus grandis* | *(661)* |

Table 3. *6,7-Dioxygenated Coumarins*

| Trivial name(s) | Year isolated | Structure | Formula | M.p. | $[\alpha]_\lambda^t$ | Solvent | Plant sources | Leading references |
|---|---|---|---|---|---|---|---|---|
| **303** | 1986 | MeO RO coumarin; R = 2'-acetyl-β-D-glucosyl | $C_{18}H_{20}O_{10}$ | 208–210 | | | *Viburnum awabuki* | (400) |
| **304** | 1986 | MeO RO coumarin; R = 6'-acetyl-β-D-glucosyl | $C_{18}H_{20}O_{10}$ | 229 | | | *Viburnum awabuki* | (400) |
| **305** | 1982 | MeO RO coumarin; R = 6'-feruloyl-β-D-glucosyl | $C_{26}H_{26}O_{12}$ | 206–208 | −110.5 | pyridine | *Haplophyllum obtusifolium* | (79) |
| **306** | 1984 1986 | MeO RO coumarin; R = 2',6'-diacetyl-β-D-glucosyl | $C_{20}H_{22}O_{11}$ | 178–179.5 184–186 | $-100^{25}$ | MeOH | *Viburnum suspensum* *Viburnum awabuki* | (335) (400) |
| **307** | 1986 | MeO RO coumarin; R = 3',6'-diacetyl-β-D-glucosyl | $C_{20}H_{22}O_{11}$ | 155–158 | | | *Viburnum awabuki* | (400) |
| **308** Lariside | 1986 | MeO RO coumarin; R = 2'-β-D-apiosyl-β-D-glucosyl | $C_{21}H_{26}O_{13}$ | 155–156 | | | *Salsola laricifolia* | (476) |

145

Table 3. (Continued)

| Trivial name(s) | Year isolated | Structure | Formula | M.p. | $[\alpha]_\lambda^t$ | Solvent | Plant sources | Leading references |
|---|---|---|---|---|---|---|---|---|
| **309** Diospyroside | 1978 | R = 6'-β-D-apiosyl-β-D-glucosyl | $C_{20}H_{24}O_{13}$ | 249–250 | $-100^{20}$ | DMSO | *Diospyros sapota* | (233) |
| **310** Hymexelsin | 1988 | R = 6'-β-D-apiosyl-β-D-glucosyl | $C_{21}H_{26}O_{13}$ | 206 | $-116^{30}$ | pyridine | *Hymenodictyon excelsum* | (522) |
| Xeroboside | 1989 | | | 238–240 | $-179^{20}$ | EtOH | *Xeromphis spinosa* | (555) |
| | 1989 | | | 195–197 | | | *Xeromphis obovata* | (587) |
| **311** Haploperoside D | 1985 | R = 2'-α-L-rhamnosyl-β-D-glucosyl | $C_{22}H_{28}O_{13}$ | 249–251 | $-37.8^{20}$ | DMF | *Haplophyllum perforatum* | (673) |
| **312** Haploperoside A | 1980 | R = 6'-α-L-rhamnosyl-β-D-glucosyl | $C_{22}H_{28}O_{13}$ | 212–213 | $-37^{22}$ | MeOH | *Haplophyllum perforatum* | (670, 673) |
| **313** Haploperoside C | 1985 | R = 2'-acetyl-6'-α-L-rhamnosyl-β-D-glucosyl | $C_{24}H_{30}O_{14}$ | 155–157 | $-27.43^{20}$ | MeOH | *Haplophyllum perforatum* | (673) |

| № | Name | Year | Structure | Formula | mp (°C) | [α] | Solvent | Source | Ref. |
|---|------|------|-----------|---------|---------|-----|---------|--------|------|
| 314 | Haploperoside B | 1980 | MeO / RO coumarin; R = 4''-acetyl-6'-α-rhamnosyl--β-D-glucosyl | $C_{24}H_{30}O_{14}$ | 89–90 | $-45^{22}$ | MeOH | *Haplophyllum perforatum* | (671–673) |
| 315 | Dauroside C | 1984 | MeO / RO coumarin; R = acetyl-6'-α-L-rhamnosyl--β-D-glucosyl | $C_{24}H_{30}O_{14}$ | 93–95 | | | *Haplophyllum dauricum* | (78) |
| 316 | Haploperoside E | 1985 | MeO / RO coumarin; R = 2',6'-bis-α-L-rhamnosyl--β-D-glucosyl | $C_{28}H_{38}O_{17}$ | 175–177 | $-56.41^{20}$ | MeOH | *Haplophyllum perforatum* | (673) |
| 317 | | 1979 | * coumarin | $C_{15}H_{16}O_5$ | 125 | $-8^{24}$ | $CHCl_3$ | *Conyza obscura* | (111) |
| 318 | Haplopinol | 1982 | coumarin | $C_{14}H_{14}O_5$ | 195–198 | | | *Haplopappus multifolius* | (164) |
| 319 | Obtusinol | 1980 | coumarin | $C_{15}H_{16}O_5$ | 97–98 | | | *Haplophyllum obtusifolium* | (20, 430) |
| 320 | Obtusinin | 1979 1980 | * coumarin | $C_{15}H_{18}O_6$ | oil 135–137 | 140.9 | $CHCl_3$ | *Conyza obscura* *Haplophyllum obtusifolium* | (111) (20, 429) |

Table 3. (Continued)

| Trivial name(s) | Year isolated | Structure | Formula | M.p. | $[\alpha]_\lambda$ | Solvent | Plant sources | Leading references |
|---|---|---|---|---|---|---|---|---|
| 321 Obtusoside | 1980 | β-D-glucosyl-O | * $C_{21}H_{28}O_{11}$ | 45–47 | 10.4 | $CHCl_3$ | Haplophyllum obtusifolium | (431) |
| 322 | 1979 | | $C_{15}H_{14}O_4$ | 118 | | | Conyza obscura | (111) |
| 323 | 1979 | | $C_{15}H_{14}O_4$ | oil | | | Conyza obscura | (111) |
| 324 Bungediol | 1982 | | * $C_{20}H_{26}O_6$ | 108–109 | $42.8^{20}$ | EtOH | Haplophyllum bungei | (3, 13) |
| 325 | 1979 | | * $C_{23}H_{30}O_6$ | 103–104 | $25.7^{20}$ | EtOH | Haplophyllum pedicellatum | (358) |
| 326 Scopofarnol | 1979 1984 | | $C_{25}H_{32}O_4$ | oil | | | Conyza obscura Artemisia persica | (111) (310) |

| No. | Name | Year | Structure | Formula | mp (°C) | $[\alpha]$ | Solvent | Source | Ref. |
|---|---|---|---|---|---|---|---|---|---|
| 327 | Scopodrimol A | 1984 | 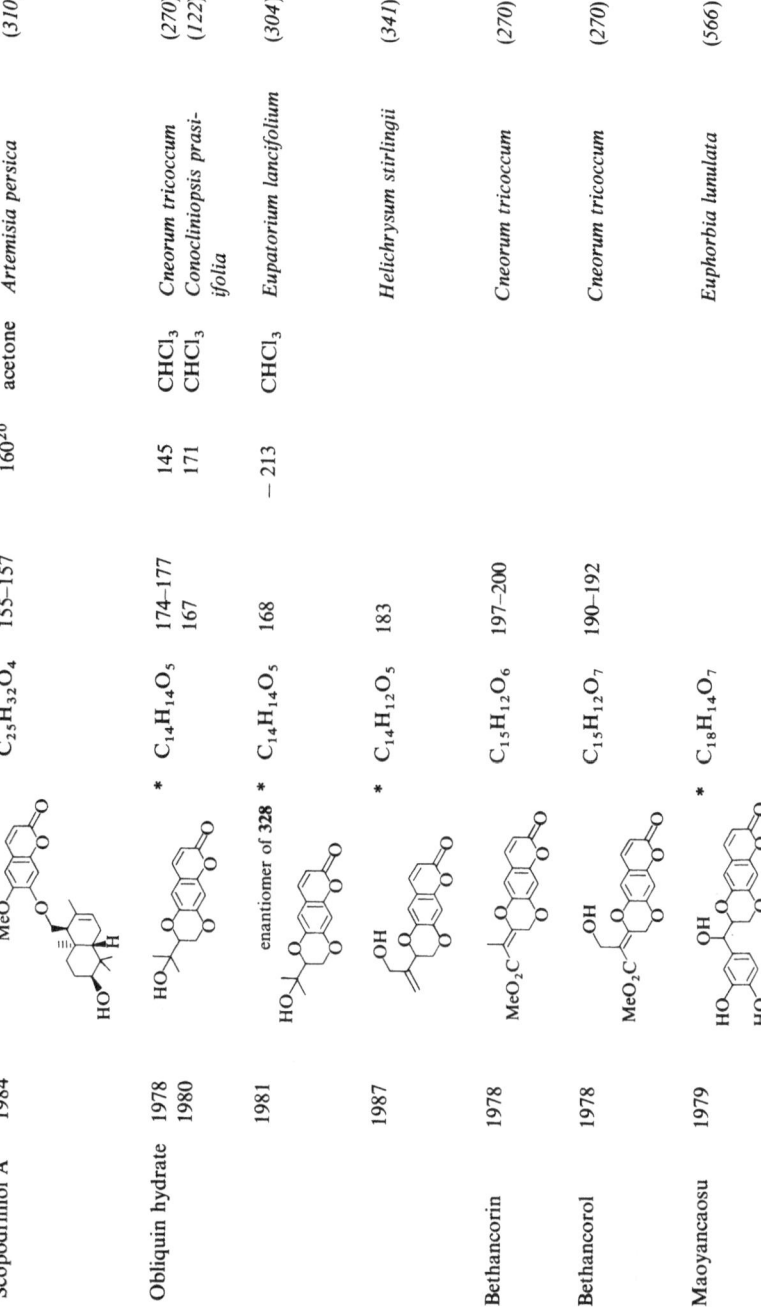 | $C_{25}H_{32}O_4$ | 155–157 | $160^{20}$ | acetone | *Artemisia persica* | (310) |
| 328 | Obliquin hydrate | 1978 1980 | | * $C_{14}H_{14}O_5$ | 174–177 167 | 145 171 | CHCl₃ CHCl₃ | *Cneorum tricoccum* *Conocliniopsis prasiifolia* | (270) (122) |
| 329 | | 1981 | enantiomer of **328** | * $C_{14}H_{14}O_5$ | 168 | $-213$ | CHCl₃ | *Eupatorium lancifolium* | (304) |
| 330 | | 1987 | | * $C_{14}H_{12}O_5$ | 183 | | | *Helichrysum stirlingii* | (341) |
| 331 | Bethancorin | 1978 | | $C_{15}H_{12}O_6$ | 197–200 | | | *Cneorum tricoccum* | (270) |
| 332 | Bethancorol | 1978 | | $C_{15}H_{12}O_7$ | 190–192 | | | *Cneorum tricoccum* | (270) |
| 333 | Maoyancaosu | 1979 | | * $C_{18}H_{14}O_7$ | | | | *Euphorbia lunulata* | (566) |

149

## Table 3. (Continued)

| | Trivial name(s) | Year isolated | Structure | Formula | M.p. | $[\alpha]_\lambda^t$ | Solvent | Plant sources | Leading references |
|---|---|---|---|---|---|---|---|---|---|
| 334 | Moluccanin | 1988 | | $C_{20}H_{18}O_8$ | 220 | | | *Aleurites moluccana* | (565) |
| 335 | | 1980 | | $C_{12}H_{10}O_5$ | 178–179 | | | *Fraxinus floribunda* | (470) |
| 336 | *O*-Methylcedrelopsin | 1983 | | $C_{16}H_{18}O_4$ | 66–68 | | | *Zanthoxylum usambarense* | (376) |
| 337 | | 1985 | | $C_{25}H_{32}O_4$ | 120–121 | | | *Brocchia cinerea* | (281) |

Table 4. 7,8-Dioxygenated Coumarins

| | Trivial name(s) | Year isolated | Structure | Formula | M.p. | $[\alpha]_\lambda^t$ | Solvent | Plant sources | Leading references |
|---|---|---|---|---|---|---|---|---|---|
| 338 | | 1984 | | $C_{14}H_{14}O_4$ | 115–117 | | | Melampodium divaricatum | (124) |
| 339 | Ferujol | 1985 | | $C_{19}H_{24}O_4$ | 68–70 | | | Ferula jaeschkeana | (590) |
| 340 | | 1979 | | $C_{11}H_{10}O_4$ | 114–116 | | | Artemisia apiacea | (581) |
| 341 | Desoxylacarol | 1985 | * | $C_{15}H_{18}O_5$ | oil | | | Artemisia armeniaca | (612) |
| 342 | | 1980 1987 | | $C_{15}H_{16}O_4$ | 94 | | | Artemisia caruifolia Coleonema calycinum | (75) (279) |
| 343 | | 1981 | | $C_{15}H_{16}O_5$ | oil | $4.6^{21}$ | $CHCl_3$ | Coleonema album | (278) |
| 344 | | 1987 | * | $C_{15}H_{18}O_6$ | oil | | | Coleonema calycinum | (279) |

151

Table 4. (Continued)

| Trivial name(s) | Year isolated | Structure | Formula | M.p. | $[\alpha]_\lambda^t$ | Solvent | Plant sources | Leading references |
|---|---|---|---|---|---|---|---|---|
| 345 Villosin | 1980 | | * $C_{20}H_{24}O_5$ | oil | 0 | | Haplophyllum villosum | (14) |
| 346 Tenudiol | 1980 | | * $C_{20}H_{26}O_6$ | oil | 0 | | Haplophyllum tenue | (14) |
| 347 | 1984 | | $C_{14}H_{14}O_4$ | 166–167 | | | Melampodium divaricatum | (124) |
| 348 Lacinartin | 1980 1986 | | $C_{15}H_{16}O_4$ | 101 | | | Artemisia apiacea Artemisia laciniata | (582) (312) |
| 349 Lacinartin epoxide | 1986 | | * $C_{15}H_{16}O_5$ | oil | $10^{20}$ | $CHCl_3$ | Artemisia laciniata | (312) |
| 350 Lacinartindiol | 1986 | | * $C_{15}H_{18}O_6$ | oil | $36^{20}$ | $CHCl_3$ | Artemisia laciniata | (312) |

| No. | Name | Year | Structure | Formula | mp (°C) | [α] | Source | Refs. |
|---|---|---|---|---|---|---|---|---|
| **351** | | 1970 | 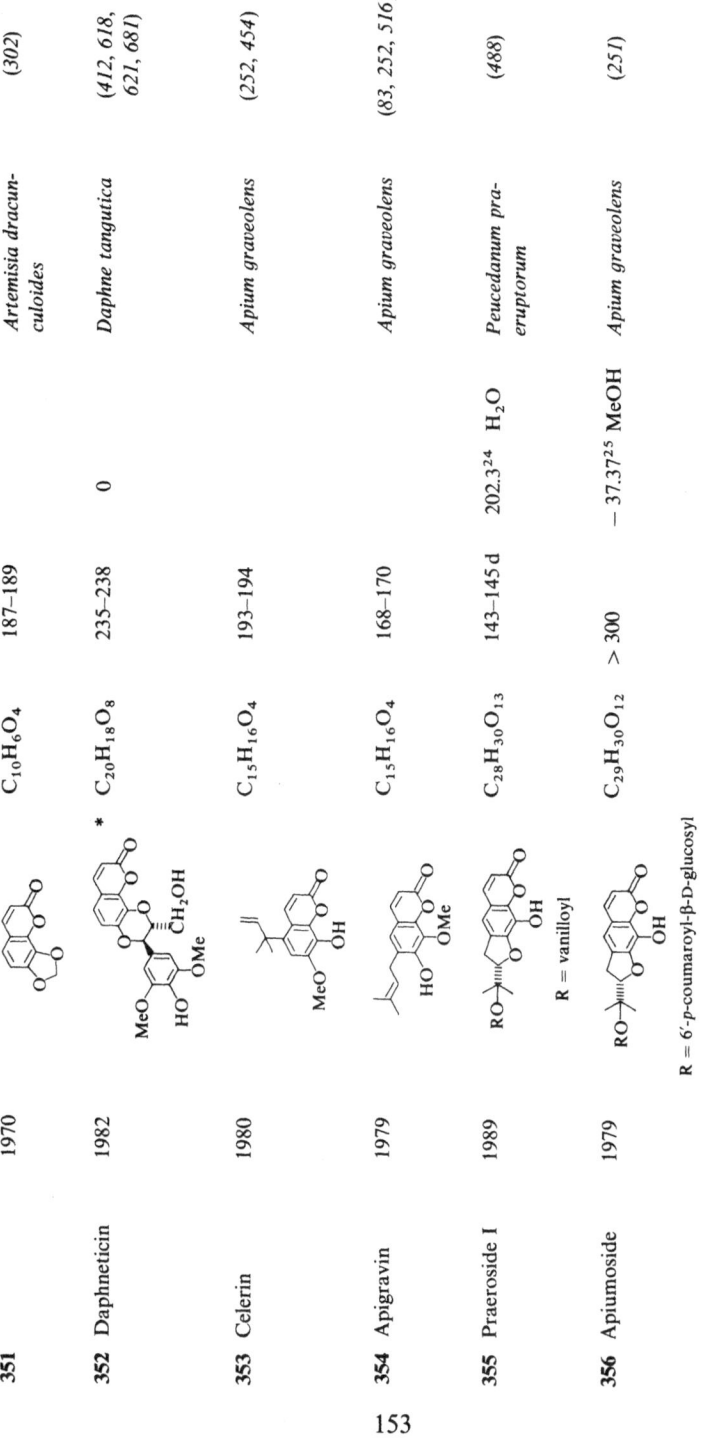 | $C_{10}H_6O_4$ | 187–189 | | *Artemisia dracunculoides* | (302) |
| **352** | Daphneticin | 1982 | | *$C_{20}H_{18}O_8$ | 235–238 | 0 | *Daphne tangutica* | (412, 618, 621, 681) |
| **353** | Celerin | 1980 | | $C_{15}H_{16}O_4$ | 193–194 | | *Apium graveolens* | (252, 454) |
| **354** | Apigravin | 1979 | | $C_{15}H_{16}O_4$ | 168–170 | | *Apium graveolens* | (83, 252, 516) |
| **355** | Praeroside I | 1989 | | $C_{28}H_{30}O_{13}$ | 143–145 d | $202.3^{24}$ $H_2O$ | *Peucedanum praeruptorum* | (488) |
| **356** | Apiumoside | 1979 | | $C_{29}H_{30}O_{12}$ | > 300 | $-37.37^{25}$ MeOH | *Apium graveolens* | (251) |

R = vanilloyl

R = 6′-p-coumaroyl-β-D-glucosyl

153

Table 4. (Continued)

| Trivial name(s) | Year isolated | Structure | Formula | M.p. | $[\alpha]^{t}_{\lambda}$ | Solvent | Plant sources | Leading references |
|---|---|---|---|---|---|---|---|---|
| **357** | 1988 | R = 6′-sinapinoyl-β-D-glucosyl | $C_{31}H_{34}O_{14}$ | 166–168 | $75^{20}$ | MeOH | *Apium graveolens* | (19) |
| **358** | 1984 | | $C_{14}H_{13}KO_8S$ | 128–130 | $-46.2^{23}$ | MeOH | *Seseli libanotis* | (409) |
| **359** Apiumetin Leptophyllidin | 1978 1978 | | $C_{14}H_{12}O_4$ | 198 200–202 | $-68.29^{19}$ $-38$ | $CHCl_3$ MeOH | *Apium graveolens* *Apium leptophyllum* | (249) (568, 571) |
| **360** | 1988 | β-D-glucosyl | $C_{20}H_{22}O_9$ | 165 | $-60^{20}$ | $CHCl_3$ | *Apium graveolens* | (19) |
| **361** Trichoclin | 1978 | | $C_{16}H_{14}O_5$ | 123.5–124 | | | *Trichocline incana* | (439) |
| **362** Heraclenol acetonide | 1977 | | $C_{19}H_{20}O_6$ | 98–99.5 | $-12.7$ | $CHCl_3$ | *Heracleum granatense* | (264) |

| No. | Name | Year | Structure | Formula | mp | [α] | Source | Ref. |
|---|---|---|---|---|---|---|---|---|
| 363 | Heraclenol 2'-O-isovalerate | 1986 | | * $C_{21}H_{24}O_7$ | gum | | Angelica archangelica | (604) |
| 364 | Heraclenol 2'-O-senecioate | 1986 | | * $C_{21}H_{22}O_7$ | gum | | Angelica archangelica | (604) |
| 365 | Dehydroindico-lactone / Wampetin / Indicolactone | 1983 / 1983 / 1984 | | * $C_{21}H_{18}O_6$ | 78 | $27.76^{22}$ CHCl$_3$ / $26.3^{25}$ CHCl$_3$ | Clausena lansium / Clausena wampi / Clausena anisata | (380) / (367) / (403) |
| 366 | Indicolactonediol | 1978 | | * $C_{21}H_{20}O_8$ | 116–117 | $13.3^{25}$ MeOH | Clausena indica | (513) |
| 367 | Anhydro-rutaretin | 1979 | | $C_{14}H_{12}O_4$ | 220–221 | | Apium leptophyllum | (567, 568) |

155

Table 4. (Continued)

| Trivial name(s) | Year isolated | Structure | Formula | M.p. | $[\alpha]_\lambda$ | Solvent | Plant sources | Leading references |
|---|---|---|---|---|---|---|---|---|
| **368** | 1980 | | $C_{16}H_{16}O_5$ | 83–85 | | | *Peucedanum ruthenicum* | (388) |
| **369** Arnottia-coumarin | 1974 | | $C_{15}H_{12}O_4$ | 140–145 | | | *Xanthoxylum arnottianum* | (324, 325) |
| **370** Seseloside | 1981 | * | $C_{20}H_{24}O_{10}$ | 257–259 | $-24.58^{20}$ | pyridine | *Seseli peucedanoides* | (57) |
| **371** Demethyl-luvangetin | 1977 | | $C_{14}H_{12}O_4$ | 195 | | | *Chloroxylon swietenia* | (100) |

Table 5. 5,6,7-Trioxygenated Coumarins

| Trivial name(s) | Year isolated | Structure | Formula | M.p. | $[\alpha]_\lambda^t$ | Solvent | Plant sources | Leading references |
|---|---|---|---|---|---|---|---|---|
| 372 Tomentin | 1988 | | $C_{11}H_{10}O_5$ | 185–186 | | | *Jatropha curcas* | (161, 448) |
| 373 | 1980 | | * $C_{14}H_{12}O_5$ | 211 | | | *Helichrysum serpyllifolium* | (120) |
| 374 | 1980 | | * $C_{15}H_{14}O_5$ | oil | | | *Helichrysum serpyllifolium* | (120) |
| 375 | 1987 | | * $C_{15}H_{14}O_6$ | oil | $- 31^{24}$ | $CHCl_3$ | *Helichrysum diosmifolium* | (341) |
| 376 | 1987 | | * $C_{15}H_{12}O_6$ | oil | | | *Helichrysum diosmifolium* | (341) |

Table 5. (Continued)

| Trivial name(s) | Year isolated | Structure | Formula | M.p. | $[\alpha]_\lambda^t$ | Solvent | Plant sources | Leading references |
|---|---|---|---|---|---|---|---|---|
| 377 Aleuritin | 1989 | | * $C_{21}H_{20}O_9$ | 238–239 | | | *Aleurites fordii* | (234) |
| 378 Murrayanone | 1988 | | $C_{17}H_{20}O_6$ | syrup | | | *Murraya paniculata* | (658) |
| 379 Murragleinin | 1984 | | * $C_{17}H_{22}O_7$ | 148–149 | − 43 | $CHCl_3$ | *Murraya gleinei* | (652) |
| 380 Heraclesol | 1978 | | $C_{17}H_{18}O_7$ | 117–118 | $30^{20}$ | MeOH | *Heracleum leskovii* | (379) |

Table 6. 5,7,8-Trioxygenated Coumarins

| Trivial name(s) | Year isolated | Structure | Formula | M.p. | $[\alpha]^t_\lambda$ | Solvent | Plant sources | Leading references |
|---|---|---|---|---|---|---|---|---|
| 381 Lacarol | 1985 | * | $C_{15}H_{18}O_6$ | 112–113 | | | Artemisia armeniaca | (612) |
| 382 Leptodactylone | 1978 | | $C_{11}H_{10}O_5$ | 149–152 | | | Leptodactylon cali-fornicum | (181) |
| 383 Methyllacarol | 1985 | * | $C_{16}H_{20}O_6$ | 104–106 | | | Artemisia laciniata | (612) |
| 384 Artanin | 1985 | | $C_{16}H_{18}O_5$ | 108–111 | | | Artemisia tanacetifolia | (612) |
| 385 | 1982 | | $C_{16}H_{18}O_5$ | | | | Toddalia aculeata | (532) |

159

Table 6. (Continued)

| Trivial name(s) | Year isolated | Structure | Formula | M.p. | $[\alpha]_\lambda^t$ | Solvent | Plant sources | Leading references |
|---|---|---|---|---|---|---|---|---|
| **386** Neoartanin | 1986 | | $C_{16}H_{18}O_5$ | 110–111 | | | Artemisia laciniata | (312) |
| **387** Prenyllacarol | 1985 | | * $C_{20}H_{26}O_6$ | 89–92 | | | Artemisia armeniaca | (612) |
| **388** Neoartaninepoxide | 1986 | | * $C_{16}H_{18}O_6$ | 145–146 | $-8^{20}$ | $CHCl_3$ | Artemisia laciniata | (312) |
| **389** Neoartanindiol | 1986 | | * $C_{16}H_{20}O_7$ | 126–128 | $30^{20}$ | $CHCl_3$ | Artemisia laciniata | (312) |

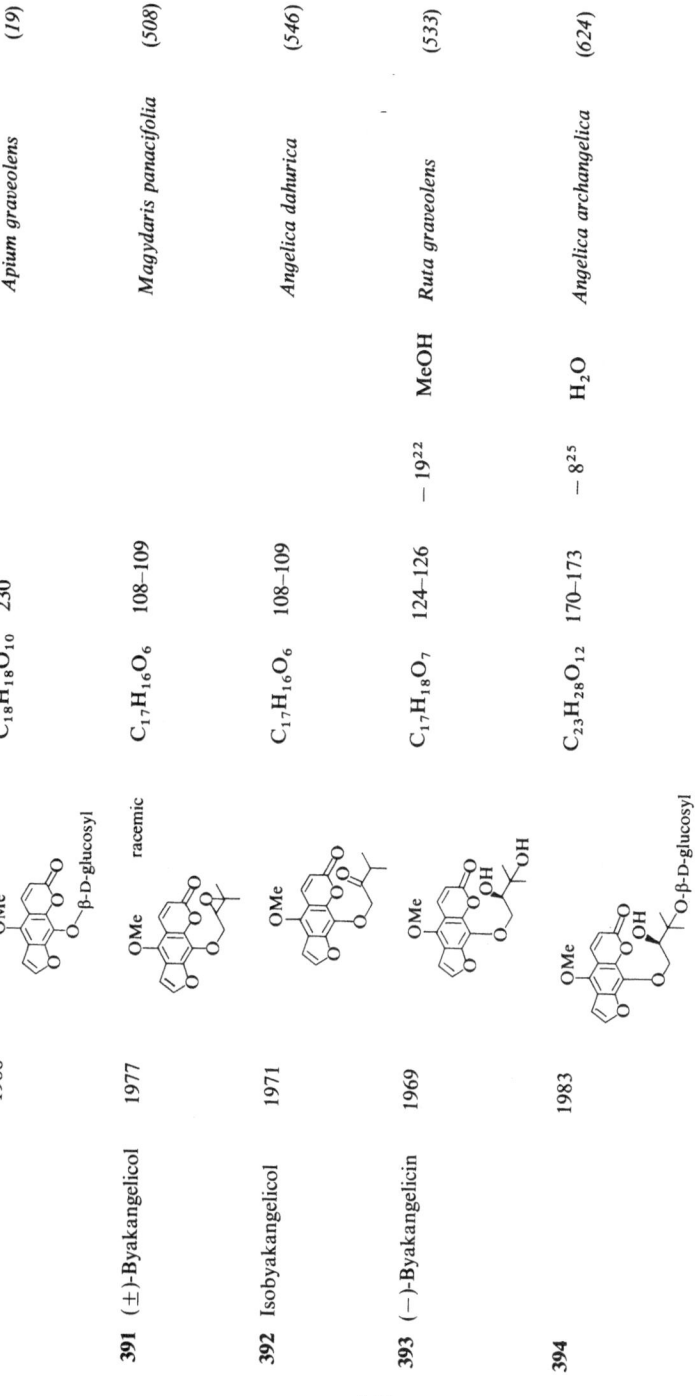

| | | | | | | | | |
|---|---|---|---|---|---|---|---|---|
| **390** | | 1988 | | $C_{18}H_{18}O_{10}$ | 230 | | *Apium graveolens* | (19) |
| **391** (±)-Byakangelicol | | 1977 | racemic | $C_{17}H_{16}O_6$ | 108–109 | | *Magydaris panacifolia* | (508) |
| **392** Isobyakangelicol | | 1971 | | $C_{17}H_{16}O_6$ | 108–109 | | *Angelica dahurica* | (546) |
| **393** (−)-Byakangelicin | | 1969 | | $C_{17}H_{18}O_7$ | 124–126 | $-19^{22}$ MeOH | *Ruta graveolens* | (533) |
| **394** | | 1983 | | $C_{23}H_{28}O_{12}$ | 170–173 | $-8^{25}$ H$_2$O | *Angelica archangelica* | (624) |

161

Table 6. (Continued)

| Trivial name(s) | Year isolated | Structure | Formula | M.p. | $[\alpha]_\lambda^t$ | Solvent | Plant sources | Leading references |
|---|---|---|---|---|---|---|---|---|
| **395** | 1983 | | $C_{23}H_{28}O_{12}$ | 111–114 | $-15^{25}$ | MeOH | *Angelica archangelica* | (624) |
| **396** | 1981 | | $C_{18}H_{20}O_7$ | 95–96 | $-17.8^{14}$ | EtOH | *Angelica dahurica* | (382) |
| **397** | 1983 | | * $C_{18}H_{20}O_7$ | 89 | $5.6^{21}$ | EtOH | *Angelica pachycarpa* | (435) |
| **398** | 1977 | | $C_{20}H_{22}O_7$ | 160–162 | $-12.9$ | $CHCl_3$ | *Heracleum granatense* | (264) |

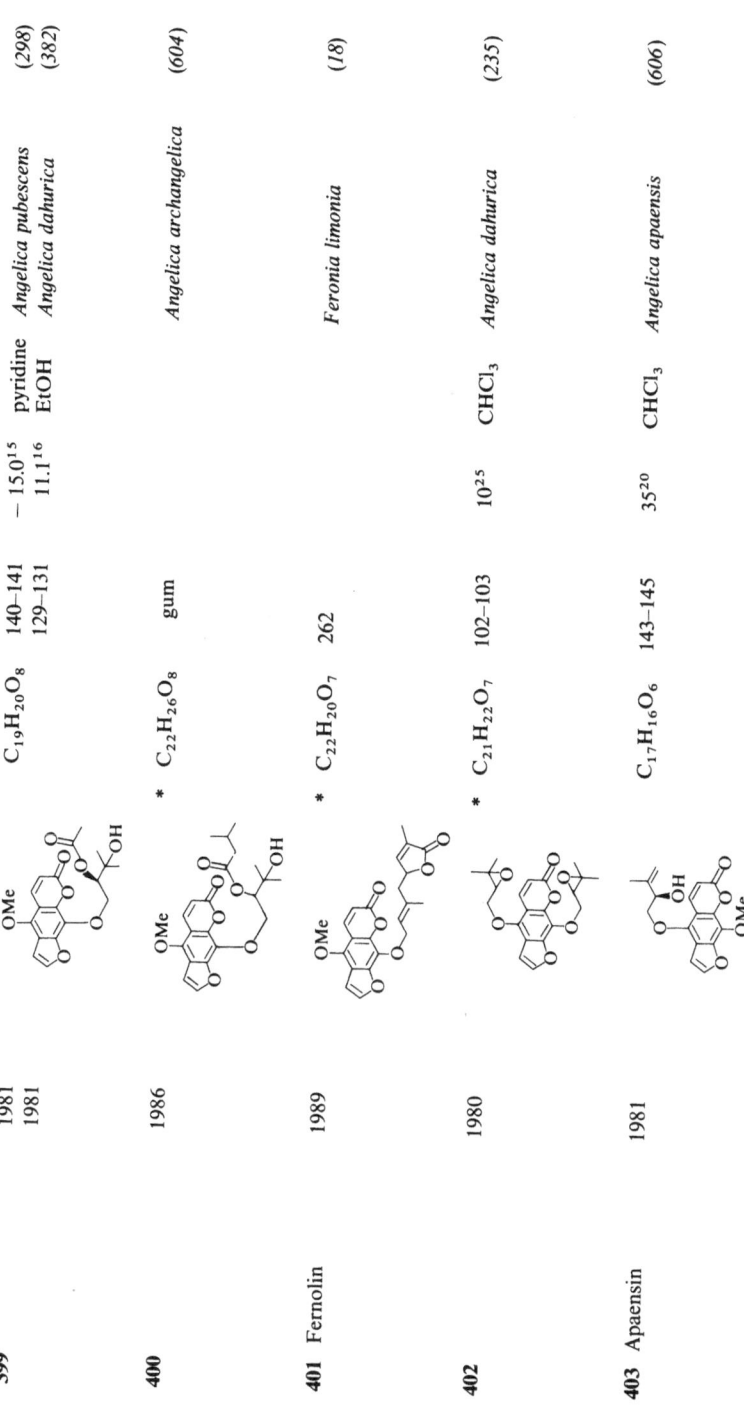

| | | | | | | |
|---|---|---|---|---|---|---|
| **399** | 1981 | | $C_{19}H_{20}O_8$ | 140–141 | $-15.0^{15}$ pyridine | *Angelica pubescens* | (298) |
| | 1981 | | | 129–131 | $11.1^{16}$ EtOH | *Angelica dahurica* | (382) |
| **400** | 1986 | | $*\ C_{22}H_{26}O_8$ | gum | | *Angelica archangelica* | (604) |
| **401** Fernolin | 1989 | | $*\ C_{22}H_{20}O_7$ | 262 | | *Feronia limonia* | (18) |
| **402** | 1980 | | $*\ C_{21}H_{22}O_7$ | 102–103 | $10^{25}$ CHCl$_3$ | *Angelica dahurica* | (235) |
| **403** Apaensin | 1981 | | $C_{17}H_{16}O_6$ | 143–145 | $35^{20}$ CHCl$_3$ | *Angelica apaensis* | (606) |

Table 6. (Continued)

| Trivial name(s) | Year isolated | Structure | Formula | M.p. | $[\alpha]_\lambda$ | Solvent | Plant sources | Leading references |
|---|---|---|---|---|---|---|---|---|
| 404 | 1983 | | $C_{23}H_{28}O_{12}$ | | $29^{25}$ | MeOH | *Angelica archangelica* | (624) |
| 405 | 1988 | * | $C_{22}H_{24}O_8$ | | $-1.6^{20}$ | $CHCl_3$ | *Peucedanum palustre* | (646) |
| 406 | 1967 | | $C_{22}H_{24}O_5$ | 74.5–75.5 | | | *Citrus medica* | (600) |
| 407 Racemosin Ceylantin | 1978 1984 | | $C_{16}H_{16}O_5$ | 125–126 126–127 | | | *Atalantia racemosa* *Atalantia ceylanica* | (25, 69, 349) (37, 450) |

Table 7. 6,7,8-Trioxygenated Coumarins

| Trivial name(s) | Year isolated | Structure | Formula | M.p. | $[\alpha]_D^t$ | Solvent | Plant sources | Leading references |
|---|---|---|---|---|---|---|---|---|
| **408** Erioside | 1980 | β-D-glucosyl-O | $C_{15}H_{16}O_{10}$ | 350d | | | Lasiosiphon eriocephalus | (92) |
| **409** Stylosin | 1983 | OR R = rhamnosylrhamnosylglucosyl | $C_{28}H_{38}O_{18}$ | 234–236 | | | Fraxinus stylosa | (288) |
| **410** | 1982 | β-D-glucosyl-O | $C_{16}H_{18}O_{10}$ | 164–166 | – 52.4 | DMF | Haplophyllum obtusifolium | (79) |
| **411** Haptusinol | 1979 | | * $C_{15}H_{18}O_6$ | 119–120 | 0 | | Haplophyllum obtusifolium | (11) |
| **412** Capensin | 1979 | | $C_{16}H_{18}O_5$ | 135–136 | | | Phyllosma capensis | (139) |
| **413** Obtusicin | 1980 | | $C_{15}H_{16}O_6$ | 89–91 | | | Haplophyllum obtusifolium | (80) |

165

Table 7. (Continued)

| Trivial name (s) | Year isolated | Structure | Formula | M.p. | $[\alpha]_\lambda$ | Solvent | Plant sources | Leading references |
|---|---|---|---|---|---|---|---|---|
| 414 Obtusin | 1979 | | * $C_{15}H_{14}O_5$ | 109–110 | $48.6^{20}$ | EtOH | Haplophyllum obtusifolium | (10) |
| 415 (−)-8-Methoxy-obliquin | 1986 | | * $C_{15}H_{14}O_5$ | gum | $-29.5_{546}$ | $CHCl_3$ | Helianthus hetero-phyllus | (303) |
| 416 Farnochrol | 1982 | | $C_{26}H_{34}O_5$ | oil | | | Artemisia vestita | (282) |
| 417 Epoxyfarnochrol | 1983 | | $C_{26}H_{34}O_6$ | oil | $-9^{20}$ | acetone | Achillea ochroleuca | (283) |
| 418 Oxofarnochrol | 1983 | | $C_{26}H_{34}O_6$ | oil | | | Achillea ochroleuca | (283) |
| 419 Deparnol | 1984 | | $C_{26}H_{34}O_6$ | oil | | | Achillea depressa | (309) |

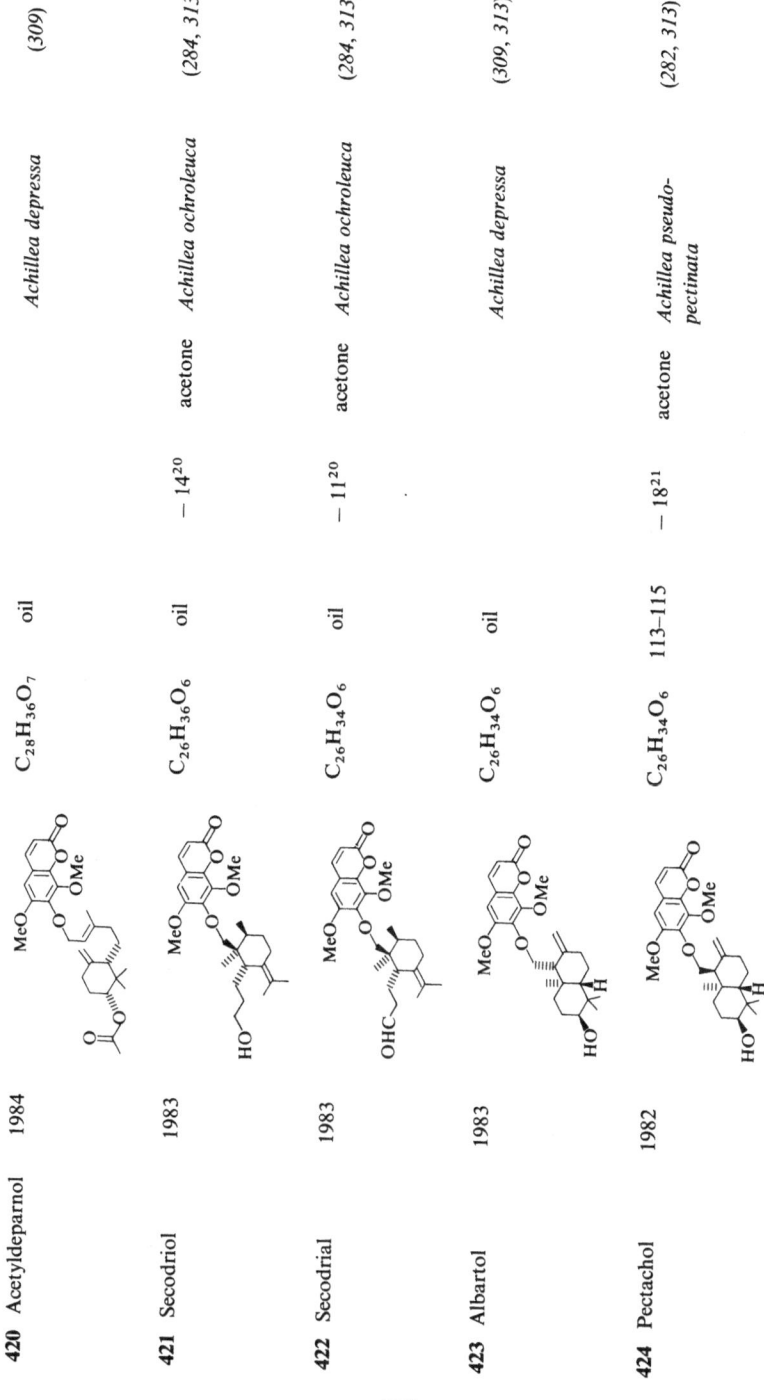

| No. | Name | Year | Structure | Formula | mp/oil | $[\alpha]$ | solvent | Source | Ref. |
|---|---|---|---|---|---|---|---|---|---|
| 420 | Acetyldeparnol | 1984 | | $C_{28}H_{36}O_7$ | oil | | | Achillea depressa | (309) |
| 421 | Secodriol | 1983 | | $C_{26}H_{36}O_6$ | oil | $-14^{20}$ | acetone | Achillea ochroleuca | (284, 313) |
| 422 | Secodrial | 1983 | | $C_{26}H_{34}O_6$ | oil | $-11^{20}$ | acetone | Achillea ochroleuca | (284, 313) |
| 423 | Albartol | 1983 | | $C_{26}H_{34}O_6$ | oil | | | Achillea depressa | (309, 313) |
| 424 | Pectachol | 1982 | | $C_{26}H_{34}O_6$ | 113–115 | $-18^{21}$ | acetone | Achillea pseudo-pectinata | (282, 313) |

167

Table 7. (Continued)

| Trivial name (s) | Year isolated | Structure | Formula | M.p. | $[\alpha]_\lambda$ | Solvent | Plant sources | Leading references |
|---|---|---|---|---|---|---|---|---|
| **425** Albartin | 1982 | | $C_{28}H_{36}O_7$ | oil | $2^{20}$ | acetone | *Artemisia alba* | (280, 313) |
| **426** Acetylpectachol | 1982 | | $C_{28}H_{36}O_7$ | 139–141 | $-7^{21}$ | acetone | *Achillea pseudo-pectinata* | (282, 313) |
| **427** Pectachol B | 1985 | | $C_{26}H_{34}O_6$ | | | | *Brocchia cinerea* | (281) |
| **428** Acetylpecta-chol B | 1985 | | $C_{28}H_{36}O_7$ | | | | *Brocchia cinerea* | (281) |

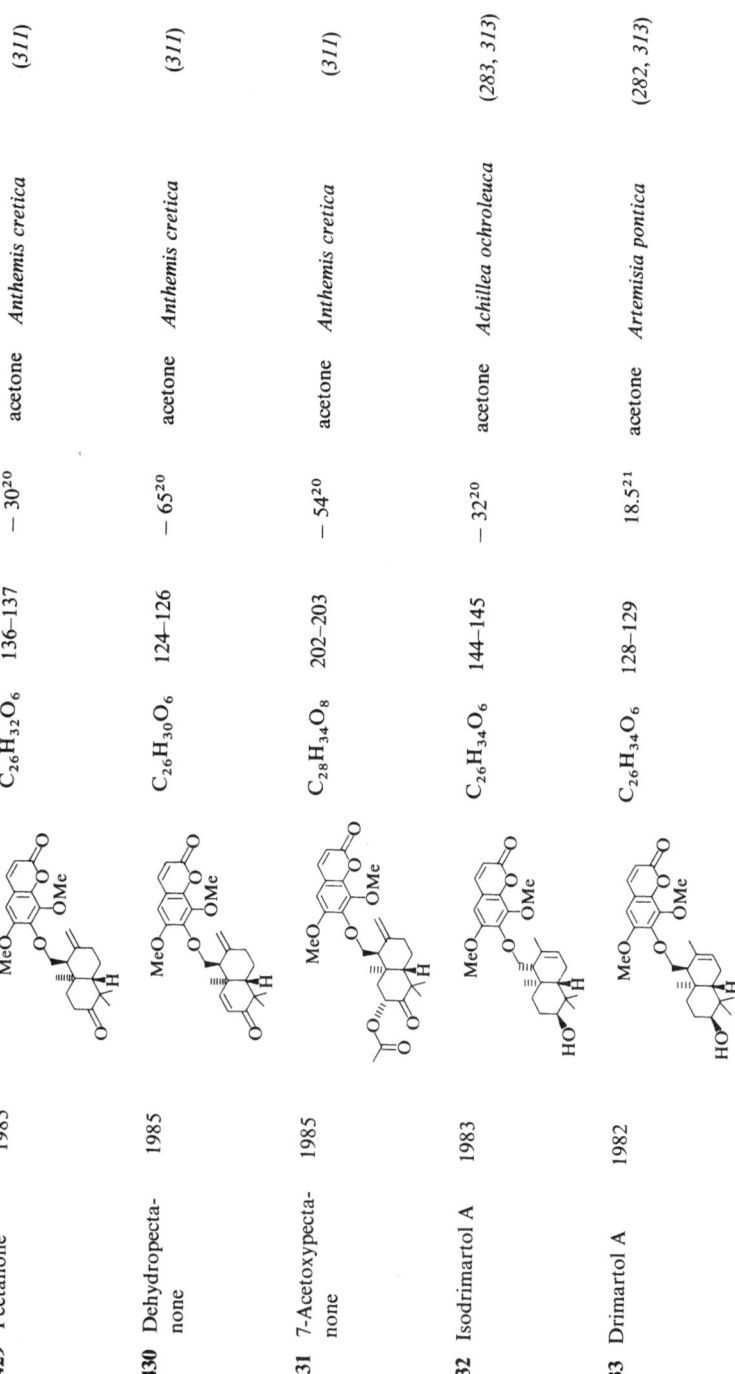

| | | | | | | | |
|---|---|---|---|---|---|---|---|
| **429** | Pectanone | 1985 | $C_{26}H_{32}O_6$ | 136–137 | − 30[20] | acetone | *Anthemis cretica* (311) |
| **430** | Dehydropecta-none | 1985 | $C_{26}H_{30}O_6$ | 124–126 | − 65[20] | acetone | *Anthemis cretica* (311) |
| **431** | 7-Acetoxypecta-none | 1985 | $C_{28}H_{34}O_8$ | 202–203 | − 54[20] | acetone | *Anthemis cretica* (311) |
| **432** | Isodrimartol A | 1983 | $C_{26}H_{34}O_6$ | 144–145 | − 32[20] | acetone | *Achillea ochroleuca* (283, 313) |
| **433** | Drimartol A | 1982 | $C_{26}H_{34}O_6$ | 128–129 | 18.5[21] | acetone | *Artemisia pontica* (282, 313) |

169

Table 7. (Continued)

| Trivial name (s) | Year isolated | Structure | Formula | M.p. | $[\alpha]_\lambda^t$ | Solvent | Plant sources | Leading references |
|---|---|---|---|---|---|---|---|---|
| **434** Acetylisodri-martol A | 1983 | | $C_{28}H_{36}O_7$ | oil | $-28^{20}$ | acetone | *Achillea ochroleuca* | (283, 313) |
| **435** Acetyldri-martol A | 1982 | | $C_{28}H_{36}O_7$ | oil | $110^{20}$ | acetone | *Achillea ochroleuca* | (280, 313) |
| **436** Drimartol B | 1982 | | $C_{26}H_{34}O_6$ | 145–146 | $-140^{21}$ | acetone | *Achillea ochroleuca* | (282, 313) |
| **437** Acetyldri-martol B | 1982 | | $C_{28}H_{36}O_7$ | 103–106 | $-145^{21}$ | acetone | *Achillea ochroleuca* | (282, 313) |

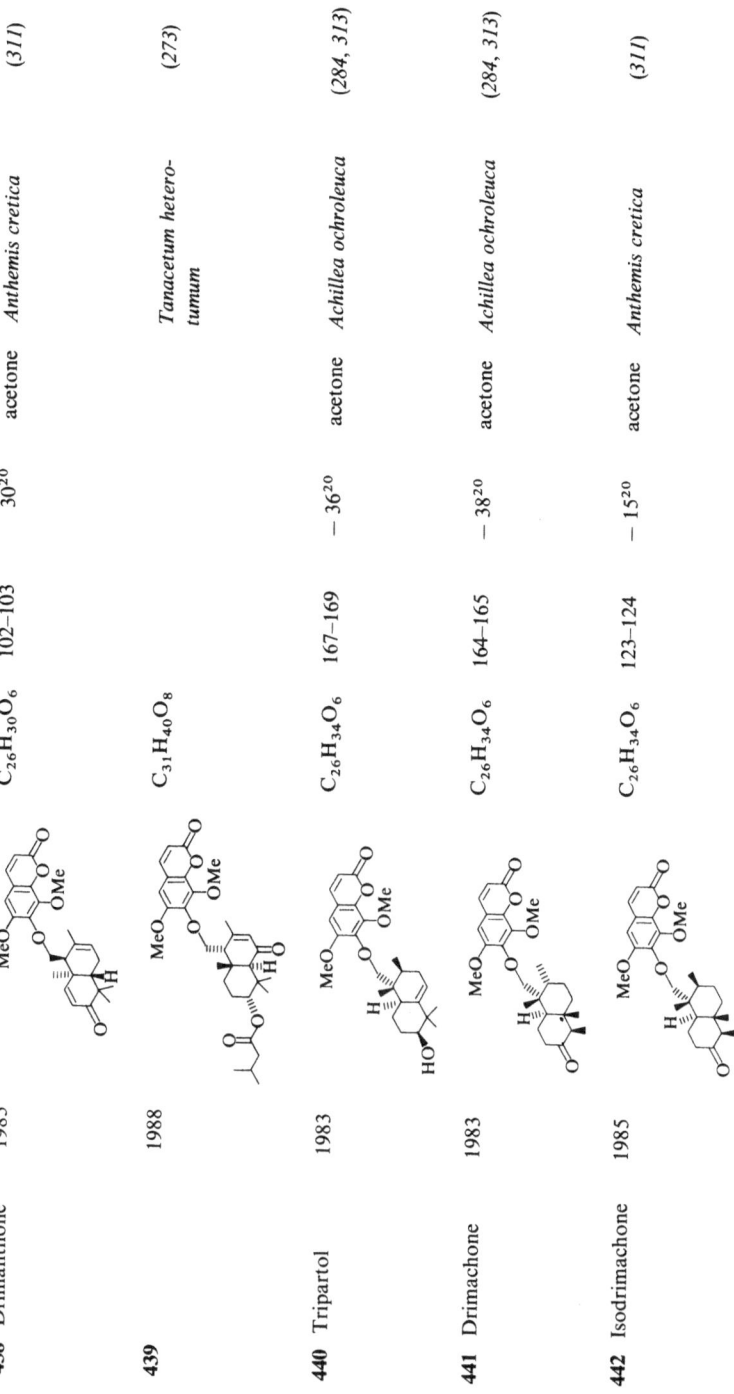

| | | | | | | | |
|---|---|---|---|---|---|---|---|
| **438** Drimanthone | 1985 | $C_{26}H_{30}O_6$ | 102–103 | $30^{20}$ | acetone | *Anthemis cretica* | *(311)* |
| **439** | 1988 | $C_{31}H_{40}O_8$ | | | | *Tanacetum hetero-tumum* | *(273)* |
| **440** Tripartol | 1983 | $C_{26}H_{34}O_6$ | 167–169 | $-36^{20}$ | acetone | *Achillea ochroleuca* | *(284, 313)* |
| **441** Drimachone | 1983 | $C_{26}H_{34}O_6$ | 164–165 | $-38^{20}$ | acetone | *Achillea ochroleuca* | *(284, 313)* |
| **442** Isodrimachone | 1985 | $C_{26}H_{34}O_6$ | 123–124 | $-15^{20}$ | acetone | *Anthemis cretica* | *(311)* |

171

Table 7. (Continued)

| Trivial name (s) | Year isolated | Structure | Formula | M.p. | $[\alpha]_\lambda^t$ | Solvent | Plant sources | Leading references |
|---|---|---|---|---|---|---|---|---|
| **443** | 1984 | | * $C_{20}H_{18}O_7$ | 245–248 | – 56 | | *Jatropha glandulifera* | (47, 411, 499, 500) |
| Jatrophin | 1988 | | | 242–244 | | | *Jatropha curcas* | (161) |
| **444** Cleomiscosin B | 1985 | | $C_{20}H_{18}O_8$ | 274 | | | *Cleome viscosa* | (396, 411, 524, 525, 620) |
| **445** Cleomiscosin D | 1988 | | * $C_{21}H_{20}O_9$ | 243–246 | | | *Cleome viscosa* | (396, 411, 617) |
| **446** Propacin | 1981 | | * $C_{20}H_{18}O_7$ | 226–228 | | | *Protium opacum* | (47, 179, 411, 616, 618) |

| No. | Name | Year | | Formula | mp | | Source | Ref. |
|---|---|---|---|---|---|---|---|---|
| **447** | Cleosandrin | 1979 | | | | | *Cleome icosandra* | (475) |
| | Cleomiscosin A | 1980 | * | $C_{20}H_{18}O_8$ | 247–249 | 0 | *Cleome viscosa* | (396, 412, 523–525, 618, 619) |
| | | 1984 | | | 250–252 | | *Soulamea soulameoides* | (45) |
| **448** | Aquillochin | 1982 | | | 220d | | *Aquilaria agallocha* | (90) |
| | Cleomiscosin C | 1985 | * | $C_{21}H_{20}O_9$ | 255 | 0 | *Cleome viscosa* | (47, 396, 411, 525, 617, 618) |
| **449** | Obtusiprenin | 1982 | | $C_{15}H_{16}O_5$ | 139–140 | | *Haplophyllum obtusifolium* | (433) |
| **450** | Obtusiprenol | 1981 | | $C_{15}H_{16}O_6$ | 106–108 | | *Haplophyllum obtusifolium* | (432) |

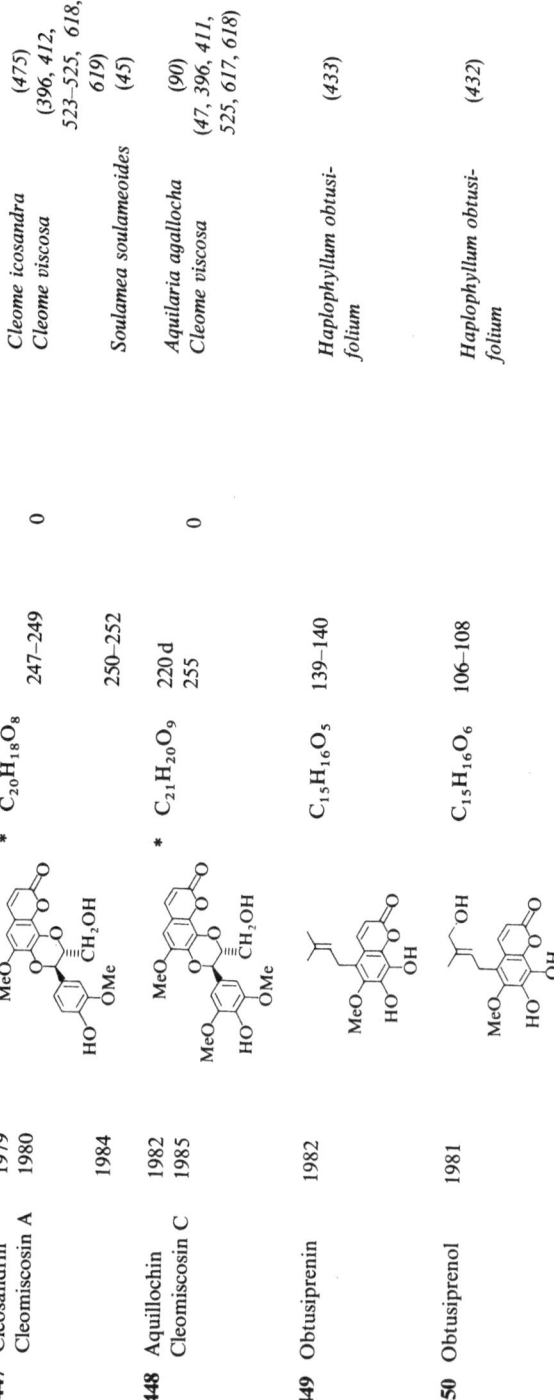

Table 8. 5,6,7,8-Tetraoxygenated Coumarins

| | Trivial name(s) | Year isolated | Structure | Formula | M.p. | $[\alpha]_\lambda^t$ | Solvent | Plant sources | Leading references |
|---|---|---|---|---|---|---|---|---|---|
| 451 | Isosabandin | 1986 | | $C_{12}H_{10}O_6$ | 128–131 | | | Artemisia lacianata | (308, 312) |

Table 9. 3-Substituted Coumarins

| | Trivial name(s) | Year isolated | Structure | Formula | M.p. | $[\alpha]_\lambda^t$ | Solvent | Plant sources | Leading references |
|---|---|---|---|---|---|---|---|---|---|
| 452 | Angustifolin | 1984 | | $C_{14}H_{14}O_3$ | | | | Ruta angustifolia | (184, 240) |
| 453 | | 1977 | | $C_{16}H_{16}O_4$ | 191–193 | | | Ruta sp. Tene 29662 | (268) |
| 454 | | 1977 | | $C_{18}H_{22}O_5$ | oil | | | Ruta sp. Tene 29662 | (268) |
| 455 | Anisocoumarin A | 1989 | | $C_{17}H_{18}O_4$ | oil | | | Clausena anisata | (479) |

174

| No. | Name | Year | Formula | mp | $[\alpha]_D$ | Solvent | Source | Ref. |
|---|---|---|---|---|---|---|---|---|
| 456 | Balsamiferone | 1979 | $C_{19}H_{22}O_3$ | 135–137 | | | Amyris balsamifera | (132, 136, 611) |
| 457 | Swieteno-coumarin I | 1980 | * $C_{20}H_{26}O_5$ | | | | Chloroxylon swietenia | (520) |
| 458 | | 1981 | $C_{19}H_{22}O_4$ | 199–200 | $23.1^{25}$ | CHCl$_3$ | Amyris elemifera | (133) |
| 459 | exo-Dehydro-chalepin | 1989 | $C_{19}H_{20}O_3$ | 100 | $7.4^{25}$ | CHCl$_3$ | Ruta graveolens | (531) |
| 460 | Anisocoumarin C | 1989 | * $C_{19}H_{22}O_5$ | oil | $15.5^{25}$ | CHCl$_3$ | Clausena anisata | (479) |
| 461 | Anisocou-marin D | 1989 | * $C_{19}H_{24}O_6$ | 210–211 | $20.5^{25}$ | MeOH | Clausena anisata | (479) |
| 462 | Elemiferone | 1985 | * $C_{16}H_{16}O_5$ | 184–185 | $39.0^{25}$ | EtOH | Amyris elemifera | (507) |
| 463 | | 1985 | * $C_{18}H_{18}O_6$ | 178.5–179.5 | $13.2^{25}$ | CHCl$_3$ | Amyris elemifera | (507) |

Table 9. (Continued)

| Trivial name(s) | Year isolated | Structure | Formula | M.p. | $[\alpha]_\lambda^t$ | Solvent | Plant sources | Leading references |
|---|---|---|---|---|---|---|---|---|
| **464** | 1985 | | * $C_{20}H_{20}O_7$ | 156–157 | $-6.5^{25}$ | CHCl$_3$ | *Amyris elemifera* | (507) |
| **465** Ulismoncadin A | 1986 | | * $C_{19}H_{22}O_4$ | 186 | $-4.6^{22}$ | CHCl$_3$ | *Helietta parvifolia* | (199) |
| **466** Shijiaocao-lactone A | 1989 | | $C_{24}H_{30}O_5$ | 95–97 | | | *Boenninghausenia sessilicarpara* | (162) |
| **467** Clausmarin A | 1978 | | * $C_{24}H_{30}O_5$ | 131 | $34^{18}$ | CHCl$_3$ | *Clausena pentaphylla* | (585) |
| **468** Clausmarin B | 1978 | | * $C_{24}H_{30}O_5$ | 80 | $-12.1^{18}$ | CHCl$_3$ | *Clausena pentaphylla* | (585) |

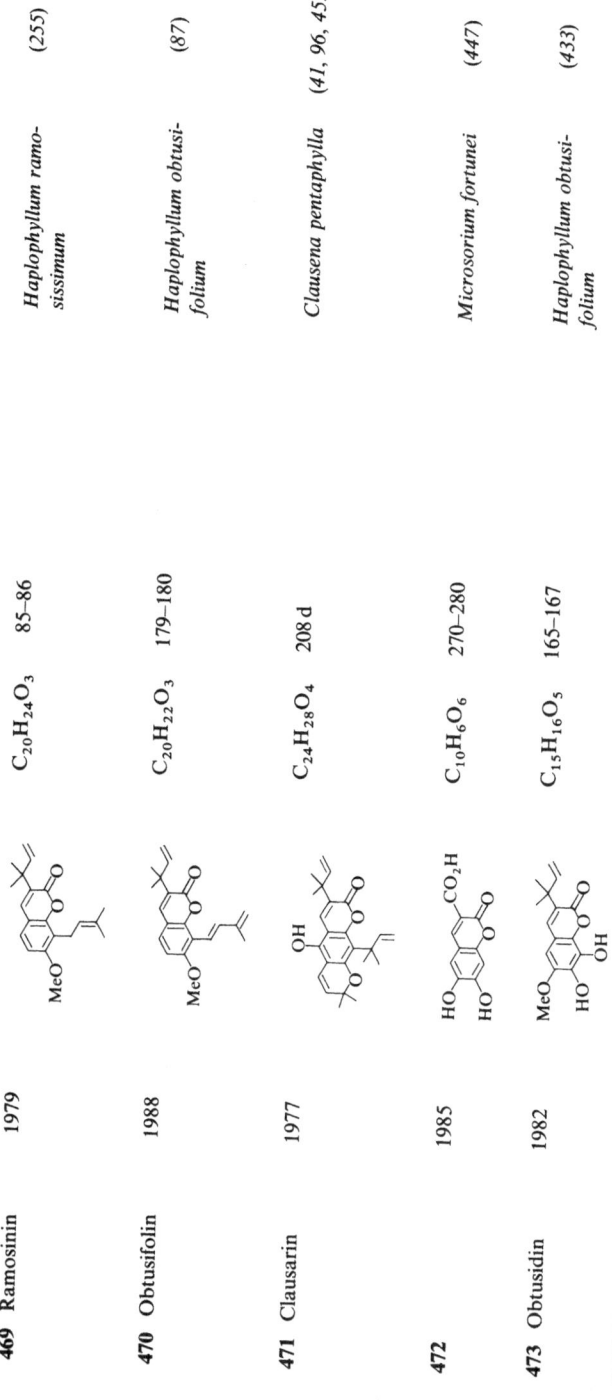

| 469 | Ramosinin | 1979 | $C_{20}H_{24}O_3$ | 85–86 | *Haplophyllum ramo-sissimum* | (255) |
| 470 | Obtusifolin | 1988 | $C_{20}H_{22}O_3$ | 179–180 | *Haplophyllum obtusi-folium* | (87) |
| 471 | Clausarin | 1977 | $C_{24}H_{28}O_4$ | 208 d | *Clausena pentaphylla* | (41, 96, 453) |
| 472 | | 1985 | $C_{10}H_6O_6$ | 270–280 | *Microsorium fortunei* | (447) |
| 473 | Obtusidin | 1982 | $C_{15}H_{16}O_5$ | 165–167 | *Haplophyllum obtusi-folium* | (433) |

Table 9.1. *3-Aryl-Substituted Coumarins*

| Trivial name(s) | Year isolated | Structure | Formula | M.p. | $[\alpha]_\lambda^t$ | Solvent | Plant sources | Leading references |
|---|---|---|---|---|---|---|---|---|
| **474** | 1974 | | $C_{16}H_{12}O_5$ | 258–260 | | | *Dalbergia oliveri* | (203) |
| **475** | 1974 | | $C_{16}H_{10}O_6$ | | | | *Dalbergia oliveri* | (203) |
| **476** Pachyrrhizin | 1945 | | $C_{19}H_{12}O_6$ | 206.5–207.5 | | | *Pachyrrhizus erosus* | (482, 517, 588) |

| No. | Name | Year | Structure | Formula | mp (°C) | | | Source | Ref. |
|---|---|---|---|---|---|---|---|---|---|
| 477 | Glycycoumarin | 1986 | | $C_{21}H_{20}O_6$ | 234–236 | | | *Glycyrrhiza uralensis* | (300) |
| 478 | Glycyrin | 1978 | | $C_{22}H_{22}O_6$ | 209–211 | | | *Glycyrrhiza spp.* | (372) |
| 479 | Licopyrano-coumarin | 1988 | * | $C_{21}H_{20}O_7$ | 137 | 14 | acetone | *Xi-bai licorice* | (299) |
| 480 | Neofolin | 1966 | | $C_{20}H_{14}O_7$ | 155.5–157.5 167–168.5 189.5–190.5 | | | *Neorautenia ficifolia* | (126) |

179

## Table 10. 4-Substituted Coumarins

| | Trivial name(s) | Year isolated | Structure | Formula | M.p. | $[\alpha]_\lambda^t$ | Solvent | Plant sources | Leading references |
|---|---|---|---|---|---|---|---|---|---|
| 481 | | 1977 | | $C_{10}H_8O_3$ | 180–181 | | | *Dalbergia volubilis* | (157) |
| 482 | | 1977 | | $C_{12}H_{10}O_4$ | 150–151 | | | *Trigonella foenumgraecum* | (94) |
| 483 | | 1987 | | $C_{13}H_{14}O_3$ | 198–200 | | | *Macrothelypteris torresiana* | (315) |
| 484 | Surangin C | 1986 | | * $C_{27}H_{36}O_6$ | oil | $16.6^{25}$ | CHCl$_3$ | *Mammea longifolia* | (418) |

| No. | Name | Year | Structure | Formula | mp | [α] | Solvent | Source | Ref. |
|---|---|---|---|---|---|---|---|---|---|
| 485 | Oblongulide | 1985 | | $C_{21}H_{22}O_5$ | 126 | | | *Calophyllum cordatooblongum* | (196) |
| 486 | Cordatolide B | 1985 | | *$C_{20}H_{22}O_5$ | 178 | −23.2 | $CHCl_3$ | *Calophyllum cordatooblongum* | (196) |
| 487 | Cordatolide A | 1985 | | *$C_{20}H_{22}O_5$ | 85 | 54.8 | $CHCl_3$ | *Calophyllum cordatooblongum* | (196) |
| 488 | | 1987 | | $C_{11}H_8O_4$ | | | | *Achillea schischkinii* | (632) |
| 489 | Rutalpinin | 1988 | | $C_{15}H_{14}O_4$ | | | | *Ruta chalepensis* | (633) |
| 490 | Troupin | 1985 | | $C_{12}H_{12}O_5$ | 178–179 | | | *Tamarix troupii* | (497) |

Table 10.1. *4-Aryl-Substituted Coumarins*

| | Trivial name(s) | Year isolated | Structure | Formula | M.p. | $[\alpha]_\lambda^t$ | Solvent | Plant sources | Leading references |
|---|---|---|---|---|---|---|---|---|---|
| **491** | Serratin | 1982 | | $C_{15}H_{10}O_4$ | 213 | | | *Passiflora serratodigitata* | (631) |
| **492** | Nivegin | 1987 | | $C_{15}H_{10}O_5$ | 262–264 | | | *Echinops niveus* | (591) |
| **493** | | 1982 | | $C_{21}H_{20}O_9$ | 166–168 | | | *Passiflora serratodigitata* | (631) |
| **494** | | 1983 | | $C_{16}H_{12}O_6$ | 208–210 | | | *Coutarea hexandra* | (527) |
| **495** | | 1988 | | $C_{17}H_{14}O_6$ | 225–226 | | | *Exostema caribaeum* | (426) |

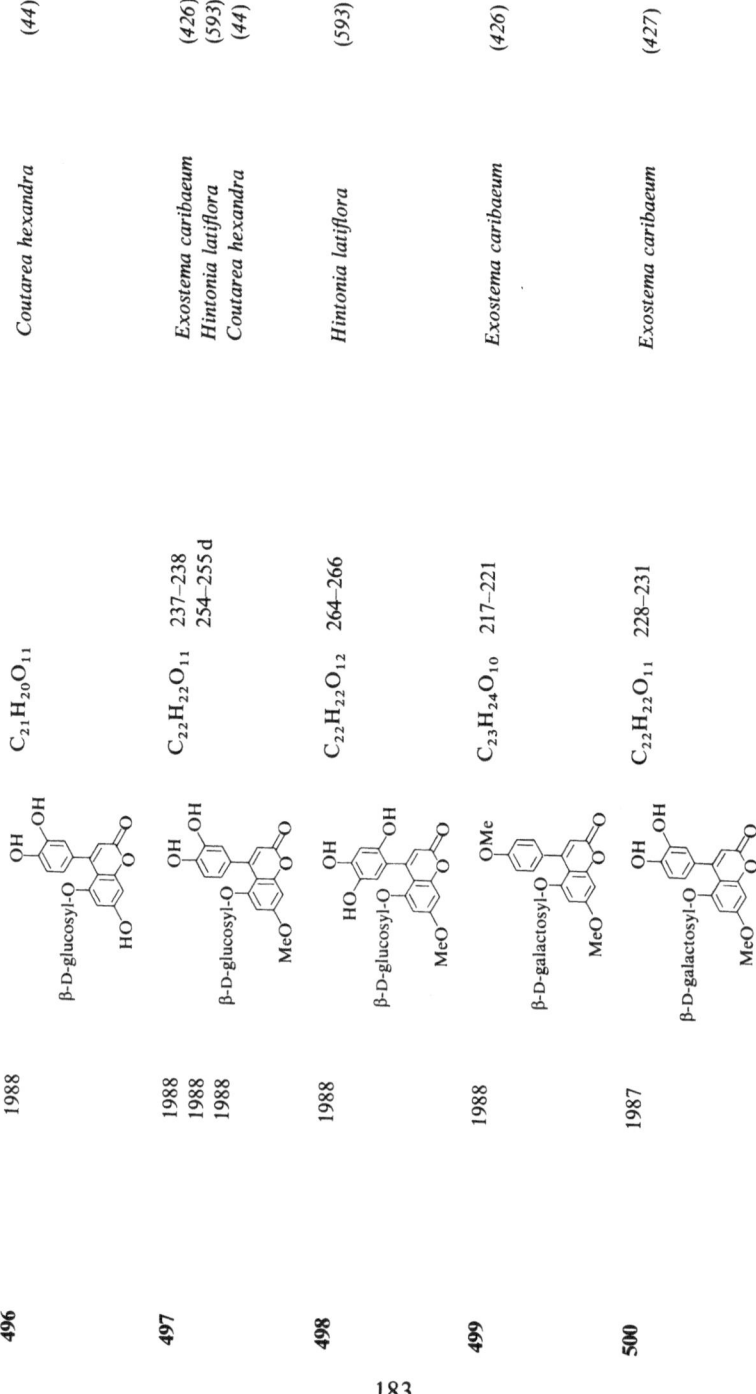

| | | | | | |
|---|---|---|---|---|---|
| **496** | 1988 | | $C_{21}H_{20}O_{11}$ | *Coutarea hexandra* | *(44)* |
| **497** | 1988<br>1988<br>1988 | | $C_{22}H_{22}O_{11}$ 237–238<br>254–255 d | *Exostema caribaeum*<br>*Hintonia latiflora*<br>*Coutarea hexandra* | *(426)*<br>*(593)*<br>*(44)* |
| **498** | 1988 | | $C_{22}H_{22}O_{12}$ 264–266 | *Hintonia latiflora* | *(593)* |
| **499** | 1988 | | $C_{23}H_{24}O_{10}$ 217–221 | *Exostema caribaeum* | *(426)* |
| **500** | 1987 | | $C_{22}H_{22}O_{11}$ 228–231 | *Exostema caribaeum* | *(427)* |

Table 10.1. (Continued)

| Trivial name(s) | Year isolated | Structure | Formula | M.p. | $[\alpha]_\lambda^t$ | Solvent | Plant sources | Leading references |
|---|---|---|---|---|---|---|---|---|
| 501 | 1988 | R = 6′-acetyl-β-D-galactosyl | $C_{24}H_{24}O_{12}$ | 215–220 | | | *Exostema caribaeum* | (426) |
| 502 | 1988 | R = 6′-β-D-apiosyl-β-D-glucosyl | $C_{26}H_{28}O_{15}$ | | | | *Coutarea hexandra* | (44) |
| 503 | 1988 | R = 6′-β-D-xylosyl-β-D-glucosyl | $C_{27}H_{30}O_{15}$ | | | | *Coutarea hexandra* | (44) |
| 504 | 1989 | | $C_{16}H_{12}O_6$ | 269–270 | | | *Coutarea hexandra* | (192, 318) |

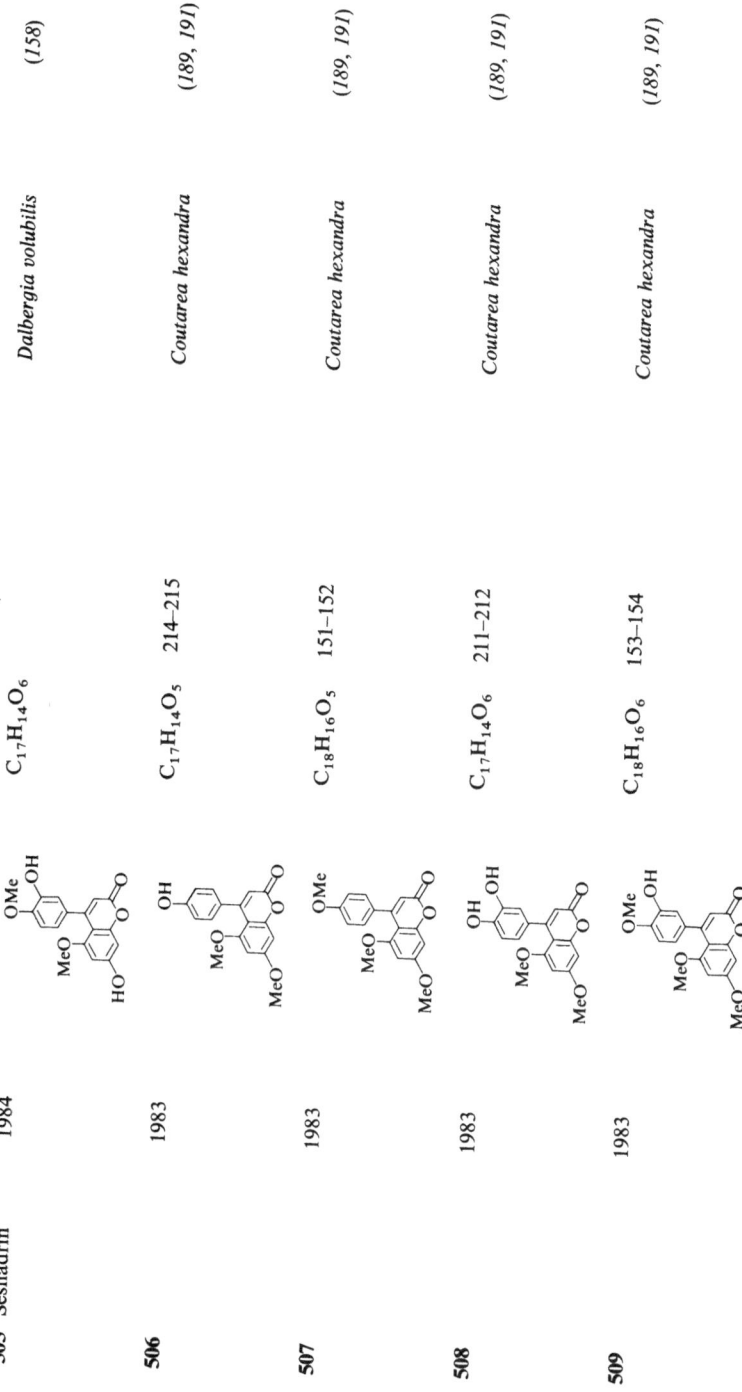

| | | | | | |
|---|---|---|---|---|---|
| **505** Seshadrin | 1984 | | $C_{17}H_{14}O_6$ | | *Dalbergia volubilis* | (158) |
| **506** | 1983 | | $C_{17}H_{14}O_5$ | 214–215 | *Coutarea hexandra* | (189, 191) |
| **507** | 1983 | | $C_{18}H_{16}O_5$ | 151–152 | *Coutarea hexandra* | (189, 191) |
| **508** | 1983 | | $C_{17}H_{14}O_6$ | 211–212 | *Coutarea hexandra* | (189, 191) |
| **509** | 1983 | | $C_{18}H_{16}O_6$ | 153–154 | *Coutarea hexandra* | (189, 191) |

Table 10.1. (Continued)

| Trivial name(s) | Year isolated | Structure | Formula | M.p. | $[\alpha]_\lambda^t$ | Solvent | Plant sources | Leading references |
|---|---|---|---|---|---|---|---|---|
| 510 | 1984 | | $C_{18}H_{14}O_6$ | 194–195 | | | *Coutarea hexandra* | (188, 189) |
| 511 | 1987 | | $C_{15}H_8O_6$ | 350d | | | *Exostema caribaeum* | (427) |
| 512 | 1984 | | $C_{16}H_{10}O_6$ | 335–342 d | | | *Coutarea latiflora* | (526) |
| 513 | 1987 | | $C_{17}H_{12}O_6$ | 273–274 | | | *Exostema caribaeum* | (427) |

186

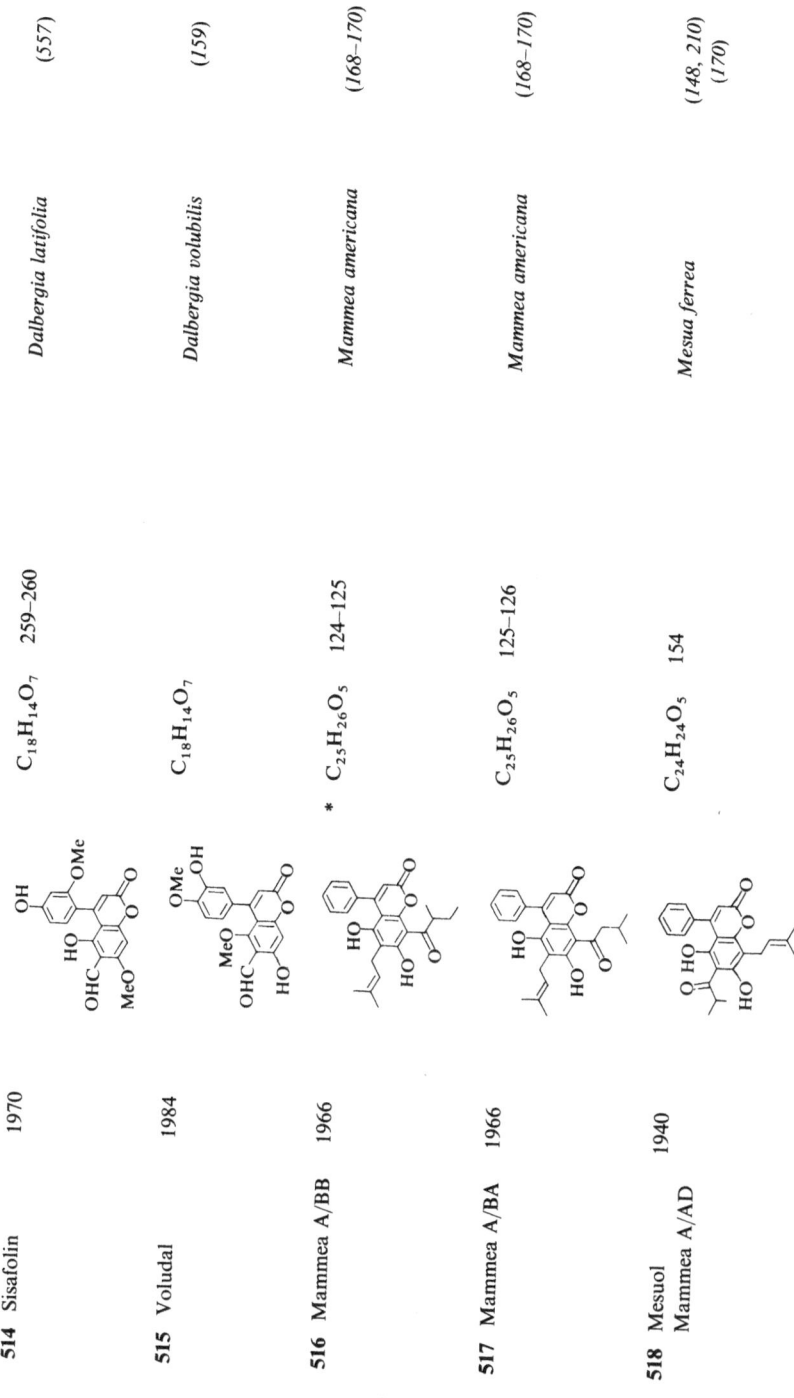

| No. | Name | Year | Formula | M.p. (°C) | Source | Ref. |
|---|---|---|---|---|---|---|
| **514** | Sisafolin | 1970 | $C_{18}H_{14}O_7$ | 259–260 | *Dalbergia latifolia* | *(557)* |
| **515** | Voludal | 1984 | $C_{18}H_{14}O_7$ | | *Dalbergia volubilis* | *(159)* |
| **516** | Mammea A/BB | 1966 | * $C_{25}H_{26}O_5$ | 124–125 | *Mammea americana* | *(168–170)* |
| **517** | Mammea A/BA | 1966 | $C_{25}H_{26}O_5$ | 125–126 | *Mammea americana* | *(168–170)* |
| **518** | Mesuol Mammea A/AD | 1940 | $C_{24}H_{24}O_5$ | 154 | *Mesua ferrea* | *(148, 210) (170)* |

187

188

Table 10.1. (Continued)

| Trivial name(s) | Year isolated | Structure | Formula | $[\alpha]_\lambda^t$ | Solvent | Plant sources | Leading references |
|---|---|---|---|---|---|---|---|
| **519** | 1981 | | $C_{24}H_{24}O_5$ | | | Ochrocarpus siamensis | (625) |
| **520** Mammea A/AB MAB 1 | 1966 1970 | | * $C_{25}H_{26}O_5$ | | | Mammea americana Mammea africana | (168–170) (140) |
| **521** Mammeisin Mammea A/AA | 1959 1966 | | $C_{25}H_{26}O_5$ | | | Mammea americana Mammea americana | (225, 227) (168–170) |
| **522** | 1975 | | * $C_{24}H_{24}O_6$ | | | Mesua thwaitesii | (62) |

| No. | Name | Year | Structure | | Formula | M.p. (°C) | Source | Ref. |
|---|---|---|---|---|---|---|---|---|
| 523 | Calaustralin | 1975 | | * | $C_{25}H_{24}O_5$ | 197–198 | *Calophyllum inophyllum* | (93) |
| 524 | Ponnalide Mammea A/BB cyclo D | 1965 | | * | $C_{25}H_{24}O_5$ | 159–160 | *Calophyllum inophyllum* | (16, 461) (170) |
| 525 | Isomammeigin | 1988 | | | $C_{25}H_{24}O_6$ | 173–175 | *Kielmeyera pumila* | (472) |
| 526 | Calophyllolide | 1951 | | | $C_{26}H_{24}O_5$ | 158–160 | *Calophyllum inophyllum* | (492, 511) |
| 527 | Mesuagin | 1969 | | | $C_{24}H_{22}O_5$ | 152–153 | *Mesua ferrea* | (59, 99, 147) |

Table 10.1. (Continued)

| Trivial name(s) | Year isolated | Structure | Formula | M.p. | $[\alpha]_\lambda^t$ | Solvent | Plant sources | Leading references |
|---|---|---|---|---|---|---|---|---|
| **528** Mesuarin | 1988 | | $C_{25}H_{24}O_5$ | 130 | | | *Mesua ferrea* | (98) |
| **529** Mammeigin | 1964 | | $C_{25}H_{24}O_5$ | 144–146 | | | *Mammea americana* | (228) |
| **530** | 1981 | | $C_{24}H_{22}O_5$ | 138–139 | | | *Ochrocarpus siamensis* | (625) |
| **531** Apetalolide | 1967 | | $C_{26}H_{24}O_5$ | 203–205 | | | *Calophyllum apetalum* | (480) |

| No. | Name | Year | Structure | | Formula | mp (°C) | $[\alpha]$ | Solvent | Source | Ref. |
|---|---|---|---|---|---|---|---|---|---|---|
| 532 | Tomentolide A | 1967 | | * | $C_{25}H_{22}O_5$ | 201–205 | | | Calophyllum tomentosum | (480) |
| 533 | Soulattrolide | 1977 | | * | $C_{25}H_{24}O_5$ | 201–202 | $-29.6^{22}$ | $CHCl_3$ | Calophyllum soulattri | (287) |
| 534 | Inophyllum B | 1972 | | * | $C_{25}H_{24}O_5$ | | $36^{20}$ | $CHCl_3$ | Calophyllum inophyllum | (365) |
| 535 | Inophyllum D | 1972 | | * | $C_{25}H_{24}O_5$ | | $35^{20}$ | $CHCl_3$ | Calophyllum inophyllum | (365) |
| 536 | Inophyllum A | 1968 | | * | $C_{25}H_{24}O_5$ | 200–202 | $43^{20}$ | acetone | Calophyllum inophyllum | (364, 365) |

191

Table 10.1. (Continued)

| | Trivial name(s) | Year isolated | Structure | Formula | M.p. | $[\alpha]_\lambda^t$ | Solvent | Plant sources | Leading references |
|---|---|---|---|---|---|---|---|---|---|
| 537 | Inophyllum C trans-(+)-Inophyllolide | 1968 1972 | | * $C_{25}H_{22}O_5$ | 188–191 | $13^{20}$ | $CHCl_3$ | Calophyllum inophyllum | (364, 365) |
| 538 | Inophyllolide | 1956 | racemic | $C_{25}H_{22}O_5$ | 186–188 | | | Calophyllum inophyllum | (511, 512) |
| 539 | Inophyllum E cis-(+)-Inophyllolide | 1968 1972 | | * $C_{25}H_{22}O_5$ | 149–150 | $70^{20}$ | $CHCl_3$ | Calophyllum inophyllum | (364, 365) |
| 540 | Nordalbergin | 1971 | | $C_{15}H_{10}O_4$ | 268–269 | | | Dalbergia sissoo | (154, 242, 446) |
| 541 | Dalbergin | 1953 | | $C_{16}H_{12}O_4$ | 210 | | | Dalbergia sissoo | (22, 34, 154, 362) |

192

| | | | $C_{16}H_{12}O_5$ | 254 | *Dalbergia stevensonii* | (35, 205) |
|---|---|---|---|---|---|---|
| **542** | Stevenin | 1973 | | | | |
| **543** | Melanettin | 1975 | $C_{16}H_{12}O_5$ | 233–234 | *Dalbergia melanoxylon* | (33, 204) |
| **544** | Isodalbergin | 1971 | $C_{16}H_{12}O_4$ | 195–196 | *Dalbergia sissoo* | (22, 446) |
| **545** | Melannein | 1966 | $C_{17}H_{14}O_6$ | 221–223 | *Dalbergia baroni* | (201, 202, 445) |
| **546** | Methyldalbergin | 1957 | $C_{17}H_{14}O_4$ | 145–146 | *Dalbergia sissoo* | (22, 34) |
| **547** | Exostemin | 1967 | $C_{18}H_{16}O_6$ | 173–174 | *Exostemma caribaeum* | (190, 444, 445, 552, 553) |

Table 10.1. (Continued)

| Trivial name(s) | Year isolated | Structure | Formula | M.p. | $[\alpha]_\lambda^t$ | Solvent | Plant sources | Leading references |
|---|---|---|---|---|---|---|---|---|
| **548** | 1989 | | $C_{17}H_{14}O_7$ | | | | *Coutarea hexandra* | (174) |
| **549** | 1987 | | $C_{19}H_{18}O_6$ | 145–146 | | | *Coutarea hexandra* | (190) |
| **550** | 1989 | | $C_{19}H_{16}O_7$ | | | | *Coutarea hexandra* | (175) |
| **551** Kuhlmannin | 1978 | | $C_{17}H_{14}O_5$ | 211 | | | *Machaerium kuhlmannii* | (489) |

Table 11. *Miscellaneous Coumarins*

| Trivial name(s) | Year isolated | Structure | Formula | M.p. | $[\alpha]_\lambda^t$ | Solvent | Plant sources | Leading references |
|---|---|---|---|---|---|---|---|---|
| **552** Trigoforin | 1982 | | $C_{12}H_{12}O_2$ | 112–113 | | | *Trigonella foenum-graecum* | *(371, 422)* |
| **553** | 1977 | | $C_9H_6O_3$ | 153–155 | | | *Alyxia lucida* | *(540)* |
| **554** | 1969 | | $C_9H_6O_3$ | 210–212 | | | *Penicillium jensenii* | *(564, 597, 599)* |
| **555** Foetidin | 1985 | | $C_{24}H_{30}O_4$ | 176–178 | $-39.8^{20}$ | EtOH | *Ferula assafoetida* | *(130)* |
| **556** 12′-Hydroxy-ferulenol | 1987 | | $C_{24}H_{30}O_4$ | gum | | | *Ferula communis* | *(635)* |
| (E)-ω-Hydroxy-ferulenol | 1988 | | | | | | *Ferula communis* | *(42)* |
| **557** ω-Hydroxyferulenol | 1987 | | $C_{24}H_{30}O_4$ | gum | | | *Ferula communis* | *(405)* |
| (Z)-ω-Hydroxy-ferulenol | 1988 | | | | | | *Ferula communis* | *(42)* |

195

Table 11. (Continued)

| Trivial name(s) | Year isolated | Structure | Formula | M.p. | $[\alpha]_\lambda^t$ | Solvent | Plant sources | Leading references |
|---|---|---|---|---|---|---|---|---|
| 558 (E)-ω-Acetoxy-ferulenol | 1988 | | $C_{26}H_{32}O_5$ | oil | | | Ferula communis | (42) |
| 559 (Z)-ω-Acetoxy-ferulenol | 1988 | | $C_{26}H_{32}O_5$ | oil | | | Ferula communis | (42) |
| 560 (E)-ω-Oxoferulenol | 1988 | | $C_{24}H_{28}O_4$ | oil | | | Ferula communis | (42) |
| 561 Ferprenin | 1988 | | * $C_{24}H_{28}O_3$ | oil | $10^{20}$ | CHCl$_3$ | Ferula communis | (43) |
| 562 (E)-ω-Hydroxy-ferprenin | 1988 | | * $C_{24}H_{28}O_4$ | oil | $0^{25}$ | | Ferula communis | (42) |
| 563 (Z)-ω-Hydroxy-ferprenin | 1988 | | * $C_{24}H_{28}O_4$ | oil | $0^{25}$ | | Ferula communis | (42) |

| No. | Name | Year | Structure | Formula | mp (°C) | [α] | Solvent | Source | Ref. |
|---|---|---|---|---|---|---|---|---|---|
| 564 | (E)-ω-Acetoxy-ferprenin | 1988 | * | $C_{26}H_{30}O_5$ | oil | $0^{25}$ | | Ferula communis | (42) |
| 565 | (Z)-ω-Acetoxy ferprenin | 1988 | * | $C_{26}H_{30}O_5$ | oil | $0^{25}$ | | Ferula communis | (42) |
| 566 | (E)-ω-Oxoferprenin | 1988 | CHO * | $C_{24}H_{26}O_4$ | oil | $0^{25}$ | | Ferula communis | (42) |
| 567 | | 1987 | Me OH | $C_{10}H_8O_3$ | 229–230 | | | Gerbera anandria | (285) |
| 568 | | 1980 | Me O-β-D-glucosyl | $C_{16}H_{18}O_8$ | 150 | $-117.5^{24}$ | MeOH | Ethulia conyzoides | (420) |
| | | 1985 | | | 152–154 | $-106.0^{20}$ | MeOH | Gerbera jamesonii | (473) |
| | | 1987 | | | 153–154 | $-109^{21}$ | MeOH | Gerbera anandria | (285) |
| 569 | | 1985 | Me O-rutinosyl | $C_{22}H_{28}O_{12}$ | 240–241 d | $-79.4^{22}$ | pyridine | Gerbera jamesonii | (473) |
| 570 | | 1987 | Me O-cellobiosyl | $C_{22}H_{28}O_{13}$ | 217–219 | $-94^{21}$ | MeOH | Gerbera anandria | (285) |

197

Table 11. (Continued)

| Trivial name(s) | Year isolated | Structure | Formula | M.p. | $[\alpha]_\lambda^1$ | Solvent | Plant sources | Leading references |
|---|---|---|---|---|---|---|---|---|
| **571** | 1987 | Me O-gentiobiosyl (structure) | $C_{22}H_{28}O_{13}$ | 155–157 | $-80^{21}$ | MeOH | *Gerbera anandria* | (285) |
| **572** | 1988 | CHO (structure) Me | $C_{17}H_{16}O_4$ | oil | | | *Mutisia orbignyana* | (680) |
| **573** | 1988 | (structure) Me | $C_{20}H_{24}O_3$ | 91 | | | *Mutisia orbignyana* | (680) |
| **574** | 1988 | OH (structure) Me | $C_{20}H_{24}O_4$ | 114 | | | *Mutisia orbignyana* | (680) |
| **575** | 1988 | OH (structure) Me | $C_{20}H_{24}O_4$ | 103 | | | *Mutisia orbignyana* | (655, 680) |

| No. | Year | Structure | Formula | | | | Species | Ref. |
|---|---|---|---|---|---|---|---|---|
| **576** | 1988 | 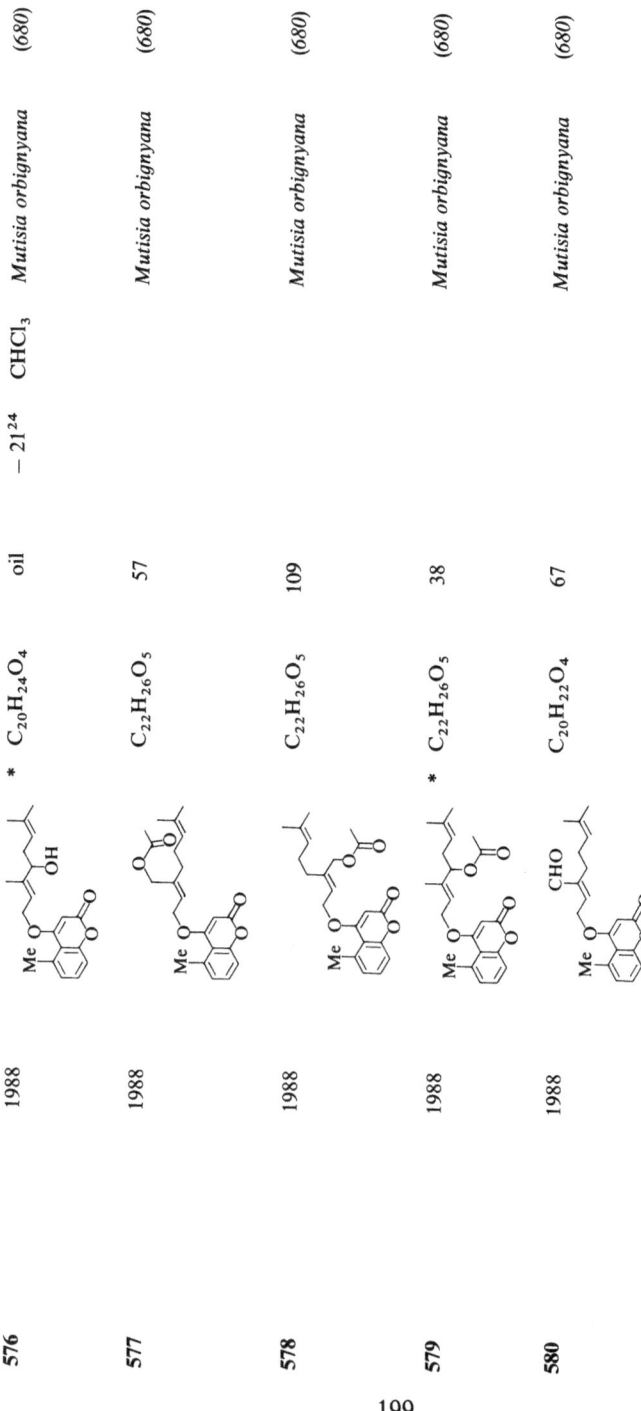 | * $C_{20}H_{24}O_4$ | oil | $-21^{24}$ | CHCl$_3$ | *Mutisia orbignyana* | *(680)* |
| **577** | 1988 | | $C_{22}H_{26}O_5$ | 57 | | | *Mutisia orbignyana* | *(680)* |
| **578** | 1988 | | $C_{22}H_{26}O_5$ | 109 | | | *Mutisia orbignyana* | *(680)* |
| **579** | 1988 | | * $C_{22}H_{26}O_5$ | 38 | | | *Mutisia orbignyana* | *(680)* |
| **580** | 1988 | | $C_{20}H_{22}O_4$ | 67 | | | *Mutisia orbignyana* | *(680)* |

199

Table 11. (Continued)

| Trivial name(s) | Year isolated | Structure | Formula | M.p. | $[\alpha]_\lambda^t$ | Solvent | Plant sources | Leading references |
|---|---|---|---|---|---|---|---|---|
| 581 | 1988 | | $C_{20}H_{22}O_4$ | 144 | | | Mutisia orbignyana | (680) |
| 582 | 1988 | | * $C_{20}H_{22}O_4$ | oil | | | Mutisia orbignyana | (655, 680) |
| 583 | 1988 | | * $C_{20}H_{24}O_5$ | | | | Mutisia orbignyana | (680) |
| 584 | 1988 | | * $C_{20}H_{24}O_5$ | | | | Mutisia orbignyana | (680) |
| 585 | 1988 | | * $C_{20}H_{24}O_5$ | | | | Mutisia orbignyana | (680) |

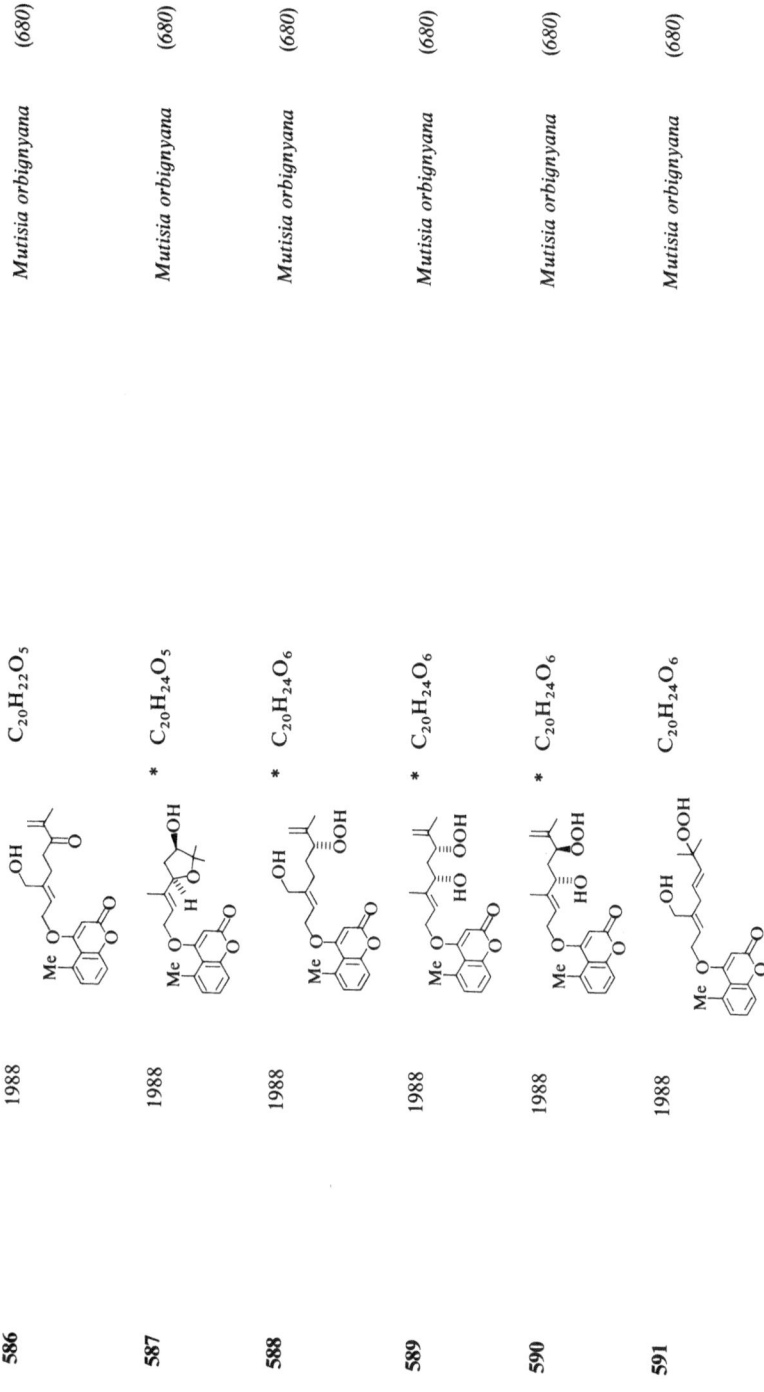

| | | | |
|---|---|---|---|
| 586 | 1988 | C$_{20}$H$_{22}$O$_5$ | *Mutisia orbignyana* (680) |
| 587 | 1988 | * C$_{20}$H$_{24}$O$_5$ | *Mutisia orbignyana* (680) |
| 588 | 1988 | * C$_{20}$H$_{24}$O$_6$ | *Mutisia orbignyana* (680) |
| 589 | 1988 | * C$_{20}$H$_{24}$O$_6$ | *Mutisia orbignyana* (680) |
| 590 | 1988 | * C$_{20}$H$_{24}$O$_6$ | *Mutisia orbignyana* (680) |
| 591 | 1988 | C$_{20}$H$_{24}$O$_6$ | *Mutisia orbignyana* (680) |

Table 11. (Continued)

| Trivial name(s) | Year isolated | Structure | Formula | M.p. | $[\alpha]_\lambda^i$ | Solvent | Plant sources | Leading references |
|---|---|---|---|---|---|---|---|---|
| **592** | 1988 | | * $C_{20}H_{24}O_6$ | | | | *Mutisia orbignyana* | (680) |
| **593** | 1988 | | $C_{20}H_{20}O_5$ | oil | | | *Mutisia orbignyana* | (680) |
| **594** Nassauvirevolutin A | 1988 | | $C_{25}H_{32}O_4$ | gum | $2.1^{24}$ | $CHCl_3$ | *Nassauvia lagascae* | (107) |
| **595** | 1989 | | $C_{25}H_{32}O_4$ | gum | $-13^{24}$ | $CHCl_3$ | *Nassauvia argentea* | (108) |
| **596** Nassauvirevolutin B | 1988 | | $C_{25}H_{30}O_4$ | gum | | | *Nassauvia lagascae* | (107) |

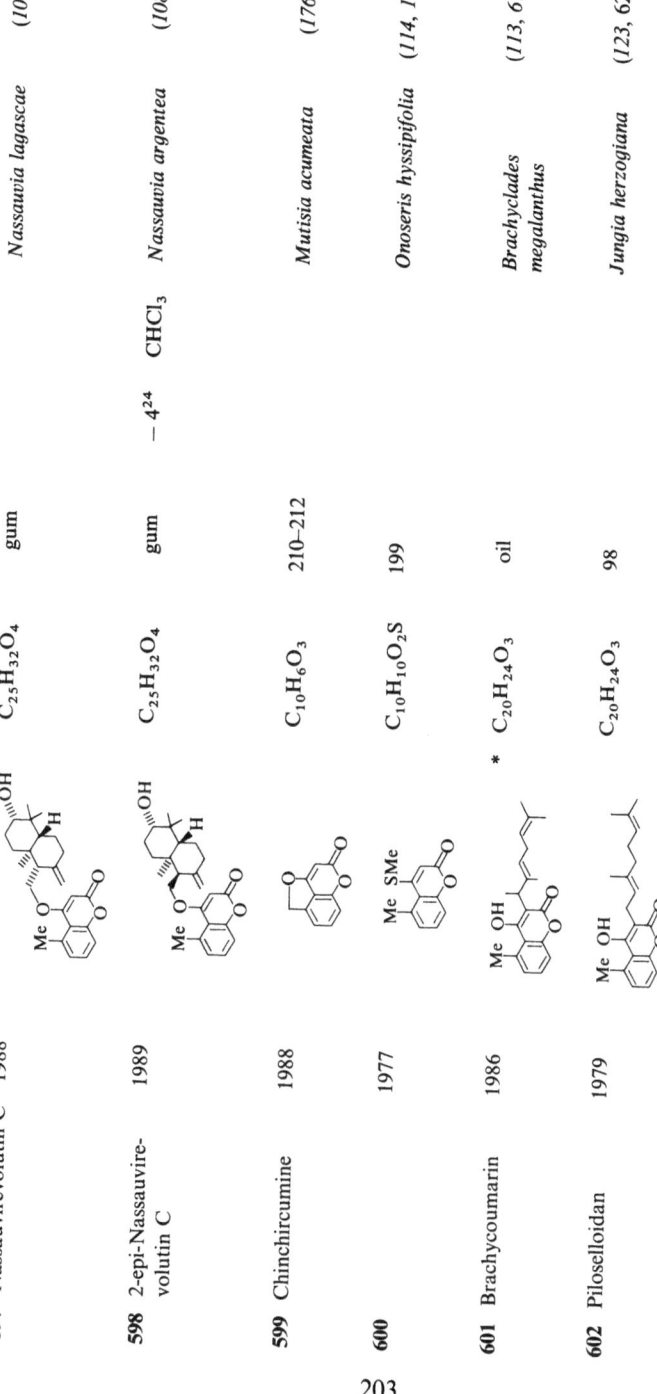

| 597 | Nassauvirevolutin C | 1988 | | $C_{25}H_{32}O_4$ | gum | | Nassauvia lagascae | (107) |
| 598 | 2-epi-Nassauvire-volutin C | 1989 | | $C_{25}H_{32}O_4$ | gum | $-4^{24}$ CHCl$_3$ | Nassauvia argentea | (108) |
| 599 | Chinchircumine | 1988 | | $C_{10}H_6O_3$ | 210–212 | | Mutisia acumeata | (176) |
| 600 | | 1977 | | $C_{10}H_{10}O_2S$ | 199 | | Onoseris hyssipifolia | (114, 116) |
| 601 | Brachycoumarin | 1986 | * | $C_{20}H_{24}O_3$ | oil | | Brachyclades megalanthus | (113, 675) |
| 602 | Piloselloidan | 1979 | | $C_{20}H_{24}O_3$ | 98 | | Jungia herzogiana | (123, 628) |

## Table 11. (Continued)

| Trivial name(s) | Year isolated | Structure | Formula | M.p. | $[\alpha]^t_\lambda$ | Solvent | Plant sources | Leading references |
|---|---|---|---|---|---|---|---|---|
| **603** | 1986 | | $C_{25}H_{32}O_3$ | oil | | | Mutisia spinosa | (675) |
| **604** | 1988 | | $C_{25}H_{30}O_4$ | oil | | | Gypothamnium pinifolium | (679) |
| **605** | 1988 | | $^* C_{25}H_{30}O_5$ | oil | | | Gypothamnium pinifolium | (679) |
| **606** 1'-Epilycoserone | 1985 | | $C_{25}H_{30}O_5$ | oil | | | Lycoseris latifolia | (112, 389) |
| **607** Lycoserone | 1985 | | $C_{25}H_{30}O_5$ | | | | Lycoseris latifolia | (112, 389) |

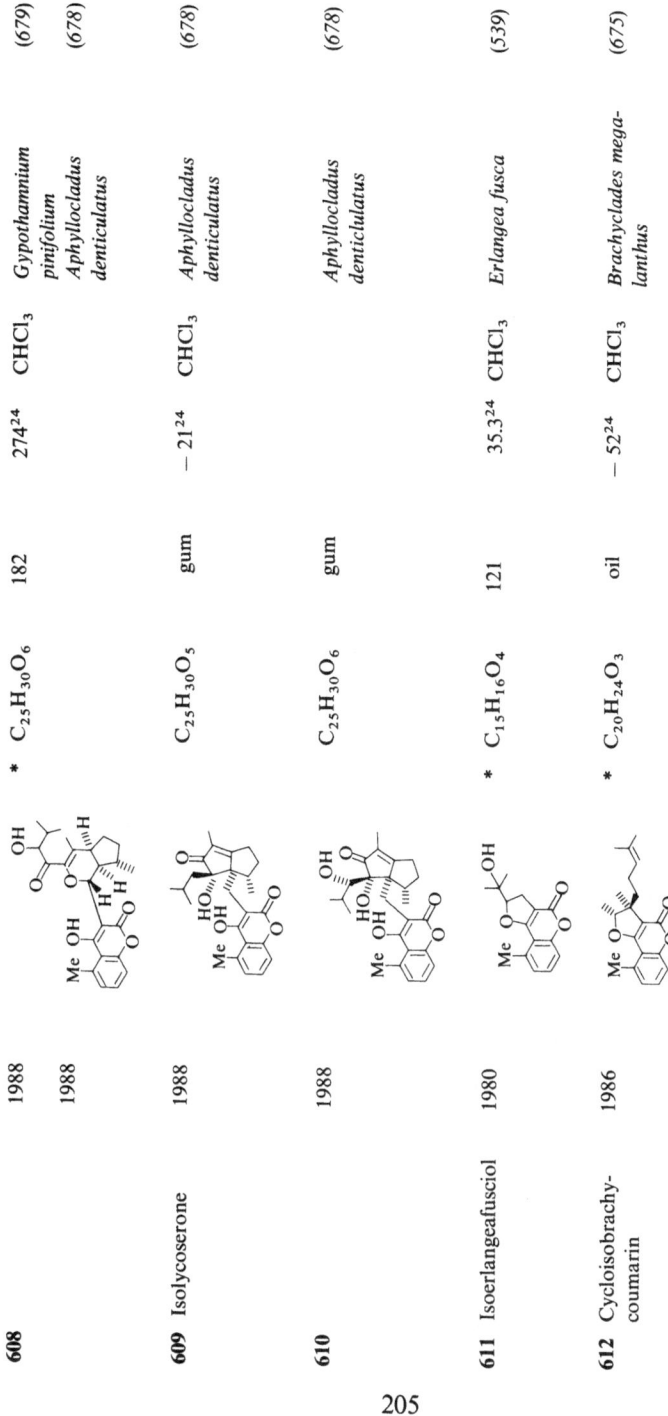

| | Year | | Formula | mp | $[\alpha]$ | Solvent | Source | Ref. |
|---|---|---|---|---|---|---|---|---|
| **608** | 1988 | | * $C_{25}H_{30}O_6$ | 182 | $274^{24}$ | $CHCl_3$ | *Gypothamnium pinifolium* | (679) |
| | 1988 | | | | | | *Aphyllocladus denticulatus* | (678) |
| **609** Isolycoserone | 1988 | | $C_{25}H_{30}O_5$ | gum | $-21^{24}$ | $CHCl_3$ | *Aphyllocladus denticulatus* | (678) |
| **610** | 1988 | | $C_{25}H_{30}O_6$ | gum | | | *Aphyllocladus denticulatus* | (678) |
| **611** Isoerlangeafusciol | 1980 | | * $C_{15}H_{16}O_4$ | 121 | $35.3^{24}$ | $CHCl_3$ | *Erlangea fusca* | (539) |
| **612** Cycloisobrachy-coumarin | 1986 | | * $C_{20}H_{24}O_3$ | oil | $-52^{24}$ | $CHCl_3$ | *Brachyclades mega-lanthus* | (675) |

206

## Table 11. (Continued)

| | Trivial name(s) | Year isolated | Structure | Formula | M.p. | $[\alpha]_\lambda^t$ | Solvent | Plant sources | Leading references |
|---|---|---|---|---|---|---|---|---|---|
| **613** | 2'-Epicycloiso-brachycoumarin | 1986 | | * $C_{20}H_{24}O_3$ | oil | | | Brachyclades megalanthus | (675) |
| **614** | Cyclobrachy-coumarin | 1986 | | * $C_{20}H_{24}O_3$ | oil | $-6^{24}$ | $CHCl_3$ | Brachyclades megalanthus | (113, 675) |
| **615** | | 1988 | | * $C_{20}H_{22}O_3$ | | | | Mutisia orbignyana | (655, 680) |
| **616** | Mutisicoumarin | 1986 | | * $C_{20}H_{24}O_4$ | 59 | | | Mutisia spinosa | (675) |
| **617** | Isotriptiliocoumarin | 1988 | | $C_{25}H_{30}O_3$ | gum | | | Triptilion benaventei | (106) |

| No. | Name | Year | Structure | Formula | mp | [α]/solvent | Source | Ref. |
|---|---|---|---|---|---|---|---|---|
| 618 | | 1988 | 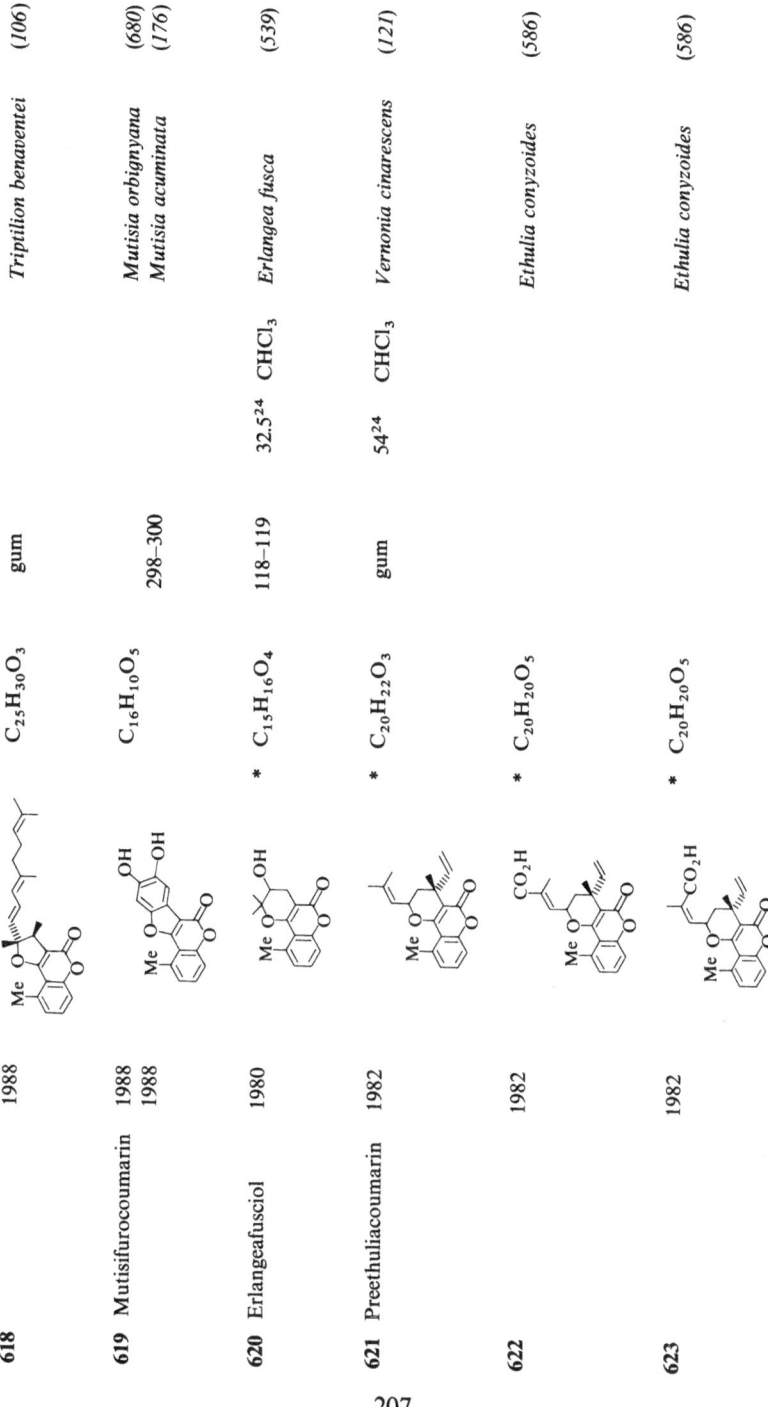 | $C_{25}H_{30}O_3$ | gum | | *Triptilion benaventei* | (106) |
| 619 | Mutisifurocoumarin | 1988 | | $C_{16}H_{10}O_5$ | 298–300 | | *Mutisia orbignyana* | (680) |
| | | 1988 | | | | | *Mutisia acuminata* | (176) |
| 620 | Erlangeafusciol | 1980 | | * $C_{15}H_{16}O_4$ | 118–119 | $32.5^{24}$ CHCl$_3$ | *Erlangea fusca* | (539) |
| 621 | Preethuliacoumarin | 1982 | | * $C_{20}H_{22}O_3$ | gum | $54^{24}$ CHCl$_3$ | *Vernonia cinarescens* | (121) |
| 622 | | 1982 | | * $C_{20}H_{20}O_5$ | | | *Ethulia conyzoides* | (586) |
| 623 | | 1982 | | * $C_{20}H_{20}O_5$ | | | *Ethulia conyzoides* | (586) |

## Table 11. (Continued)

| Trivial name(s) | Year isolated | Structure | Formula | M.p. | $[\alpha]_\lambda^t$ | Solvent | Plant sources | Leading references |
|---|---|---|---|---|---|---|---|---|
| **624** Ethuliacoumarin | 1977 | | * $C_{20}H_{22}O_5$ | 61 | $28^{24}$ | $CHCl_3$ | *Ethulia conyzoides* | *(118)* |
| **625** Isoethulia-coumarin A | 1980 | | * $C_{20}H_{22}O_5$ | 156–157 | $-20.4^{24}$ | $CHCl_3$ | *Ethulia conyzoides* | *(60)* |
| **626** Isoethulia-coumarin B | 1980 | | * $C_{20}H_{22}O_5$ | 267–268 | $47.1^{24}$ | $CHCl_3$ | *Ethulia conyzoides* | *(60)* |
| **627** | 1982 | | * $C_{20}H_{20}O_5$ | 193–194 | | | *Ethulia conyzoides* | *(586)* |
| **628** Cycloethulia-coumarin | 1977 | | * $C_{20}H_{20}O_5$ | 129 | $21^{24}$ | $CHCl_3$ | *Ethulia conyzoides* | *(118)* |

*References, pp. 283–316*

208

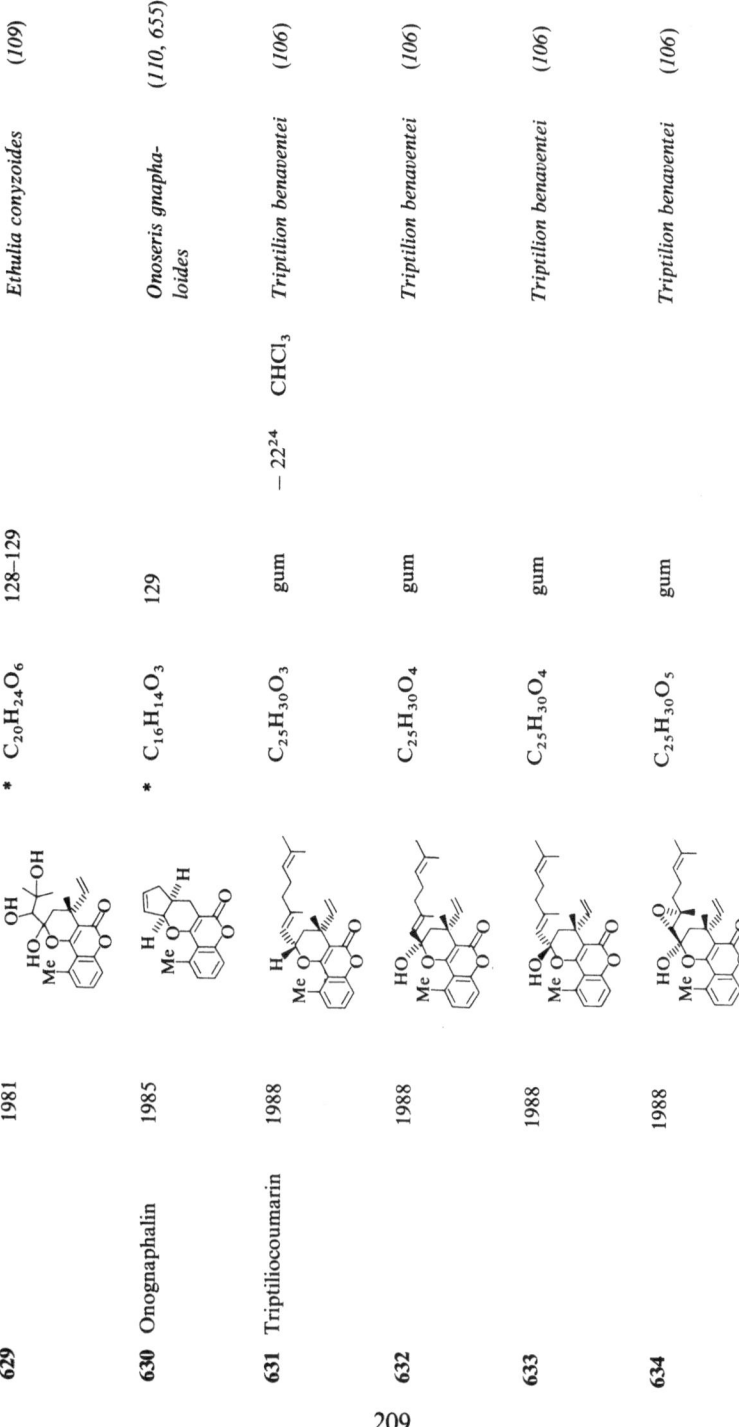

| | | | | | | |
|---|---|---|---|---|---|---|
| **629** | 1981 | | * $C_{20}H_{24}O_6$ | 128–129 | | *Ethulia conyzoides* | (109) |
| **630** Onognaphalin | 1985 | | * $C_{16}H_{14}O_3$ | 129 | | *Onoseris gnapha-loides* | (110, 655) |
| **631** Triptiliocoumarin | 1988 | | $C_{25}H_{30}O_3$ | gum | $-22^{24}$ CHCl$_3$ | *Triptilion benaventei* | (106) |
| **632** | 1988 | | $C_{25}H_{30}O_4$ | gum | | *Triptilion benaventei* | (106) |
| **633** | 1988 | | $C_{25}H_{30}O_4$ | gum | | *Triptilion benaventei* | (106) |
| **634** | 1988 | | $C_{25}H_{30}O_5$ | gum | | *Triptilion benaventei* | (106) |

209

## Table 11. (Continued)

| Trivial name(s) | Year isolated | Structure | Formula | M.p. | $[\alpha]_\lambda^t$ | Solvent | Plant sources | Leading references |
|---|---|---|---|---|---|---|---|---|
| 635 Triptispinocoumarin | 1988 | | $C_{14}H_{12}O_3$ | gum | | | *Triptilion spinosum* | *(106, 655)* |
| 636 Isotriptospino-coumarin | 1988 | | $C_{14}H_{12}O_3$ | oil | | | *Nassauvia magel-lanica* | *(107)* |
| 637 Bothrioclinin | 1977 | | $C_{15}H_{14}O_3$ | oil | | | *Bothriocline laxa* | *(119)* |
| 638 | 1988 | | $C_{25}H_{28}O_3$ | gum | | | *Triptilion benaventei* | *(106)* |
| 639 Isoethulia-coumarin C | 1980 | | * $C_{20}H_{24}O_6$ | 128–129 | | | *Ethulia conyzoides* | *(109)* |

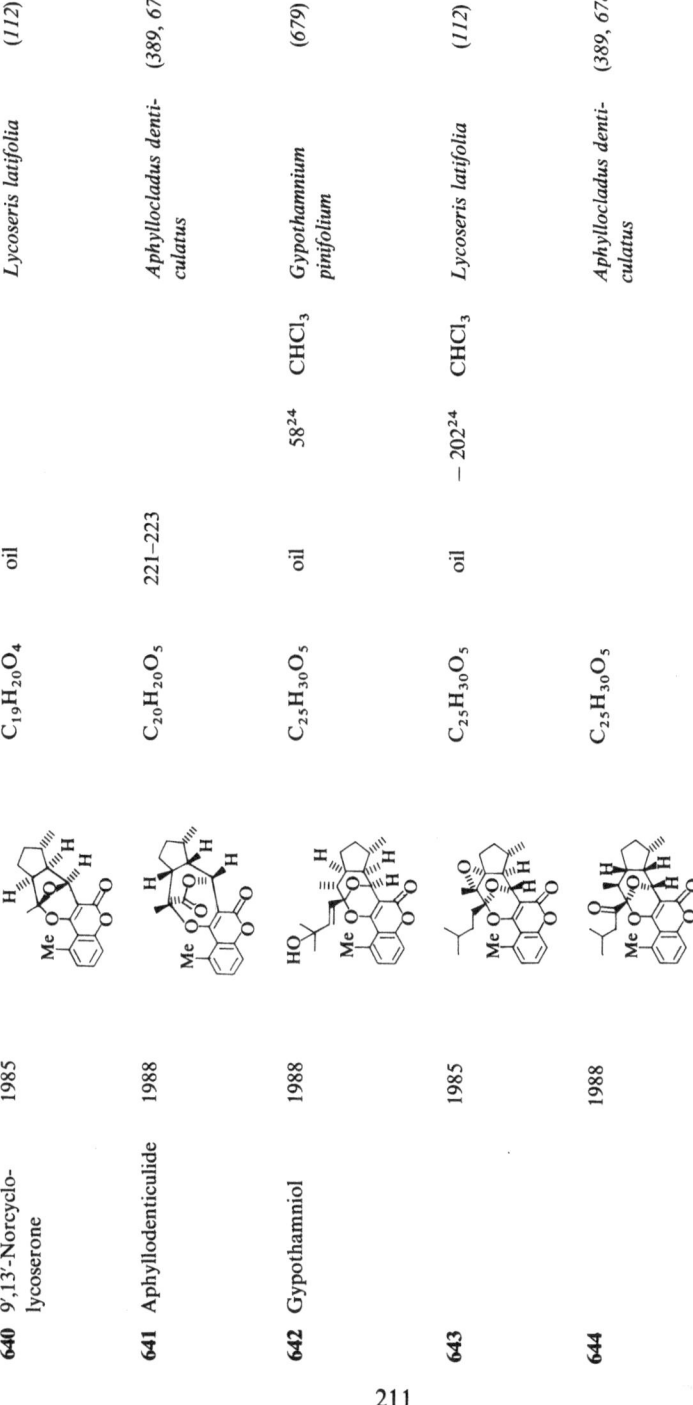

| | | | | | | |
|---|---|---|---|---|---|---|
| **640** 9′,13′-Norcyclo-lycoserone | 1985 | | $C_{19}H_{20}O_4$ | oil | | *Lycoseris latifolia* (112) |
| **641** Aphyllodenticulide | 1988 | | $C_{20}H_{20}O_5$ | 221–223 | | *Aphyllocladus denti-culatus* (389, 678) |
| **642** Gypothamniol | 1988 | | $C_{25}H_{30}O_5$ | oil | $58^{24}$ CHCl$_3$ | *Gypothamnium pinifolium* (679) |
| **643** | 1985 | | $C_{25}H_{30}O_5$ | oil | $-202^{24}$ CHCl$_3$ | *Lycoseris latifolia* (112) |
| **644** | 1988 | | $C_{25}H_{30}O_5$ | | | *Aphyllocladus denti-culatus* (389, 678) |

Table 11. (Continued)

| Trivial name(s) | Year isolated | Structure | Formula | M.p. | $[\alpha]_\lambda^t$ | Solvent | Plant sources | Leading references |
|---|---|---|---|---|---|---|---|---|
| **645** Cyclolycoserone | 1985 | | $C_{25}H_{30}O_5$ | oil | $63^{24}$ | $CHCl_3$ | Lycoseris latifolia | (112, 389) |
| **646** 10′,11′-Dehydro-cyclolycoserone | 1988 | | $C_{25}H_{28}O_5$ | gum | | | Aphyllocladus denti-culatus | (678) |
| **647** 10′-Hydroxycyclo-lycoserone | 1988 | | * $C_{25}H_{30}O_6$ | gum | | | Aphyllocladus denti-culatus | (678) |
| **648** 1′-Epi-6′,7′-dehydro-cyclolycoserone | 1985 | | $C_{25}H_{28}O_5$ | oil | $-55^{24}$ | $CHCl_3$ | Lycoseris latifolia | (112) |

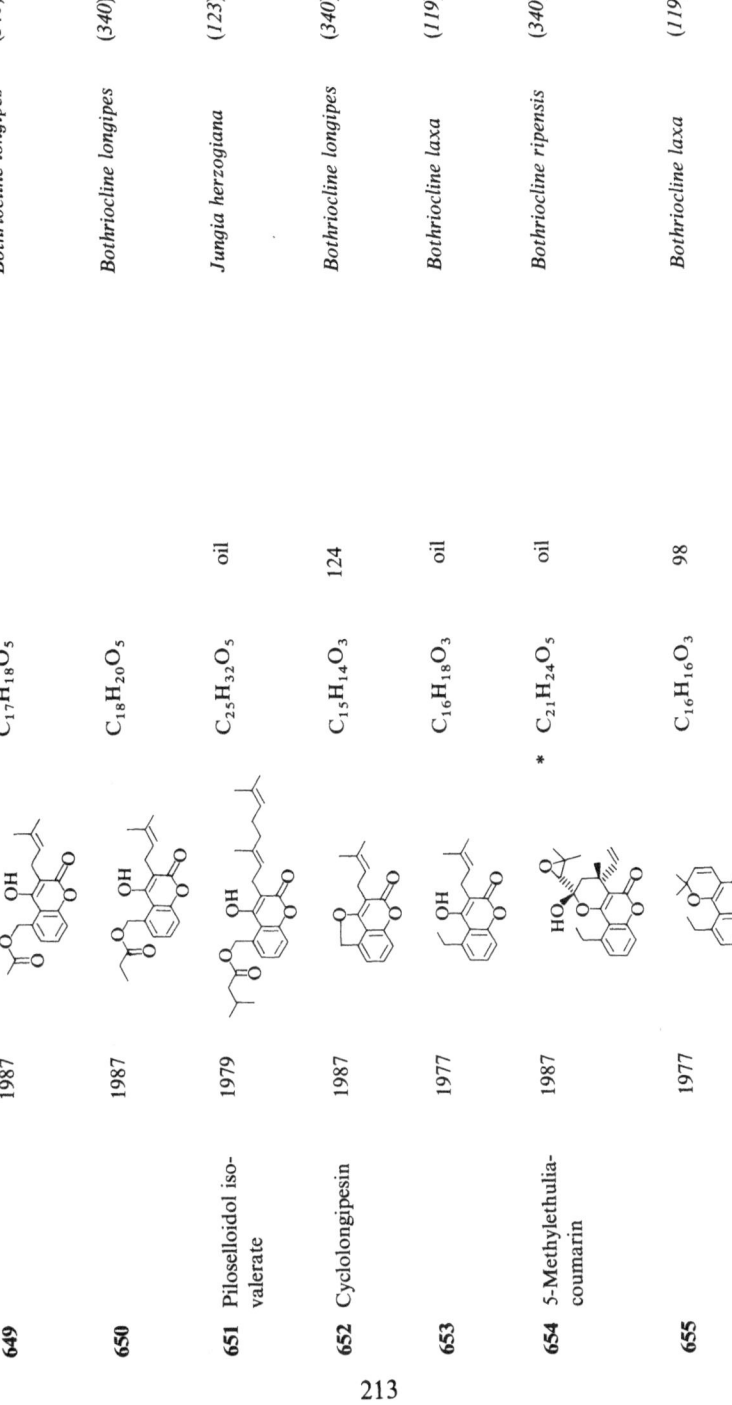

| | | | | |
|---|---|---|---|---|
| **649** | | 1987 | $C_{17}H_{18}O_5$ | *Bothriocline longipes* (340) |
| **650** | | 1987 | $C_{18}H_{20}O_5$ | *Bothriocline longipes* (340) |
| **651** | Piloselloidol iso-valerate | 1979 | $C_{25}H_{32}O_5$ — oil | *Jungia herzogiana* (123) |
| **652** | Cyclolongipesin | 1987 | $C_{15}H_{14}O_3$ — 124 | *Bothriocline longipes* (340) |
| **653** | | 1977 | $C_{16}H_{18}O_3$ — oil | *Bothriocline laxa* (119) |
| **654** | 5-Methylethulia-coumarin | 1987 | * $C_{21}H_{24}O_5$ — oil | *Bothriocline ripensis* (340) |
| **655** | | 1977 | $C_{16}H_{16}O_3$ — 98 | *Bothriocline laxa* (119) |

*Table 11.* (Continued)

| Trivial name(s) | Year isolated | Structure | Formula | M.p. | $[\alpha]_\lambda^t$ | Solvent | Plant sources | Leading references |
|---|---|---|---|---|---|---|---|---|
| **656** 9-Methyllongipesin | 1987 | | $C_{16}H_{18}O_4$ | | | | *Bothriocline longipes* | *(340)* |
| **657** | 1987 | | * $C_{18}H_{20}O_5$ | | | | *Bothriocline longipes* | *(340)* |
| **658** | 1987 | | * $C_{19}H_{22}O_5$ | | | | *Bothriocline longipes* | *(340)* |
| **659** | 1987 | | * $C_{16}H_{16}O_3$ | 85 | | | *Bothriocline longipes* | *(340)* |
| **660** | 1987 | | * $C_{18}H_{18}O_5$ | oil | | | *Bothriocline longipes* | *(340)* |
| **661** | 1987 | | * $C_{19}H_{20}O_5$ | oil | | | *Bothriocline longipes* | *(340)* |

| No. | Name | Year | Structure | Formula | mp | $[\alpha]$ | Source | Ref. |
|---|---|---|---|---|---|---|---|---|
| 662 | Ventilatone A | 1985 | | * $C_{17}H_{12}O_6$ | 284 | $133^{25}$ $CHCl_3$ | Ventilago calyculata | (295) |
| 663 | | 1977 | | $C_9H_6O_3$ | 220–222 | | Alyxia lucida | (540) |
| 664 | | 1987 | | $C_{11}H_{10}O_3$ | 137–138 | | Edgeworthia gardneri | (23, 153) |
| 665 | Pygmaeoherin | 1988 | | $C_{17}H_{18}O_4$ | 198–200 | | Pygmaeopremna herbacea | (436) |
| 666 | | 1987 | | $C_{13}H_{14}O_3$ | 121–122 | | Edgeworthia gardneri | (23, 153) |
| 667 | Necatorin | 1983 | | $C_{15}H_{18}N_2O_3$ | 220–225 | | Lactarias necator | (607, 608) |
| 668 | | 1977 | | $C_9H_6O_3$ | 157–160 | | Alyxia lucida | (540) |

215

Table 11. (Continued)

| Trivial name(s) | Year isolated | Structure | Formula | M.p. | $[\alpha]_\lambda^t$ | Solvent | Plant sources | Leading references |
|---|---|---|---|---|---|---|---|---|
| 669 | 1964 | (structure, OMe) | $C_{10}H_8O_3$ | | | | *Triticum sativum* | (217) |
| 670 | 1980 | (structure, Me OMe OMe) | $C_{12}H_{12}O_4$ | 89–90 | | | *Fraxinus floribunda* | (470, 495) |
| | 1977 | | | 85 | | | *Perezia multiflora* | (32, 116) |
| 671 | 1979 | (structure, OH, HO) | $C_9H_6O_4$ | 249 | | | *Euphorbia terracina* | (419) |
| 672 | 1979 | (structure, OH, MeO) | $C_{10}H_8O_4$ | 225 | | | *Euphorbia paralias* | (419) |
| 673 | 1987 | (structure, OMe, Me, HO) | $C_{14}H_{16}O_4$ | 197–199 | | | *Macrothelypteris torresiana* | (315) |
| 674 | 1989 | (structure, Me O-β-D-glucosyl, HO) | $C_{16}H_{18}O_9$ | 202–203 | $-96.4^{20}$ | MeOH | *Gerbera jamesonii* | (474) |
| 675 | 1980 | (structure, Me OMe, HO) | $C_{11}H_{10}O_4$ | 267–268 | | | *Gerbera jamesonii* | (292) |

216

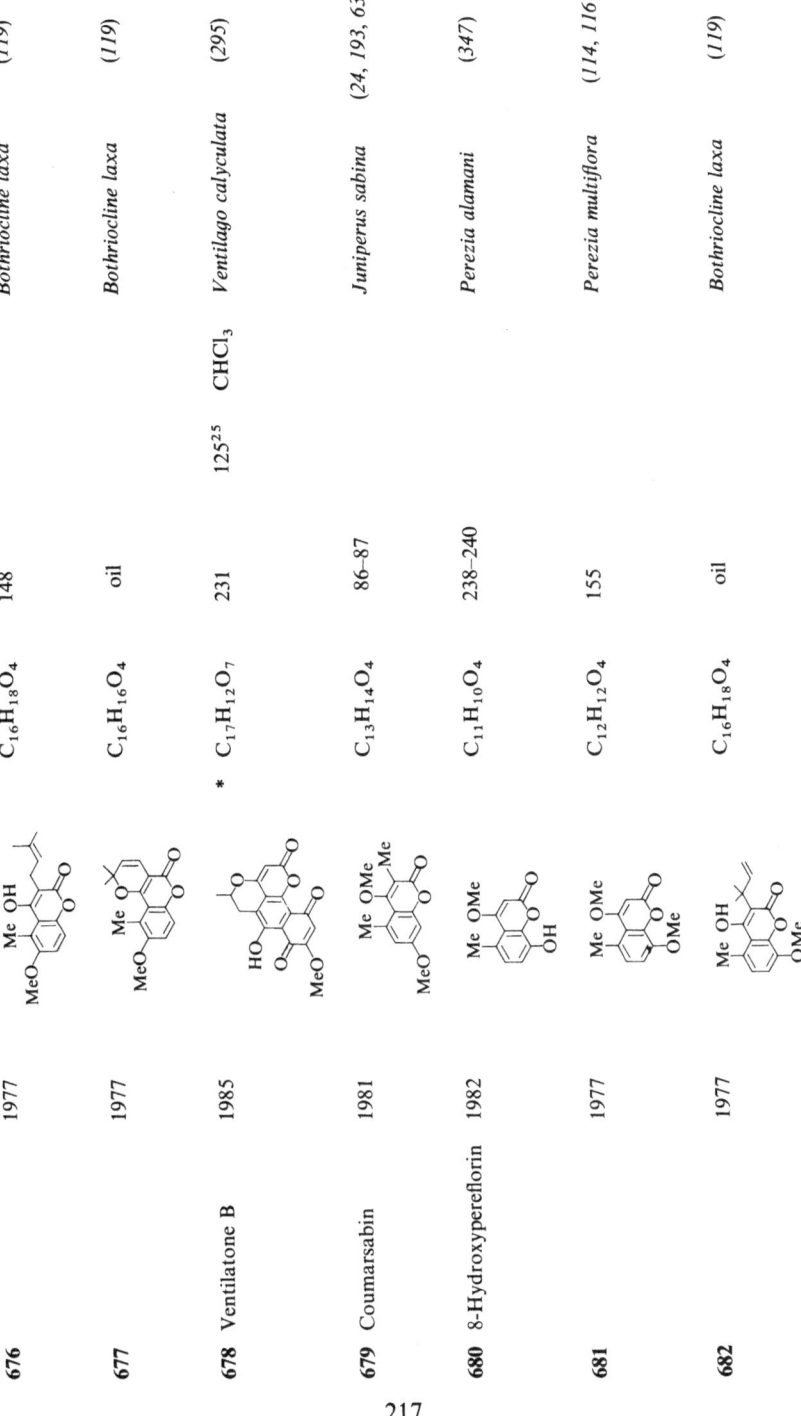

| 676 | 1977 | | C$_{16}$H$_{18}$O$_4$ | 148 | | *Bothriocline laxa* | (119) |
| 677 | 1977 | | C$_{16}$H$_{16}$O$_4$ | oil | | *Bothriocline laxa* | (119) |
| 678 Ventilatone B | 1985 | * C$_{17}$H$_{12}$O$_7$ | 231 | 125$^{25}$ | CHCl$_3$ | *Ventilago calyculata* | (295) |
| 679 Coumarsabin | 1981 | | C$_{13}$H$_{14}$O$_4$ | 86–87 | | *Juniperus sabina* | (24, 193, 639) |
| 680 8-Hydroxyperreflorin | 1982 | | C$_{11}$H$_{10}$O$_4$ | 238–240 | | *Perezia alamani* | (347) |
| 681 | 1977 | | C$_{12}$H$_{12}$O$_4$ | 155 | | *Perezia multiflora* | (114, 116) |
| 682 | 1977 | | C$_{16}$H$_{18}$O$_4$ | oil | | *Bothriocline laxa* | (119) |

Table 11. (Continued)

| Trivial name(s) | Year isolated | Structure | Formula | M.p. | $[\alpha]^i_\lambda$ | Solvent | Plant sources | Leading references |
|---|---|---|---|---|---|---|---|---|
| **683** | 1977 | | $C_{16}H_{18}O_4$ | oil | | | *Bothriocline laxa* | *(119)* |
| **684** | 1977 | | * $C_{16}H_{18}O_4$ | 168 | | | *Erlangea rogersii* | *(114, 115)* |
| **685** | 1988 | | * $C_{25}H_{30}O_4$ | gum | | | *Triptilion spinosum* | *(106)* |
| **686** | 1977 | | $C_{15}H_{14}O_4$ | > 200 d | | | *Bothriocline laxa* | *(119)* |
| **687** Heliclactone | 1988 | | * $C_{15}H_{16}O_4$ | 285 | | | *Helicteres angusti-folia* | *(649)* |
| **688** | 1982 | | $C_{10}H_8O_4$ | 184–187 | | | *Gomortega keule* | *(222)* |

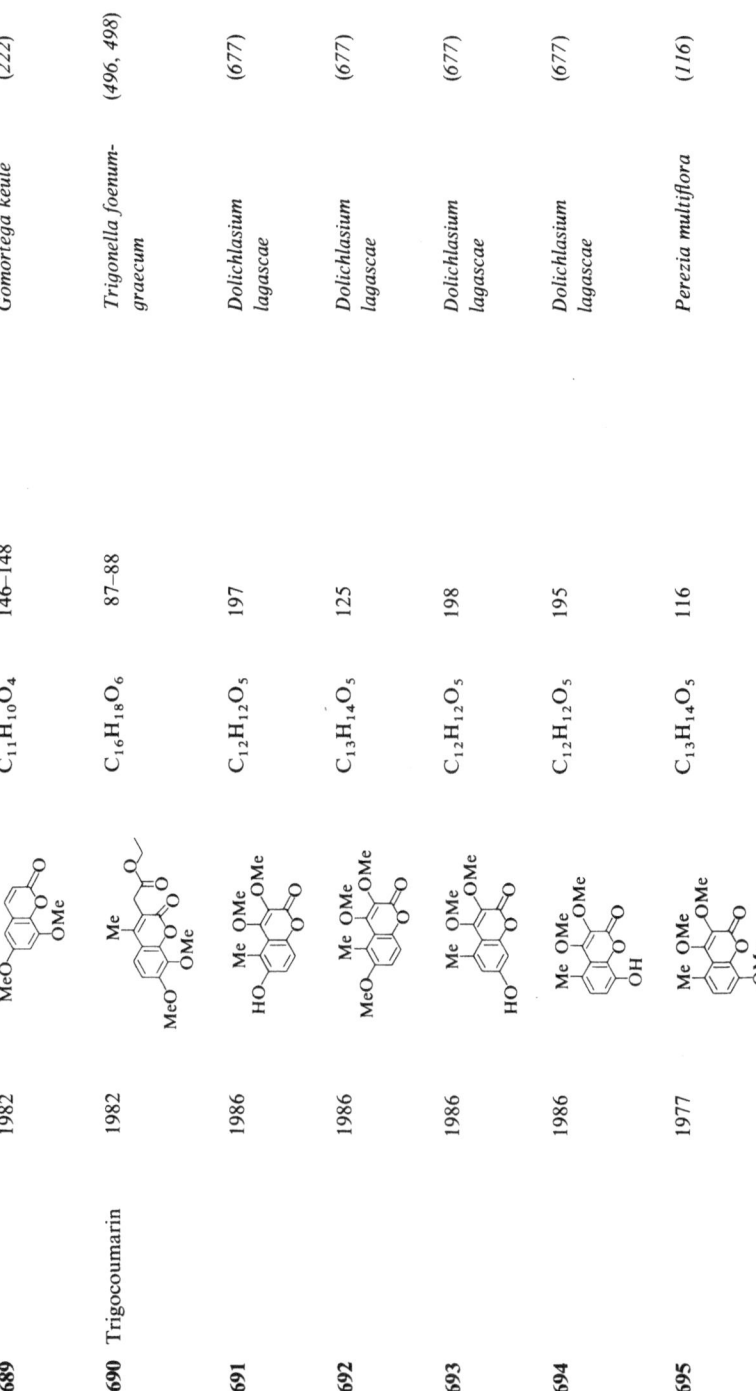

| No. | Name | Year | Formula | M.p. | Source | Ref. |
|---|---|---|---|---|---|---|
| **689** | | 1982 | $C_{11}H_{10}O_4$ | 146–148 | *Gomortega keule* | (222) |
| **690** | Trigocoumarin | 1982 | $C_{16}H_{18}O_6$ | 87–88 | *Trigonella foenum-graecum* | (496, 498) |
| **691** | | 1986 | $C_{12}H_{12}O_5$ | 197 | *Dolichlasium lagascae* | (677) |
| **692** | | 1986 | $C_{13}H_{14}O_5$ | 125 | *Dolichlasium lagascae* | (677) |
| **693** | | 1986 | $C_{12}H_{12}O_5$ | 198 | *Dolichlasium lagascae* | (677) |
| **694** | | 1986 | $C_{12}H_{12}O_5$ | 195 | *Dolichlasium lagascae* | (677) |
| **695** | | 1977 | $C_{13}H_{14}O_5$ | 116 | *Perezia multiflora* | (116) |

Table 11. (Continued)

| Trivial name(s) | Year isolated | Structure | Formula | M.p. | $[\alpha]_\lambda^t$ | Solvent | Plant sources | Leading references |
|---|---|---|---|---|---|---|---|---|
| **696** | 1984 | (OHC, OMe, OMe, OMe; chromone) | $C_{13}H_{12}O_6$ | 185–187 | | | *Perezia coerulescens* | (40) |
| **697** | 1983 | (OH, HO, HO; coumarin) | $C_9H_6O_5$ | 272–274 | | | *Dasycladus vermicularis* | (437) |
| **698** | 1978 | (OMe, HO, OH; coumarin) | $C_{10}H_8O_5$ | 221–222 | | | *Haplophyllum schelkovnikovii* | (8) |
| **699** 8-Methoxy-coumarsabin | 1981 | (Me, OMe, Me, MeO, OMe; coumarin) | $C_{14}H_{16}O_5$ | 125–126 | | | *Juniperus sabina* | (24, 193) |
| **700** Pereflorin B | 1978 | (OHC, OMe, OMe, HO, OMe; chromone) | $C_{13}H_{12}O_7$ | 199–200 | | | *Perezia multiflora* | (348) |

Table 11.1. 3-Aryl Oxygenated Coumarins

| Trivial name(s) | Year isolated | Structure | Formula | M.p. | [α]$_\lambda^t$ | Solvent | Plant sources | Leading references |
|---|---|---|---|---|---|---|---|---|
| **701** Derrusnin | 1969 | | $C_{19}H_{16}O_7$ | 187 | | | Derris robusta | (215) |
| **702** Glabrescin | 1977 | * | $C_{23}H_{20}O_7$ | 112–113 | | | Derris glabrescens | (187) |
| **703** Robustic acid | 1942 | | $C_{22}H_{20}O_6$ | 208–210 | | | Derris robusta | (215, 297, 337, 344) |
| **704** Robustin | 1969 | | $C_{22}H_{18}O_7$ | 206–207 | | | Derris robusta | (215, 338) |
| **705** Methyl robustate | 1969 | | $C_{23}H_{22}O_6$ | 195–196 | | | Derris robusta | (215) |
| **706** | 1969 | | $C_{23}H_{20}O_7$ | 210–211 | | | Derris robusta | (215) |

221

Table 11.1. (Continued)

| | Trivial name(s) | Year isolated | Structure | Formula | M.p. | $[\alpha]_\lambda^t$ | Solvent | Plant sources | Leading references |
|---|---|---|---|---|---|---|---|---|---|
| 707 | | 1986 | | $C_{22}H_{18}O_7$ | 202–204 | | | *Derris spruceana* | *(247)* |
| 708 | | 1986 | | $C_{23}H_{20}O_7$ | 195 | | | *Derris spruceana* | *(247)* |
| 709 | Scandenin | 1943 | | $C_{26}H_{26}O_6$ | 230–232 | | | *Derris scandens* | *(165, 345, 501)* |
| 710 | Lonchocarpic acid | 1934 | | $C_{26}H_{26}O_6$ | 210–212 | | | *Lonchocarpus sp.* | *(223, 345, 346, 501)* |

Table 11.2. Coumestans

| Trivial name(s) | Year isolated | Structure | Formula | M.p. | $[\alpha]_\lambda^t$ | Solvent | Plant sources | Leading references |
|---|---|---|---|---|---|---|---|---|
| 711 Lonchocarpenin | 1969 | | $C_{27}H_{28}O_6$ | 181–182 | | | *Derris scandens* | *(223)* |
| 712 Thonningine B | 1983 | | $C_{23}H_{20}O_7$ | | | | *Milletia thonningii* | *(366)* |
| 713 Thonningine A | 1983 | | $C_{23}H_{18}O_8$ | 205–208 | | | *Milletia thonningii* | *(366)* |
| 714 Coumestrol · | 1957 | | $C_{15}H_8O_5$ | 385d | | | *Medicago sativa* | *(101, 103, 218, 355)* |
| 715 | 1965 | | $C_{16}H_{10}O_5$ | 331–332 | | | *Medicago sativa* | *(102)* |

223

*Table 11.2.* (Continued)

| | Trivial name(s) | Year isolated | Structure | Formula | M.p. | $[\alpha]_\lambda^t$ | Solvent | Plant sources | Leading references |
|---|---|---|---|---|---|---|---|---|---|
| **716** | Isosojagol | 1984 | | $C_{20}H_{16}O_5$ | | | | *Phaseolus aureus* | *(491)* |
| **717** | Sojagol | 1968 | | $C_{20}H_{16}O_5$ | 284–286 | | | *Soja hispida* | *(682)* |
| **718** | Sophoracoume- stan A | 1981 | | $C_{20}H_{14}O_5$ | > 300 | | | *Sophora franchetiana* | *(377)* |
| **719** | | 1966 | | $C_{16}H_{10}O_6$ | 329–329.5 d | | | *Medicago sativa* | *(104)* |
| **720** | | 1966 | | $C_{17}H_{12}O_6$ | 306 | | | *Medicago sativa* | *(595)* |

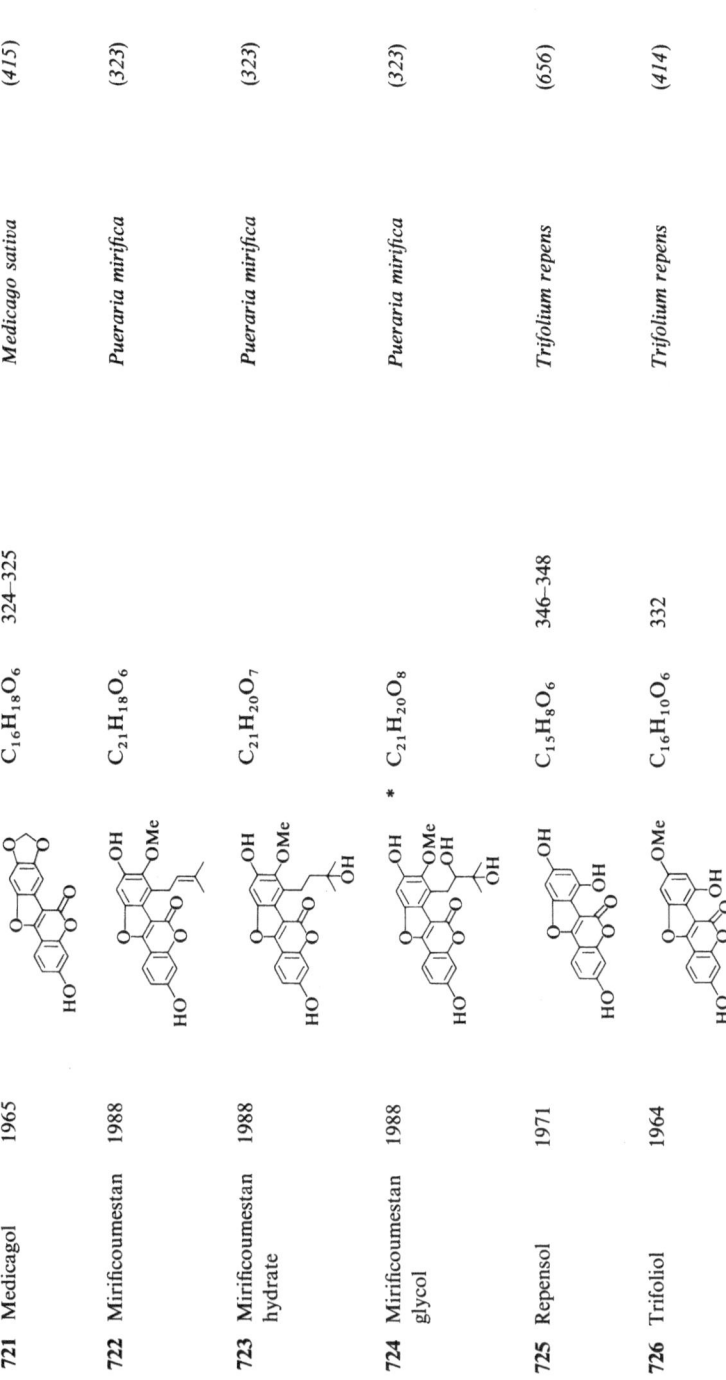

| No. | Name | Year | Structure | Formula | mp | Source | Ref. |
|---|---|---|---|---|---|---|---|
| 721 | Medicagol | 1965 | | $C_{16}H_{18}O_6$ | 324–325 | *Medicago sativa* | *(415)* |
| 722 | Mirificoumestan | 1988 | | $C_{21}H_{18}O_6$ | | *Pueraria mirifica* | *(323)* |
| 723 | Mirificoumestan hydrate | 1988 | | $C_{21}H_{20}O_7$ | | *Pueraria mirifica* | *(323)* |
| 724 | Mirificoumestan glycol | 1988 | | * $C_{21}H_{20}O_8$ | | *Pueraria mirifica* | *(323)* |
| 725 | Repensol | 1971 | | $C_{15}H_8O_6$ | 346–348 | *Trifolium repens* | *(656)* |
| 726 | Trifoliol | 1964 | | $C_{16}H_{10}O_6$ | 332 | *Trifolium repens* | *(414)* |

Table 11.2. (Continued)

| Trivial name(s) | Year isolated | Structure | Formula | M.p. | $[\alpha]_\lambda^t$ | Solvent | Plant sources | Leading references |
|---|---|---|---|---|---|---|---|---|
| 727 Wairol | 1980 | OMe, OMe, HO | $C_{17}H_{12}O_6$ | 292–294 | | | Medicago sativa | (105, 577) |
| 728 Coumestrin | 1984 | OH, RO, R = β-D-glucosyl | $C_{21}H_{18}O_{10}$ | oil | | | Glycine max | (410) |
| 729 Tuberostan | 1985 | MeO | $C_{21}H_{16}O_5$ | 228 | | | Pueraria tuberosa | (514) |
| 730 Flemichap-parin C | 1973 | MeO | $C_{17}H_{10}O_6$ | 272 | | | Flemingia chappar | (17) |
| 731 Psoralidin | 1948 | OH, HO | $C_{20}H_{16}O_5$ | 290–292 | | | Psoralea corylifolia | (150, 370) |
| 732 Psoralidin oxide | 1980 | * OH, HO | $C_{20}H_{16}O_6$ | | | | Psoralea corylifolia | (290) |

226

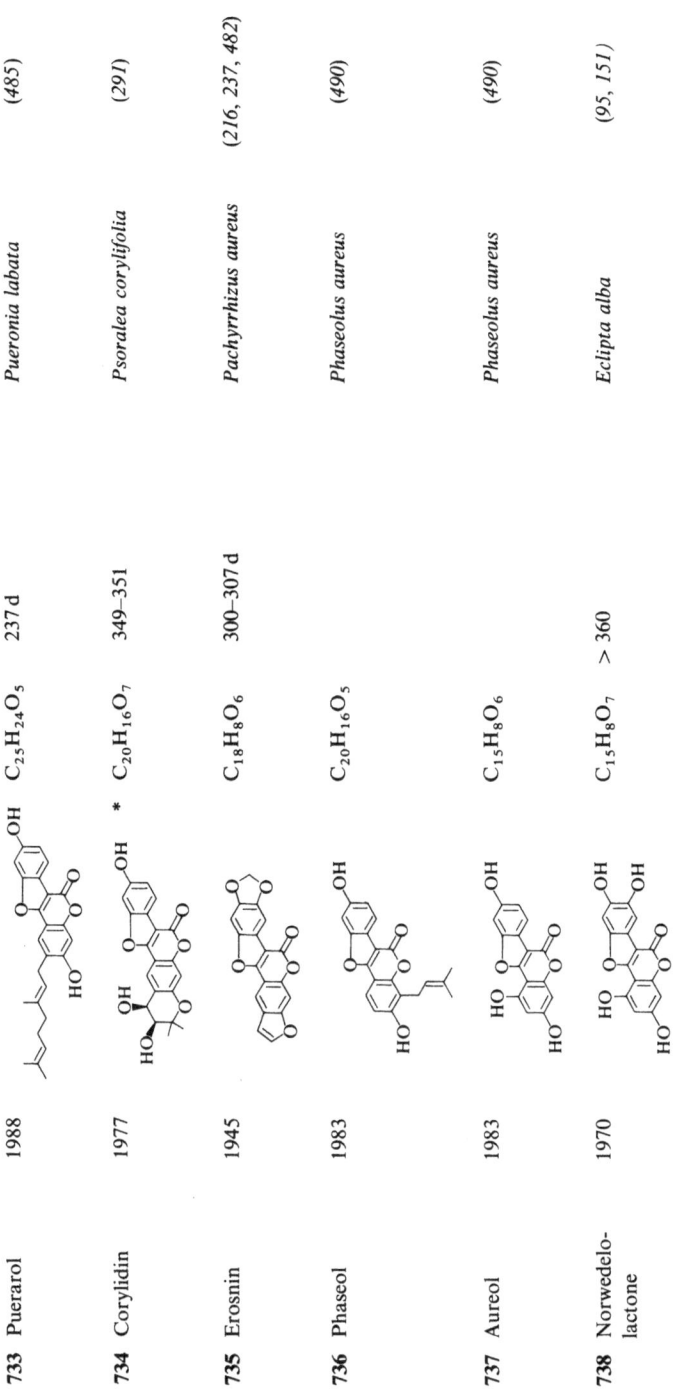

| | | | | | | |
|---|---|---|---|---|---|---|
| **733** Puerarol | 1988 | | $C_{25}H_{24}O_5$ | 237 d | *Pueronia labata* | (485) |
| **734** Corylidin | 1977 | * | $C_{20}H_{16}O_7$ | 349–351 | *Psoralea corylifolia* | (291) |
| **735** Erosnin | 1945 | | $C_{18}H_8O_6$ | 300–307 d | *Pachyrrhizus aureus* | (216, 237, 482) |
| **736** Phascol | 1983 | | $C_{20}H_{16}O_5$ | | *Phaseolus aureus* | (490) |
| **737** Aureol | 1983 | | $C_{15}H_8O_6$ | | *Phaseolus aureus* | (490) |
| **738** Norwedelo-lactone | 1970 | | $C_{15}H_8O_7$ | > 360 | *Eclipta alba* | (95, 151) |

227

Table 11.2. (Continued)

| Trivial name(s) | Year isolated | Structure | Formula | M.p. | $[\alpha]_\lambda^t$ | Solvent | Plant sources | Leading references |
|---|---|---|---|---|---|---|---|---|
| 739 | 1970 | (structure) R = β-D-glucosyl | $C_{21}H_{18}O_{12}$ | > 340 | | | Eclipta alba | (95) |
| 740 Wedelolactone | 1956 | (structure) | $C_{16}H_{10}O_7$ | 327–330 d | | | Wedelia calendulacea | (274, 275, 650) |
| 741 Glycyrol | 1969 1989 | (structure) | $C_{21}H_{18}O_6$ | 243.5–245 | | | Glycyrrhiza spp. Glycyrrhiza uralensis | (548) (583) |
| 742 | 1969 1989 | (structure) | $C_{22}H_{20}O_6$ | 259–260.5 | | | Glycyrrhiza spp. Glycyrrhiza uralensis | (548) (583) |
| 743 Isoglycyrol | 1969 1989 | (structure) | $C_{21}H_{18}O_6$ | 298–300 d | | | Glycyrrhiza spp. Glycyrrhiza uralensis | (548) (583) |
| 744 Gancaonin F | 1989 | (structure) | $C_{21}H_{16}O_6$ | 290–291 | | | Glycyrrhiza sp. | (236) |

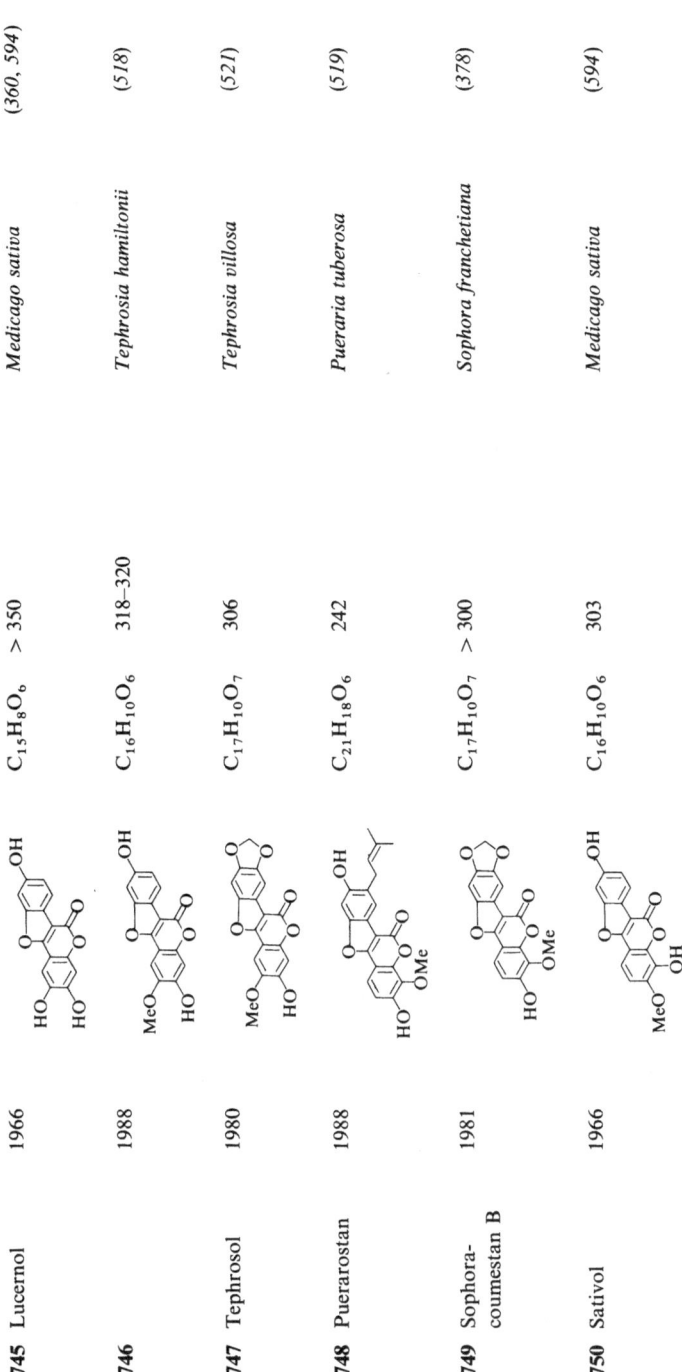

| No. | Name | Year | Formula | mp | Species | Ref. |
|---|---|---|---|---|---|---|
| 745 | Lucernol | 1966 | $C_{15}H_8O_6$ | > 350 | *Medicago sativa* | (360, 594) |
| 746 | | 1988 | $C_{16}H_{10}O_6$ | 318–320 | *Tephrosia hamiltonii* | (518) |
| 747 | Tephrosol | 1980 | $C_{17}H_{10}O_7$ | 306 | *Tephrosia villosa* | (521) |
| 748 | Puerarostan | 1988 | $C_{21}H_{18}O_6$ | 242 | *Pueraria tuberosa* | (519) |
| 749 | Sophora-coumestan B | 1981 | $C_{17}H_{10}O_7$ | > 300 | *Sophora franchetiana* | (378) |
| 750 | Sativol | 1966 | $C_{16}H_{10}O_6$ | 303 | *Medicago sativa* | (594) |

Table 12. *Biscoumarins*

| | Trivial name(s) | Year isolated | Structure | Formula | M.p. | $[\alpha]_\lambda^i$ | Solvent | Plant sources | Leading references |
|---|---|---|---|---|---|---|---|---|---|
| **751** | Bicoumol | 1967 | | $C_{18}H_{10}O_6$ | 293–294 | | | *Trifolium repens* | (596) |
| **752** | Matsukazelactone | 1964 | | $C_{20}H_{14}O_6$ | 267–268 | | | *Boenninghausenia albiflora* | (440, 441) |
| **753** | Bhubaneswin | 1984 | | $C_{19}H_{12}O_6$ | 320 | | | *Boenninghausenia albiflora* | (77) |
| **754** | Jayantinin | 1989 | | $C_{20}H_{14}O_6$ | 255–256 | | | *Boenninghausenia albiflora* | (350) |
| **755** | Ipomopsin | 1984 | | $C_{20}H_{14}O_8$ | > 310 | | | *Ipomopsis aggregata* | (46) |

| | | | | | | | |
|---|---|---|---|---|---|---|---|
| **756** | Euphorbetin | 1971 | | $C_{18}H_{10}O_8$ | > 320 | | *Euphorbia lathyris* | (211, 213, 570) |
| **757** | Isoeuphorbetin | 1973 | | $C_{18}H_{10}O_8$ | > 300 | | *Euphorbia lathyris* | (212, 213) |
| **758** | Desertorin A | 1987 | | $C_{22}H_{18}O_8$ | > 300 | | *Emericella desertorum* | (483) |
| **759** | Desertorin B | 1987 | | $C_{23}H_{20}O_8$ | > 300 | | *Emericella desertorum* | (483) |
| **760** | Desertorin C | 1987 | | * $C_{24}H_{22}O_8$ | 235–237 | 16.8 CHCl$_3$ | *Emericella desertorum* | (483, 534) |

## Table 12. (Continued)

| Trivial name(s) | Year isolated | Structure | Formula | M.p. | $[\alpha]_\lambda^t$ | Solvent | Plant sources | Leading references |
|---|---|---|---|---|---|---|---|---|
| **761** Orlandin | 1979 | | $C_{22}H_{18}O_8$ | 285 d | | | *Aspergillus niger* | (173) |
| **762** Desmethylkotanin | 1971 | | * $C_{23}H_{20}O_8$ | > 315 | $-13.3^{25}$ | $CHCl_3$ | *Aspergillus glaucus* | (129) |
| **763** Kotanin | 1971 | | * $C_{24}H_{22}O_8$ | > 315 | $33.1^{25}$ | $CHCl_3$ | *Aspergillus glaucus* | (129) |
| **764** Edgeworin | 1989 | | $C_{18}H_{10}O_6$ | 284–296 d | | | *Edgeworthia chrysantha* | (55) |

| No. | Name | Year | Structure | Formula | M.p. | [α] / Solvent | Source | Refs. |
|---|---|---|---|---|---|---|---|---|
| 765 | Lasiocephalin | 1971 | | $C_{19}H_{12}O_6$ | 215 | | Lasiosiphon eriocephalus | (97, 178, 645) |
| 766 | Fatagarin | 1975 | | $C_{19}H_{12}O_6$ | 233–234 | | Ruta oreojasme | (271) |
| 767 | Edgeworthin Demethyldaphno-retin | 1974 1986 | | $C_{18}H_{10}O_7$ | 280–282 d | | Edgeworthia gardneri Daphne gnidiodes | (421) (634) |
| 768 | Daphnoretin | 1963 | | $C_{19}H_{12}O_7$ | 244–247 | | Daphne mezereum | (374, 628) |
| 769 | Daphnorin | 1963 | R = β-D-glucosyl | $C_{25}H_{22}O_{12}$ | 202–204 | $-78^{20}$ $H_2O$ | Daphne mezereum | (629) |
| 770 | Rutarensin | 1988 | R = 6-(3-hydroxy-3-methylglutaryl)--β-D-glucosyl | $C_{31}H_{30}O_{17}$ | 220 d | | Ruta chalepensis | (230) |
| 771 | | 1968 | | $C_{20}H_{14}O_7$ | 240 | | Ruta graveolens | (373, 506, 530) |

233

Table 12. (Continued)

| | Trivial name(s) | Year isolated | Structure | Formula | M.p. | $[\alpha]_\lambda^t$ | Solvent | Plant sources | Leading references |
|---|---|---|---|---|---|---|---|---|---|
| 772 | Acetyldaphnoretin | 1986 | | $C_{21}H_{14}O_8$ | 230–232 | | | *Edgeworthia gardneri* | (145) |
| 773 | Oreojasmin | 1975 | | $C_{20}H_{14}O_7$ | 238–239 | | | *Ruta oreojasme* | (271) |
| 774 | Lasioerin | 1978 | | $C_{18}H_8O_5$ | > 351 d | | | *Lasiosiphon eriocephalus* | (561) |
| 775 | Gnidiacoumarin | 1975 | | $C_{18}H_8O_5$ | 355–365 d | | | *Gnidia lamprantha* | (399) |
| 776 | Eriocephaloside | 1981 | R = rhamnosyl | $C_{24}H_{18}O_{10}$ | 350 d | | | *Lasiosiphon eriocephalus* | (91) |
| 777 | Dicoumarol | 1941 | | $C_{19}H_{12}O_6$ | 288–289 | | | *Melilotus alba* | (138, 317, 598) |

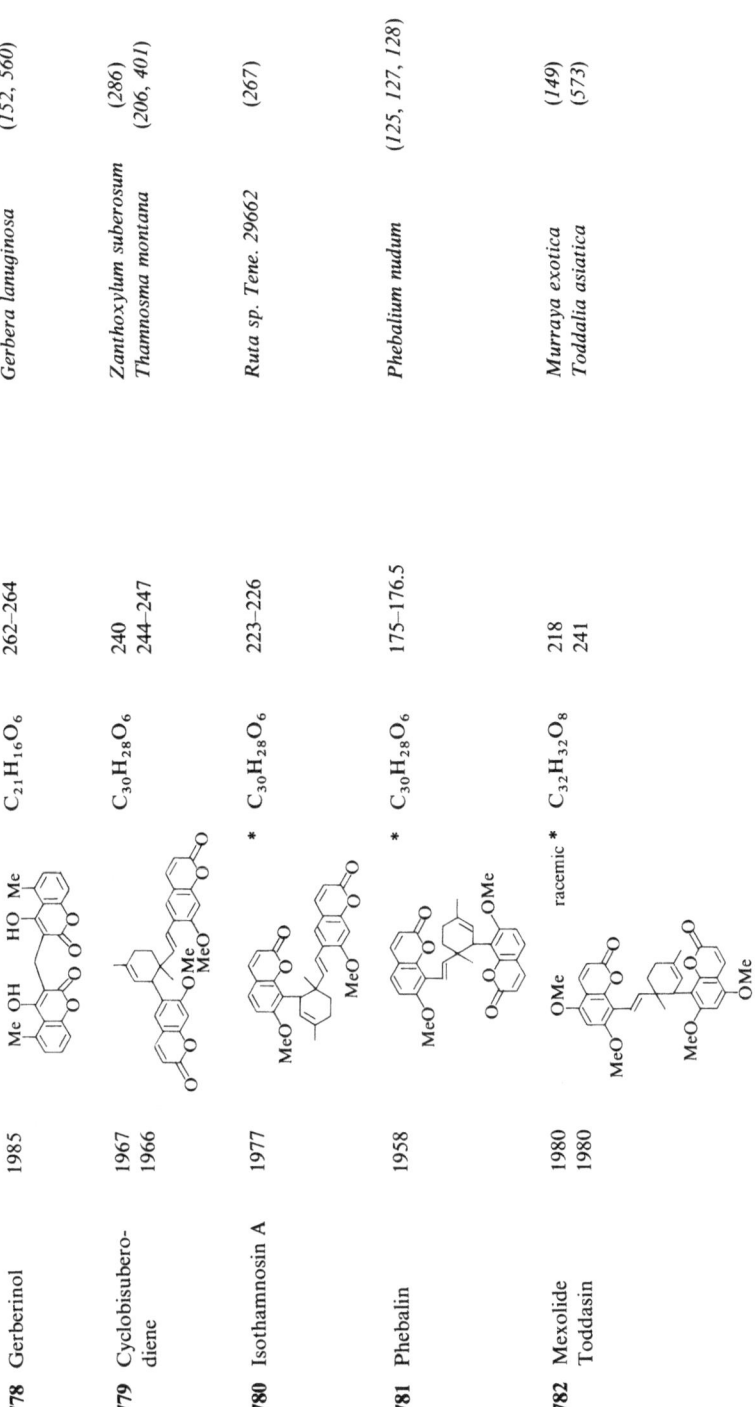

| | | | | | |
|---|---|---|---|---|---|
| **778** Gerberinol | 1985 | $C_{21}H_{16}O_6$ | 262–264 | *Gerbera lanuginosa* | (152, 560) |
| **779** Cyclobisubero-diene | 1967 1966 | $C_{30}H_{28}O_6$ | 240 244–247 | *Zanthoxylum suberosum* *Thamnosma montana* | (286) (206, 401) |
| **780** Isothamnosin A | 1977 | * $C_{30}H_{28}O_6$ | 223–226 | *Ruta sp. Tene. 29662* | (267) |
| **781** Phebalin | 1958 | * $C_{30}H_{28}O_6$ | 175–176.5 | *Phebalium nudum* | (125, 127, 128) |
| **782** Mexolide Toddasin | 1980 1980 | racemic * $C_{32}H_{32}O_8$ | 218 241 | *Murraya exotica* *Toddalia asiatica* | (149) (573) |

235

## Table 12. (Continued)

| | Trivial name(s) | Year isolated | Structure | Formula | M.p. | $[\alpha]_\lambda^t$ | Solvent | Plant sources | Leading references |
|---|---|---|---|---|---|---|---|---|---|
| 783 | Candicanin | 1971 | * | $C_{32}H_{28}O_{10}$ | 152–153 | | | Heracleum candicans | (63–65) |
| 784 | Ismailin | 1983 | | $C_{31}H_{18}O_9$ | 284–285 d | | | Diospyros ismailii | (342) |

## Table 13. Triscoumarins

| | Trivial name(s) | Year isolated | Structure | Formula | M.p. | $[\alpha]_\lambda^t$ | Solvent | Plant sources | Leading references |
|---|---|---|---|---|---|---|---|---|---|
| 785 | Edgeworoside A | 1989 | R = α-L-rhamnosyl | $C_{33}H_{24}O_{13}$ | 208–209 | | | Edgeworthia chrysantha | (55) |

*Amendments/Additions to Entries in Table 1 in Reference (448)*

| | Trivial Name(s) | Structure | Formula | Amendment/addition | Leading reference |
|---|---|---|---|---|---|
| 786 (8) | 6′,7′-Epoxyaurapten | | $C_{19}H_{22}O_4$ | Synthesis | (50, 51) |
| 787 (9) | 3′,6′-Epoxycycloaurapten | | $C_{19}H_{22}O_4$ | Synthesis | (50, 51) |
| 788 (15) | *trans*-Diversin | | $C_{19}H_{20}O_4$ | Given a trivial name | (562) |
| 789 (16) | Diversin | | $C_{19}H_{20}O_4$ | Revised stereochemistry | (562) |
| 790 (17) | Auraptenal | | $C_{19}H_{20}O_4$ | Given a trivial name | (563) |
| 791 (18) | Marmin | | $C_{19}H_{24}O_5$ | Synthesis | (50, 51) |

Amendments/Additions to Entries in Table 1 (Continued)

| Trivial Name(s) | Structure | Formula | Amendment/addition | Leading reference |
|---|---|---|---|---|
| 792 (22) Geiparvarin | | $C_{19}H_{18}O_5$ | Synthesis | (71, 160, 336, 343, 361, 547, 549, 630) |
| 793 (23) Capnolactone | | * $C_{19}H_{18}O_5$ | Given a trivial name | (563) |
| 794 (24) Oxycapnolactone | | * $C_{19}H_{18}O_6$ | Given a trivial name | (563) |
| 795 (28) Karatavicinol | | $C_{24}H_{32}O_5$ | Isolated with m.p. 72–74° Stereochemistry | (643) (131) |
| 796 (29) Reoselin | R = β-sophorosyl | $C_{36}H_{52}O_{15}$ | Structure revision | (131) |
| 797 (32) Farnesiferol C | | $C_{24}H_{30}O_4$ | Synthesis of antipode | (443) |

| No. | Name | Structure | Formula | Stereochemistry | Ref. |
|---|---|---|---|---|---|
| 798 (35) | Kopeolin | | $C_{24}H_{32}O_5$ | Stereochemistry | (464) |
| 799 (36) | Kopeoside | R = β-D-glucosyl | $C_{30}H_{42}O_{10}$ | Stereochemistry | (464) |
| 800 (37) | Feropolol | * | $C_{24}H_{34}O_6$ | Stereochemistry | (545) |
| 801 (38) | Feropolin | * | $C_{26}H_{36}O_7$ | Stereochemistry | (545) |
| 802 (39) | Feropolone | * | $C_{24}H_{32}O_6$ | Stereochemistry | (545) |

*Amendments/Additions to Entries in Table 1* (Continued)

| Trivial Name(s) | Structure | Formula | Amendment/addition | Leading reference |
|---|---|---|---|---|
| **803** (40) Karatavic acid | | $C_{24}H_{28}O_5$ | Revised structure<br>Synthesis | (493)<br>(481) |
| **804** (41) Galbanic acid | | $C_{24}H_{30}O_5$ | Revised structure<br>Stereochemistry | (58)<br>(468) |
| **805** (42) Methyl galbanate | | $C_{25}H_{32}O_5$ | Revised structure<br>Stereochemistry | (58)<br>(468) |
| **806** (49) Colladonin | | $C_{24}H_{30}O_4$ | Revised stereochemistry | (544) |

| | | | | |
|---|---|---|---|---|
| 807 (50) | Colladin | | $C_{26}H_{32}O_5$ | Revised stereochemistry (544) |
| 808 (51) | Badrakemin | | $C_{24}H_{30}O_4$ | Revised stereochemistry (544) |
| 809 (52) | Badrakemin acetate | | $C_{26}H_{32}O_5$ | Revised stereochemistry (544) |
| 810 (53) | Badrakemone | | $C_{24}H_{28}O_4$ | Revised stereochemistry (544) |
| 811 (54) | Feselol | | $C_{24}H_{30}O_4$ | Revised stereochemistry (466) |

*Amendments/Additions to Entries in Table 1* (Continued)

| Trivial Name(s) | Structure | Formula | Amendment/addition | Leading reference |
|---|---|---|---|---|
| **812** (55) Conferol | | $C_{24}H_{30}O_4$ | Revised stereochemistry | (544) |
| **813** (56) Conferol acetate | | $C_{26}H_{32}O_5$ | Revised stereochemistry | (544) |
| **814** (57) Conferone | | $C_{24}H_{28}O_4$ | Revised stereochemistry | (544) |
| **815** (58) Feropolidin | | $C_{24}H_{30}O_4$ | Stereochemistry | (545) |

*References, pp. 283–316*

| | | | | |
|---|---|---|---|---|
| 816 (59) | Mogoltacin | $C_{24}H_{30}O_4$ | Revised stereochemistry | (544) |
| 817 (60) | Mogoltin | $C_{24}H_{30}O_5$ | Stereochemistry | (368) |
| 818 (61) | Mogoltavin | $C_{26}H_{32}O_6$ | Stereochemistry | (368) |
| 819 (62) | Mogoltavinin | $C_{29}H_{36}O_6$ | Stereochemistry | (368) |
| 820 (63) | Conferin | $C_{26}H_{30}O_6$ | Revised stereochemistry | (544) |

Amendments/Additions to Entries in Table 1 (Continued)

| | Trivial Name(s) | Structure | Formula | Amendment/addition | Leading reference |
|---|---|---|---|---|---|
| **821**<br>**(64)** | Conferdione | | $C_{24}H_{26}O_5$ | Revised stereochemistry | (544) |
| **822**<br>**(65)** | Nevskin | | $C_{24}H_{32}O_5$ | Revised stereochemistry | (477, 542) |
| **823**<br>**(65)** | Isosamarcandin | | $C_{24}H_{32}O_5$ | Revised stereochemistry | (477) |
| **824**<br>**(66)** | Isosamarcandin angelate | | $C_{29}H_{38}O_6$ | Revised stereochemistry | (477) |

| | | | | |
|---|---|---|---|---|
| **825** (67) | Samarcandin | | $C_{24}H_{32}O_5$ | Revised stereochemistry (477, 502) |
| **826** (68) | Samarcandin acetate | | $C_{26}H_{34}O_6$ | Revised stereochemistry (477, 502) |
| **827** (69) | Samarcandone | | $C_{24}H_{30}O_5$ | Revised stereochemistry (477) |
| **828** (70) | Colladocin | | $C_{26}H_{34}O_6$ | Revised stereochemistry (542) |

*Amendments/Additions to Entries in Table 1 (Continued)*

| Trivial Name(s) | Structure | Formula | Amendment/addition | Leading reference |
|---|---|---|---|---|
| **829** (71) Mogoltavidin | | $C_{24}H_{32}O_5$ | Revised stereochemistry | (477, 544) |
| **830** (72) Mogoltavicin | | $C_{26}H_{34}O_6$ | Revised stereochemistry | (477, 544) |
| **831** (73) Kellerin | | $C_{24}H_{32}O_5$ | Stereochemistry | (463, 465, 502) |

*Amendments/Additions to Entries in Table 1.1 in Reference (448)*

| | Trivial name(s) | Structure | Formula | Amendment/addition | Leading reference |
|---|---|---|---|---|---|
| **832** (82) | Peuruthenicin | | $C_{11}H_8O_5$ | Synthesis | (26) |
| **833** (82) | Officinalin | | $C_{11}H_8O_5$ | Structure revision required | (26) |
| **834** (83) | Officinalin acetate | | $C_{13}H_{10}O_6$ | Structure revision required | (26) |
| **835** (84) | Officinalin isobutyrate | | $C_{15}H_{14}O_6$ | Structure revision required | (26) |
| **836** (86) | Demethylsuberosin | | $C_{14}H_{14}O_3$ | Synthesis | (134, 135) |
| **837** (87) | Suberosin | | $C_{15}H_{16}O_3$ | Synthesis | (135) |

*Amendments/Additions to Entries in Table 1.1 (Continued)*

| | Trivial name(s) | Structure | Formula | Amendment/addition | Leading reference |
|---|---|---|---|---|---|
| 838 (90) | Geijerin | | $C_{15}H_{16}O_4$ | Synthesis | (137) |
| 839 (92) | Dehydrogeijerin | | $C_{15}H_{14}O_4$ | Synthesis | (137) |
| 840 (93) | Peucedanol | | $C_{14}H_{16}O_5$ | Stereochemistry | (407) |
| 841 (95) | Ulopterol Pubesinol | | * $C_{15}H_{18}O_5$ | New trivial name given | RS Kapil, personal communication |
| 842 (98) | Angelol Angelol A | | $C_{20}H_{24}O_7$ | Structure revised New trivial name given | (53, 54) |
| 843 (99) | Micromelin | | $C_{15}H_{12}O_6$ | Preferred conformation assigned | (404, 614) |

| | | | | | |
|---|---|---|---|---|---|
| 844 (100) | Ostruthin | | $C_{19}H_{22}O_3$ | Synthesis | (135) |
| 845 (102) | Ammirin | | $C_{14}H_{12}O_3$ | Synthesis | (651) |
| 846 (118) | Prandiol | * | $C_{14}H_{14}O_5$ | M.p. 172–173 from *Apium graveolens* | (254) |
| 847 (120) | Smirniorin | | $C_{18}H_{18}O_7$ | Stereochemistry | (408) |
| 848 (121) | Smirnioridin | | $C_{21}H_{22}O_7$ | Stereochemistry | (408) |
| 849 (122) | Psoralen | | $C_{11}H_6O_3$ | Synthesis | (301, 529) |
| 850 (124) | Prangosin | | $C_{14}H_{13}NO_3$ | Synthesis | (558) |
| 851 (141) | Xanthyletin | | $C_{14}H_{12}O_3$ | Synthesis | (651) |

*Amendments/Additions to Entries in Table 1.2 in Reference (448)*

| | Trivial name(s) | Structure | Formula | Amendment/addition | Leading reference |
|---|---|---|---|---|---|
| **852** (144) | Ramosin | | $C_{19}H_{22}O_3$ | Given a trivial name | (255) |
| **853** (148) | Auraptenol | | $C_{15}H_{16}O_4$ | Stereochemistry | (73) |
| **854** (149) | Arnottinin | | $C_{14}H_{14}O_4$ | Synthesis | (70) |
| **855** (151) | Citrusal | | $C_{15}H_{16}O_4$ | Given a trivial name | (563) |
| **856** (154) | Meranzin hydrate | | $C_{15}H_{18}O_5$ | Stereochemistry | |

| 857 | *trans*-Dehydroosthol | | $C_{15}H_{14}O_3$ | Synthesis | (460) |
| (158) | | | | | |
| 858 | Murralongin | | $C_{15}H_{14}O_4$ | Structure revised | (320) |
| (160) | | | | | |
| 859 | Murrangatin | | $C_{15}H_{16}O_5$ | Stereochemistry | (328) |
| (164) | | | | | |
| 860 | Apterin | | $C_{20}H_{24}O_{10}$ | Optical rotation sign reversed | (580) |
| (185) | | | | | |
| 861 | Cniforin A | | $C_{20}H_{22}O_7$ | Given a trivial name | (52) |
| (187) | | | | | |
| 862 | Isopeucenidin | | $C_{21}H_{22}O_7$ | Optical rotation sign reversed | (363) |
| (189) | | | | | |

Amendments/Additions to Entries in Table 1.2 (Continued)

| Trivial name(s) | Structure | Formula | Amendment/addition | Leading reference |
|---|---|---|---|---|
| **863** Angelicin (**199**) | | $C_{11}H_6O_3$ | Synthesis | (529, 664) |
| **864** Oroselol (**200**) | | $C_{14}H_{12}O_4$ | Synthesis | (207) |
| **865** Oroselone (**202**) | | $C_{14}H_{10}O_3$ | Synthesis | (559) |
| **866** Lomatin (**203**) Libanotin A | | $C_{14}H_{14}O_4$ | New trivial name given but incorrect structure assigned | (648) |

| | | | | | |
|---|---|---|---|---|---|
| **867** (215) | Laserpitin | | $C_{19}H_{20}O_6$ | Given a trivial name | (563) |
| **868** (230) | (+)-Anomalin (+)-Praeruptorin **B** | | $C_{24}H_{26}O_7$ | New trivial name given | (258) |
| **869** (234) | Oxyanomalin | | * $C_{24}H_{26}O_8$ | Given a trivial name | (258) |
| **870** (239) | Dioxyanomalin | | * $C_{24}H_{26}O_9$ | Given a trivial name | (258) |

*Amendments/Additions to Entries in Table 2 in Reference (448)*

| Trivial name(s) | Structure | Formula | Amendment/addition | Leading reference |
|---|---|---|---|---|
| **871** (272) Aurantiumal | | $C_{21}H_{22}O_5$ | Given a trivial name Shown to be an artefact | (563) (231) |
| **872** (275) Eriobrucinol | | $C_{19}H_{20}O_4$ | Synthesis | (172) |
| **873** (276) Hydroxyeriobrucinol | | $C_{19}H_{20}O_5$ | Stereochemistry | (260) |
| **874** (277) Deoxybruceol | | $C_{19}H_{20}O_4$ | Crystal structure | (259) |
| **875** (278) Bruceol | | $C_{19}H_{20}O_5$ | Stereochemistry | (84, 259) |

| No. | Name | Structure | Formula | Notes | Ref. |
|---|---|---|---|---|---|
| 876 (280) | Xanthoxyletin | | $C_{15}H_{14}O_4$ | Synthesis | (452) |
| 877 (282) | Coumurrayin | | $C_{16}H_{18}O_4$ | Synthesis | (452) |
| 878 (284) | Isosibiricin | | $C_{16}H_{18}O_5$ | Given a trivial name | (563) |
| 879 (290) | | | $C_{16}H_{14}O_4$ | Synthesis  Revision of m.p. to 231–235 | (457) |
| 880 (292) | Furopinnarin | | $C_{17}H_{16}O_4$ | Synthesis | (457) |

*Amendments/Additions to Entries in Table 2 (Continued)*

| | Trivial name(s) | Structure | Formula | Amendment/addition | Leading reference |
|---|---|---|---|---|---|
| **881** (293) | Hortiolone | | $C_{19}H_{18}O_4$ | Structure confirmation<br>Synthesis | (186)<br>(456) |
| **882** (300) | Nordentatin | | $C_{19}H_{20}O_4$ | Synthesis | (453) |
| **883** (301) | Trachyphyllin | | $C_{19}H_{20}O_4$ | Synthesis | (452) |
| **884** (302) | Dentatin | | $C_{20}H_{22}O_4$ | Synthesis | (453) |

*Amendments/Additions to Entries in Table 3 in Reference (448)*

| | Trivial name(s) | Structure | Formula | Amendment/addition | Leading reference |
|---|---|---|---|---|---|
| **885** (322) | Obliquetol | | $C_{14}H_{14}O_4$ | Synthesis | (451) |
| **886** (323) | Obliquetin | | $C_{15}H_{16}O_4$ | Synthesis | (451) |
| **887** (328) | Nieshoutin | | * $C_{15}H_{16}O_4$ | Synthesis | (451) |

*Amendments/Additions to Entries in Table 3* (Continued)

| Trivial name(s) | Structure | Formula | Amendment/addition | Leading reference |
|---|---|---|---|---|
| **888** (329) Heratomol | | $C_{11}H_6O_4$ | Synthesis | (30) |
| **889** (330) Sphondin | | $C_{12}H_8O_4$ | Synthesis | (664) |
| **890** (331) Heratomin | | $C_{16}H_{14}O_4$ | Synthesis | (30, 664) |

*Amendments/Additions to Entries in Table 4 in Reference (448)*

| | Trivial name(s) | Structure | Formula | Amendment/addition | Leading reference |
|---|---|---|---|---|---|
| **891** (346) | Rutaretin Leptophyllin Racemol | | $C_{14}H_{14}O_5$ | Stereochemistry New trivial name given New trivial name given | (327) (568, 571) (575) |
| **892** (347) | Isorutarin Leptophylloside | | $C_{20}H_{24}O_{10}$ | Stereochemistry New trivial name given | (327) (568, 571) |
| **893** (348) | Rutarin | | $C_{20}H_{24}O_{10}$ | Stereochemistry | (327) |
| **894** (349) | Rutaretin methyl ether | | $C_{15}H_{16}O_5$ | Stereochemistry | (327) |
| **895** (352) | Imperatorin Marmelide | | $C_{16}H_{14}O_4$ | New trivial name given | (146, 455) |
| **896** (372) | Benahorin | | $C_{17}H_{16}O_4$ | Synthesis | (455) |

Amendments/Additions to Entries in Table 5 in Reference (448)

| Trivial name(s) | Structure | Formula | Amendment/addition | Leading reference |
|---|---|---|---|---|
| **897** (377) Isofraxetin | | $C_{10}H_8O_5$ | Synthesis | (28) |
| **898** (378) Tomentin | | $C_{11}H_{10}O_5$ | Isolated as a natural product | (161) |
| **899** (380) Umckalin | | $C_{11}H_{10}O_5$ | Synthesis | (28) |
| **900** (382) | | $C_{15}H_{14}O_5$ | Structure revision is required | (29, 416, 417) |
| **901** (383) Tomenin | | $C_{17}H_{20}O_{10}$ | Synthesis | (29) |
| **902** (387) | | $C_{21}H_{26}O_5$ | Synthesis | (28) |

*Amendments/Additions to Entries in Table 6 in Reference (448)*

| | Trivial name(s) | Structure | Formula | Amendment/addition | Leading reference |
|---|---|---|---|---|---|
| **903** (391) | | | $C_{11}H_{10}O_5$ | Synthesis necessitates the structural revision to this formula | (27) |
| **904** (392) | | | $C_{12}H_{12}O_5$ | Synthesis | (27) |

*Amendments/Additions to Entries in Table 8 in Reference (448)*

| | Trivial name(s) | Structure | Formula | Amendment/addition | Leading reference |
|---|---|---|---|---|---|
| **908** | Artelin | | $C_{13}H_{14}O_6$ | Previous identification believed to be erroneous | (36) |
| (427) | | | | Now isolated from *Sapium sebiferum* | (668) |

Amendments/Additions to Entries in Table 7 in Reference (448)

| Trivial name(s) | Structure | Formula | Amendment/addition | Leading reference |
|---|---|---|---|---|
| **905** (416) Dracunculin | | $C_{11}H_8O_5$ | Given a trivial name Synthesis Identical with a wrongly assigned coumarin from *Artemisia vulgaris* | (563) (142) (459, 602) |
| **906** (420) Obtusifol | | * $C_{15}H_{16}O_6$ | Structure now revised | (9) |
| **907** (424) Isofraxoside | | $C_{16}H_{18}O_{10}$ | Structure assigned to a supposed new coumarin from *Haplophyllum obtusifolium* | (79) |

*Amendments/Additions to Entries in Table 9 in Reference (448)*

| | Trivial name(s) | Structure | Formula | Amendment/ addition | Leading reference |
|---|---|---|---|---|---|
| **909** (430) | Gravelliferone | | $C_{19}H_{22}O_3$ | Synthesis | (136, 425, 575) |
| **910** (433) | Chalepin | | $C_{19}H_{22}O_4$ | Synthesis Stereochemistry | (425, 575) (133) |
| **911** (434) | Rutamarin | | $C_{21}H_{24}O_5$ | Synthesis Stereochemistry | (425) (133) |
| **912** (437) | | | $C_{19}H_{20}O_3$ | Synthesis | (239, 575, 576) |
| **913** (438) | | | $C_{19}H_{20}O_3$ | Synthesis | (611) |
| **914** (439) | | | $C_{24}H_{28}O_3$ | Synthesis | (575) |

*Amendments/Additions to Entries in Table 10 in Reference (448)*

| Trivial name(s) | Structure | Formula | Amendment/addition | Leading reference |
|---|---|---|---|---|
| **915** (445) | | * $C_{19}H_{24}O_5$ | Synthesis | (170) |
| **916** (446) | | $C_{21}H_{26}O_5$ | Synthesis | (170) |
| **917** (447) | | $C_{21}H_{26}O_5$ | Synthesis | (170) |
| **918** (448) | | $C_{22}H_{28}O_5$ | Synthesis<br>Stereochemistry and synthesis | (170)<br>(83) |
|  |  |  | Mammea B/BC |  |
|  |  |  | Mammea B/BD |  |
|  |  |  | Mammea B/BB |  |

*References, pp. 283–316*

**919** Ferruol A  
**(449)** Mammea D/BB

$C_{23}H_{30}O_5$  Synthesis  (171)

**920** Mammea C/BB  
**(450)**

$C_{24}H_{32}O_5$  Synthesis  (170)

**921** Surangin A  
**(451)**

$C_{27}H_{36}O_5$  Synthesis  (170)

**922** Mammea B/BA  
**(452)**

$C_{22}H_{28}O_5$  Synthesis  (170)

*Amendments/Additions to Entries in Table 10* (Continued)

| Trivial name(s) | Structure | Formula | Amendment/addition | Leading reference |
|---|---|---|---|---|
| **923** (453) Mammea B/AC | | $C_{21}H_{26}O_5$ | Synthesis | (170) |
| **924** (454) Mammea B/AB | | * $C_{22}H_{28}O_5$ | Synthesis | (170) |
| **925** (455) Mammea B/AA | | $C_{22}H_{28}O_5$ | Synthesis | (170) |
| **926** (456) Mammea B/BC cyclo F | | * $C_{21}H_{26}O_6$ | Synthesis | (170) |

| | | | | |
|---|---|---|---|---|
| 927 (457) | Mammea B/BB cyclo F | | * $C_{22}H_{28}O_6$ | Synthesis | (170) |
| 928 (458) | Mammea B/BA cyclo F | | * $C_{22}H_{28}O_6$ | Synthesis | (170) |
| 929 (462) | Mammea B/AC cyclo F | | * $C_{21}H_{26}O_6$ | Synthesis | (170) |
| 930 (463) | Mammea B/AB cyclo F | | * $C_{22}H_{28}O_6$ | Synthesis | (170) |
| 931 (464) | Mammea B/AA cyclo F | | * $C_{22}H_{28}O_6$ | Synthesis | (170) |

*Amendments/Additions to Entries in Table 10* (Continued)

| Trivial name(s) | Structure | Formula | Amendment/addition | Leading reference |
|---|---|---|---|---|
| **932** (**465**) | | * $C_{21}H_{26}O_6$ | Synthesis | (170) |
| Mammea B/BC cyclo E | | | | |
| **933** (**466**) | | * $C_{22}H_{28}O_6$ | Synthesis | (170) |
| Mammea B/BB cyclo E | | | | |
| **934** (**467**) | | * $C_{22}H_{28}O_6$ | Synthesis | (170) |
| Mammea B/BA cyclo E | | | | |
| **935** (**471**) | | * $C_{22}H_{26}O_5$ | Synthesis | (170) |
| Mammea B/AB cyclo D | | | | |

| | | | | |
|---|---|---|---|---|
| 936 (472) | Mammea C/AB cyclo D | Synthesis | * $C_{24}H_{30}O_5$ | (170) |
| 937 (478) | Mammea E/BB | Synthesis | * $C_{24}H_{30}O_7$ | (171) |
| 938 (479) | Surangin B | Synthesis | * $C_{29}H_{38}O_7$ | (83, 171) |
| 939 (480) | Mammea E/BA | Synthesis | * $C_{24}H_{30}O_7$ | (171) |

Amendments/Additions to Entries in Table 11 in Reference (448)

| | Trivial name(s) | Structure | Formula | Amendment/addition | Leading reference |
|---|---|---|---|---|---|
| 940 (482) | Ekersenin | Me OMe | $C_{11}H_{10}O_3$ | Given a trivial name | (31, 486) |
| | Pereflorin | | | Synthesis | |
| | | | | Given a trivial name | (116) |
| 941 (485) | Isogerberacoumarin | racemic Me | $C_{15}H_{16}O_3$ | Synthesis | (163) |
| 942 (486) | | OMe | $C_{10}H_8O_3$ | Synthesis | (26) |
| 943 (499) | | Me OMe HO OMe | $C_{12}H_{12}O_5$ | Synthesis | (21) |

# Formula Index

| Formula | Compound number | Formula | Compound number |
|---|---|---|---|
| $C_9H_6O_3$ | (553), (554), (663), (668) | $C_{14}H_{13}KO_9S$ | (234) |
| $C_9H_6O_4$ | (244), (671) | $C_{14}H_{13}NO_3$ | (850) |
| $C_9H_6O_5$ | (697) | $C_{14}H_{14}O_3$ | (130), (452), (836) |
| | | $C_{14}H_{14}O_4$ | (1), (3), (76), (80), (81), (131), |
| $C_{10}H_6O_3$ | (599) | | (145), (158), (246), (338), |
| $C_{10}H_6O_4$ | (73), (351) | | (347), (854), (866), (885) |
| $C_{10}H_6O_6$ | (472) | $C_{14}H_{14}O_5$ | (77), (123), (205), (253), (318), |
| $C_{10}H_8O_3$ | (70), (481), (567), (669), (942) | | (328), (329), (846), (891) |
| $C_{10}H_8O_4$ | (245), (672), (688) | $C_{14}H_{16}O_4$ | (78), (673) |
| $C_{10}H_8O_5$ | (698), (897) | $C_{14}H_{16}O_5$ | (2), (163), (699), (840) |
| $C_{10}H_{10}O_2S$ | (600) | | |
| | | $C_{15}H_8O_5$ | (714) |
| $C_{11}H_6O_3$ | (849), (863) | $C_{15}H_8O_6$ | (511), (725), (737), (745) |
| $C_{11}H_6O_4$ | (286), (888) | $C_{15}H_8O_7$ | (738) |
| $C_{11}H_8O_4$ | (68), (75), (143), (488) | $C_{15}H_{10}O_4$ | (491), (540) |
| $C_{11}H_8O_5$ | (74), (832), (833), (905) | $C_{15}H_{10}O_5$ | (492) |
| $C_{11}H_{10}O_3$ | (71), (664), (940) | $C_{15}H_{12}O_4$ | (288), (369) |
| $C_{11}H_{10}O_4$ | (72), (142), (340), (675), (680), | $C_{15}H_{12}O_5$ | (150) |
| | (689) | $C_{15}H_{12}O_6$ | (331), (376), (843) |
| $C_{11}H_{10}O_5$ | (372), (382), (898), (899), (903) | $C_{15}H_{12}O_7$ | (332) |
| | | $C_{15}H_{14}O_3$ | (94), (172), (637), (652), (857) |
| $C_{12}H_8O_4$ | (889) | $C_{15}H_{14}O_4$ | (173), (293), (322), (323), |
| $C_{12}H_{10}O_4$ | (144), (482) | | (489), (686), (839), (858), (876) |
| $C_{12}H_{10}O_5$ | (247), (335) | $C_{15}H_{14}O_5$ | (4), (107), (193), (248), (268), |
| $C_{12}H_{10}O_6$ | (451) | | (270), (374), (414), (415), (900) |
| $C_{12}H_{12}O_2$ | (552) | $C_{15}H_{14}O_6$ | (108), (109), (375), (835) |
| $C_{12}H_{12}O_4$ | (670), (681) | $C_{15}H_{16}O_3$ | (837), (941) |
| $C_{12}H_{12}O_5$ | (269), (490), (691), (693), | $C_{15}H_{16}O_4$ | (82), (83), (84), (88), (159), |
| | (694), (904), (943) | | (272), (342), (353), (354), |
| | | | (611), (620), (687), (838), |
| $C_{13}H_{10}O_6$ | (834) | | (853), (855), (886), (887) |
| $C_{13}H_{12}O_6$ | (696) | $C_{15}H_{16}O_5$ | (134), (151), (175), (176), |
| $C_{13}H_{12}O_7$ | (700) | | (177), (183), (184), (186), |
| $C_{13}H_{14}O_3$ | (483), (666) | | (187), (188), (211), (212), |
| $C_{13}H_{14}O_4$ | (679) | | (275), (317), (319), (343), |
| $C_{13}H_{14}O_5$ | (692), (695) | | (349), (449), (473), (859), (894) |
| $C_{13}H_{14}O_6$ | (908) | $C_{15}H_{16}O_6$ | (289), (413), (450), (906) |
| | | $C_{15}H_{16}O_9$ | (252) |
| $C_{14}H_{10}O_3$ | (129), (865) | $C_{15}H_{16}O_{10}$ | (408) |
| $C_{14}H_{10}O_4$ | (206) | $C_{15}H_{17}ClO_4$ | (171) |
| $C_{14}H_{12}O_3$ | (635), (636), (845), (851) | $C_{15}H_{18}N_2O_3$ | (667) |
| $C_{14}H_{12}O_4$ | (128), (146), (147), (292), | $C_{15}H_{18}O_4$ | (79) |
| | (359), (367), (371), (864) | $C_{15}H_{18}O_5$ | (89), (162), (341), (841), (856) |
| $C_{14}H_{12}O_5$ | (330), (373) | $C_{15}H_{18}O_6$ | (95), (320), (344), (360), (381), |
| $C_{14}H_{13}KO_8S$ | (209), (358) | | (411) |

| Formula | Compound number | Formula | Compound number |
|---------|-----------------|---------|-----------------|
| $C_{16}H_{10}O_5$ | (619), (715) | $C_{17}H_{20}O_{10}$ | (901) |
| $C_{16}H_{10}O_6$ | (475), (512), (719), (726), (746), (750) | $C_{17}H_{22}O_5$ | (170) |
|  |  | $C_{17}H_{22}O_6$ | (251), (283) |
| $C_{16}H_{10}O_7$ | (740) | $C_{17}H_{22}O_7$ | (379) |
| $C_{16}H_{12}O_4$ | (541), (544) |  |  |
| $C_{16}H_{12}O_5$ | (474), (542), (543) | $C_{18}H_8O_5$ | (774), (775) |
| $C_{16}H_{12}O_6$ | (494), (504) | $C_{18}H_8O_6$ | (735) |
| $C_{16}H_{14}O_3$ | (240), (630) | $C_{18}H_{10}O_6$ | (764), (751) |
| $C_{16}H_{14}O_4$ | (241), (242), (287), (294), (301), (879), (890), (895) | $C_{18}H_{10}O_7$ | (767) |
|  |  | $C_{18}H_{10}O_8$ | (756), (757) |
| $C_{16}H_{14}O_5$ | (361) | $C_{18}H_{14}O_6$ | (510) |
| $C_{16}H_{16}O_3$ | (655), (659) | $C_{18}H_{14}O_7$ | (333), (514), (515) |
| $C_{16}H_{16}O_4$ | (280), (453), (677) | $C_{18}H_{16}O_5$ | (507) |
| $C_{16}H_{16}O_5$ | (119), (243), (368), (407), (462) | $C_{18}H_{16}O_6$ | (509), (547) |
| $C_{16}H_{18}O_3$ | (653) | $C_{18}H_{18}O_5$ | (660) |
| $C_{16}H_{18}O_4$ | (85), (86), (199), (271), (336), (656), (676), (682), (683), (684), (877) | $C_{18}H_{18}O_5S$ | (200) |
|  |  | $C_{18}H_{18}O_6$ | (463) |
|  |  | $C_{18}H_{18}O_7$ | (259), (847) |
| $C_{16}H_{18}O_5$ | (178), (249), (250), (273), (276), (278), (384), (385), (386), (412), (878) | $C_{18}H_{18}O_{10}$ | (390) |
|  |  | $C_{18}H_{20}O_5$ | (650), (657) |
|  |  | $C_{18}H_{20}O_6$ | (220), (256) |
| $C_{16}H_{18}O_6$ | (168), (279), (285), (290), (388), (690), (721) | $C_{18}H_{20}O_7$ | (396), (397) |
|  |  | $C_{18}H_{20}O_{10}$ | (303), (304) |
| $C_{16}H_{18}O_8$ | (568) | $C_{18}H_{22}O_5$ | (454) |
| $C_{16}H_{18}O_9$ | (674) |  |  |
| $C_{16}H_{18}O_{10}$ | (410), (907) | $C_{19}H_{12}O_6$ | (476), (753), (765), (766), (777) |
| $C_{16}H_{20}O_6$ | (281), (383) | $C_{19}H_{12}O_7$ | (768) |
| $C_{16}H_{20}O_7$ | (389) | $C_{19}H_{16}O_7$ | (550), (701) |
|  |  | $C_{19}H_{18}O_4$ | (881) |
| $C_{17}H_{10}O_6$ | (730) | $C_{19}H_{18}O_5$ | (792), (793) |
| $C_{17}H_{10}O_7$ | (747), (749) | $C_{19}H_{18}O_6$ | (235), (236), (237), (549), (794) |
| $C_{17}H_{12}O_4$ | (69) | $C_{19}H_{20}O_3$ | (459), (912), (913) |
| $C_{17}H_{12}O_6$ | (513), (662), (720), (727) | $C_{19}H_{20}O_4$ | (174), (302), (640), (788), (789), (790), (872), (874), (882), (883) |
| $C_{17}H_{12}O_7$ | (678) |  |  |
| $C_{17}H_{14}O_4$ | (546) |  |  |
| $C_{17}H_{14}O_5$ | (506), (551) | $C_{19}H_{20}O_5$ | (121), (122), (132), (141), (661), (873), (875) |
| $C_{17}H_{14}O_6$ | (495), (505), (508), (545) |  |  |
| $C_{17}H_{14}O_7$ | (548) | $C_{19}H_{20}O_6$ | (127), (135), (139), (201), (215), (228), (229), (230), (258), (362), (867) |
| $C_{17}H_{16}O_4$ | (295), (572), (880), (896) |  |  |
| $C_{17}H_{16}O_5$ | (192), (296), (297) |  |  |
| $C_{17}H_{16}O_6$ | (194), (391), (392), (403) | $C_{19}H_{20}O_8$ | (399) |
| $C_{17}H_{16}O_7$ | (110) | $C_{19}H_{22}O_3$ | (238), (456), (844), (852), (909) |
| $C_{17}H_{18}O_4$ | (455), (665) | $C_{19}H_{22}O_4$ | (5), (9), (153), (156), (160), (239), (458), (465), (786), (787), (910) |
| $C_{17}H_{18}O_5$ | (148), (649) |  |  |
| $C_{17}H_{18}O_6$ | (181), (189), (298) |  |  |
| $C_{17}H_{18}O_7$ | (380), (393) | $C_{19}H_{22}O_5$ | (11), (120), (460), (658) |
| $C_{17}H_{20}O_4$ | (87) | $C_{19}H_{22}O_6$ | (93), (213), (214), (222), (223), (224) |
| $C_{17}H_{20}O_5$ | (179) |  |  |
| $C_{17}H_{20}O_6$ | (378) | $C_{19}H_{24}O_4$ | (339) |

| Formula | Compound number |
|---------|-----------------|
| $C_{19}H_{24}O_5$ | (154), (165), (791), (915) |
| $C_{19}H_{24}O_6$ | (12), (13), (161), (461) |
| | |
| $C_{20}H_{14}O_5$ | (718) |
| $C_{20}H_{14}O_6$ | (752), (754) |
| $C_{20}H_{14}O_7$ | (480), (771), (773) |
| $C_{20}H_{14}O_8$ | (755) |
| $C_{20}H_{16}O_5$ | (716), (717), (731), (736) |
| $C_{20}H_{16}O_6$ | (732) |
| $C_{20}H_{16}O_7$ | (734) |
| $C_{20}H_{18}O_7$ | (443), (446) |
| $C_{20}H_{18}O_8$ | (334), (352), (444), (447) |
| $C_{20}H_{20}O_5$ | (593), (622), (623), (627), (628), (641) |
| $C_{20}H_{20}O_7$ | (464) |
| $C_{20}H_{22}O_3$ | (470), (615), (621) |
| $C_{20}H_{22}O_4$ | (580), (581), (582), (884) |
| $C_{20}H_{22}O_5$ | (284), (486), (487), (586), (624), (625), (626) |
| $C_{20}H_{22}O_6$ | (190) |
| $C_{20}H_{22}O_7$ | (218), (221), (398), (861) |
| $C_{20}H_{22}O_9$ | (360) |
| $C_{20}H_{22}O_{11}$ | (306), (307) |
| $C_{20}H_{24}O_3$ | (195), (469), (573), (601), (602), (612), (613), (614) |
| $C_{20}H_{24}O_4$ | (574), (575), (576), (616) |
| $C_{20}H_{24}O_5$ | (274), (277), (345), (583), (584), (585), (587) |
| $C_{20}H_{24}O_6$ | (185), (191), (588), (589), (590), (591), (592), (629), (639) |
| $C_{20}H_{24}O_7$ | (100), (104), (105), (106), (291), (842) |
| $C_{20}H_{24}O_9$ | (196), (208), (210) |
| $C_{20}H_{24}O_{10}$ | (124), (125), (126), (216), (217), (255), (370), (860), (892), (893) |
| $C_{20}H_{24}O_{12}$ | (65) |
| $C_{20}H_{24}O_{13}$ | (309) |
| $C_{20}H_{26}O_5$ | (457) |
| $C_{20}H_{26}O_6$ | (169), (282), (324), (346), (387) |
| $C_{20}H_{26}O_7$ | (96), (97), (98), (99), (102), (103) |
| $C_{20}H_{26}O_{10}$ | (90), (91), (92), (164), (166) |
| | |
| $C_{21}H_{14}O_8$ | (772) |
| $C_{21}H_{16}O_5$ | (729) |
| $C_{21}H_{16}O_6$ | (744), (778) |
| $C_{21}H_{18}O_5$ | (133) |

| Formula | Compound number |
|---------|-----------------|
| $C_{21}H_{18}O_6$ | (266), (365), (722), (741), (743), (748) |
| $C_{21}H_{18}O_7$ | (267) |
| $C_{21}H_{18}O_{10}$ | (728) |
| $C_{21}H_{18}O_{12}$ | (739) |
| $C_{21}H_{19}NO_5$ | (149) |
| $C_{21}H_{20}O_4$ | (265) |
| $C_{21}H_{20}O_6$ | (477) |
| $C_{21}H_{20}O_7$ | (479), (723) |
| $C_{21}H_{20}O_8$ | (366), (724) |
| $C_{21}H_{20}O_9$ | (377), (445), (449), (493) |
| $C_{21}H_{20}O_{11}$ | (496) |
| $C_{21}H_{22}O_5$ | (261), (263), (264), (485), (871) |
| $C_{21}H_{22}O_6$ | (262) |
| $C_{21}H_{22}O_7$ | (136), (138), (140), (225), (231), (364), (402), (848), (862) |
| $C_{21}H_{24}O_5$ | (6), (654), (911) |
| $C_{21}H_{24}O_7$ | (137), (260), (363) |
| $C_{21}H_{26}O_5$ | (902), (916), (917), (923) |
| $C_{21}H_{26}O_6$ | (926), (929), (932) |
| $C_{21}H_{26}O_{13}$ | (308), (310) |
| $C_{21}H_{28}O_{10}$ | (167) |
| $C_{21}H_{28}O_{11}$ | (321) |
| | |
| $C_{22}H_{16}O_6$ | (111) |
| $C_{22}H_{18}O_7$ | (704), (707) |
| $C_{22}H_{18}O_8$ | (758), (761) |
| $C_{22}H_{20}O_6$ | (703), (742) |
| $C_{22}H_{20}O_7$ | (401) |
| $C_{22}H_{22}O_6$ | (478) |
| $C_{22}H_{22}O_{11}$ | (497), (500) |
| $C_{22}H_{22}O_{12}$ | (498) |
| $C_{22}H_{24}O_5$ | (406) |
| $C_{22}H_{24}O_8$ | (405) |
| $C_{22}H_{26}O_5$ | (577), (578), (579), (935) |
| $C_{22}H_{26}O_8$ | (400) |
| $C_{22}H_{26}O_{11}$ | (257) |
| $C_{22}H_{28}O_5$ | (918), (922), (924), (925) |
| $C_{22}H_{28}O_6$ | (927), (928), (930), (931), (933), (934) |
| $C_{22}H_{28}O_{12}$ | (569) |
| $C_{22}H_{28}O_{13}$ | (311), (312), (570), (571) |
| | |
| $C_{23}H_{18}O_8$ | (713) |
| $C_{23}H_{20}O_7$ | (219), (702), (706), (708), (712) |
| $C_{23}H_{20}O_8$ | (759), (762) |
| $C_{23}H_{22}O_6$ | (705) |
| $C_{23}H_{24}O_7$ | (226) |
| $C_{23}H_{24}O_{10}$ | (499) |

| Formula | Compound number |
| --- | --- |
| $C_{23}H_{26}O_7$ | (202) |
| $C_{23}H_{28}O_{12}$ | (394), (395), (404) |
| $C_{23}H_{28}O_{13}$ | (66) |
| $C_{23}H_{30}O_5$ | (919) |
| $C_{23}H_{30}O_6$ | (325) |
| $C_{24}H_{18}O_{10}$ | (776) |
| $C_{24}H_{22}O_5$ | (527), (530) |
| $C_{24}H_{22}O_8$ | (760), (763) |
| $C_{24}H_{24}O_5$ | (518), (519), (733) |
| $C_{24}H_{24}O_6$ | (522) |
| $C_{24}H_{24}O_{12}$ | (501) |
| $C_{24}H_{26}O_4$ | (566) |
| $C_{24}H_{26}O_5$ | (821) |
| $C_{24}H_{26}O_7$ | (203), (233), (868) |
| $C_{24}H_{26}O_8$ | (869) |
| $C_{24}H_{26}O_9$ | (232), (870) |
| $C_{24}H_{28}O_3$ | (561), (914) |
| $C_{24}H_{28}O_4$ | (23), (36), (37), (42), (299), (300), (471), (560), (562), (563), (810), (814) |
| $C_{24}H_{28}O_5$ | (43), (44), (46), (47), (803) |
| $C_{24}H_{28}O_7$ | (227) |
| $C_{24}H_{28}O_8$ | (207) |
| $C_{24}H_{30}O_3$ | (155) |
| $C_{24}H_{30}O_4$ | (14), (19), (20), (21), (30), (32), (39), (40), (61), (555), (556), (557), (797), (806), (808), (811), (812), (815), (816) |
| $C_{24}H_{30}O_5$ | (7), (8), (15), (17), (25), (27), (32), (53), (60), (63), (64), (157), (466), (467), (468), (804), (817), (827), (936) |
| $C_{24}H_{30}O_6$ | (10), (254) |
| $C_{24}H_{30}O_7$ | (937), (939) |
| $C_{24}H_{30}O_{14}$ | (313), (314), (315) |
| $C_{24}H_{32}O_4$ | (28), (62) |
| $C_{24}H_{32}O_5$ | (24), (50), (52), (55), (56), (59), (795), (796), (822), (823), (825), (829), (831), (920) |
| $C_{24}H_{32}O_6$ | (54), (802) |
| $C_{24}H_{34}O_6$ | (26), (800) |
| $C_{25}H_{22}O_5$ | (532), (537), (538), (539) |
| $C_{25}H_{22}O_{12}$ | (769) |
| $C_{25}H_{24}O_5$ | (523), (524), (528), (529), (533), (534), (535), (536) |
| $C_{25}H_{24}O_6$ | (525) |
| $C_{25}H_{26}O_5$ | (516), (517), (520), (521) |
| $C_{25}H_{28}O_3$ | (638) |
| $C_{25}H_{28}O_5$ | (646), (648) |
| $C_{25}H_{30}O_3$ | (617), (618), (631) |
| $C_{25}H_{30}O_4$ | (596), (604), (632), (633), (685) |
| $C_{25}H_{30}O_5$ | (605), (606), (607), (609), (634), (642), (643), (644), (645) |
| $C_{25}H_{30}O_6$ | (608), (610), (647) |
| $C_{25}H_{32}O_3$ | (603) |
| $C_{25}H_{32}O_4$ | (326), (327), (337), (594), (595), (597), (598) |
| $C_{25}H_{32}O_5$ | (651), (805) |
| $C_{25}H_{32}O_{13}$ | (115), (197) |
| $C_{26}H_{24}O_5$ | (526), (531) |
| $C_{26}H_{26}O_6$ | (709), (710) |
| $C_{26}H_{26}O_{12}$ | (305) |
| $C_{26}H_{28}O_{15}$ | (502) |
| $C_{26}H_{30}O_5$ | (564), (565) |
| $C_{26}H_{30}O_6$ | (45), (430), (438), (820) |
| $C_{26}H_{32}O_5$ | (22), (558), (559), (807), (809), (813) |
| $C_{26}H_{32}O_6$ | (16), (35), (429), (818) |
| $C_{26}H_{34}O_5$ | (29), (416) |
| $C_{26}H_{34}O_6$ | (18), (51), (57), (417), (418), (419), (422), (423), (424), (427), (432), (433), (436), (440), (441), (442), (826), (828), (830) |
| $C_{26}H_{34}O_{12}$ | (101) |
| $C_{26}H_{34}O_{13}$ | (152) |
| $C_{26}H_{34}O_{14}$ | (116), (117), (118), (198) |
| $C_{26}H_{36}O_6$ | (421) |
| $C_{26}H_{36}O_7$ | (801) |
| $C_{27}H_{28}O_6$ | (711) |
| $C_{27}H_{30}O_{15}$ | (503) |
| $C_{27}H_{36}O_5$ | (921) |
| $C_{27}H_{36}O_6$ | (484) |
| $C_{28}H_{26}O_8$ | (204) |
| $C_{28}H_{30}O_{13}$ | (355) |
| $C_{28}H_{34}O_8$ | (431) |
| $C_{28}H_{36}O_7$ | (420), (425), (426), (428), (434), (435), (437) |
| $C_{28}H_{38}O_{17}$ | (316) |
| $C_{28}H_{38}O_{18}$ | (409) |

| Formula | Compound number | Formula | Compound number |
|---|---|---|---|
| $C_{29}H_{30}O_5$ | (41) | $C_{31}H_{18}O_9$ | (784) |
| $C_{29}H_{30}O_{12}$ | (356) | $C_{31}H_{29}NO_8$ | (112), (113) |
| $C_{29}H_{34}O_6$ | (49) | $C_{31}H_{30}O_{17}$ | (770) |
| $C_{29}H_{36}O_6$ | (38), (819) | $C_{31}H_{34}O_{14}$ | (357) |
| $C_{29}H_{38}O_5$ | (31) | $C_{31}H_{40}O_8$ | (439) |
| $C_{29}H_{38}O_6$ | (824) | $C_{31}H_{46}O_6$ | (182) |
| $C_{29}H_{38}O_7$ | (938) | | |
| $C_{29}H_{40}O_6$ | (58) | $C_{32}H_{28}O_{10}$ | (783) |
| | | $C_{32}H_{32}O_8$ | (782) |
| $C_{30}H_{27}NO_8$ | (114) | $C_{33}H_{24}O_{13}$ | (785) |
| $C_{30}H_{28}O_6$ | (779), (780), (781) | | |
| $C_{30}H_{32}O_{14}$ | (67) | $C_{34}H_{32}O_{12}$ | (180) |
| $C_{30}H_{38}O_{10}$ | (48) | | |
| $C_{30}H_{40}O_{10}$ | (33) | $C_{36}H_{50}O_{15}$ | (34) |
| $C_{30}H_{42}O_{10}$ | (799) | $C_{36}H_{52}O_{15}$ | (796) |

# Trivial Name Index

| Name | Compound number | Name | Compound number |
|---|---|---|---|
| (E)-ω-Acetoxyferprenin | (564) | Angelicin | (863) |
| (Z)-ω-Acetoxyferprenin | (565) | Angelin | (284) |
| (E)-ω-Acetoxyferulenol | (558) | Angelol | (842) |
| (Z)-ω-Acetoxyferulenol | (559) | Angelol A | (842) |
| 7-Acetoxypectanone | (431) | Angelol B | (100) |
| Acetyldaphnoretin | (772) | Angelol C | (97) |
| Acetyldeparnol | (420) | Angelol D | (104) |
| Acetyldihydromelin A | (110) | Angelol E | (98) |
| Acetyldrimartol A | (435) | Angelol F | (96) |
| Acetyldrimartol B | (437) | Angelol G | (106) |
| Acetylisodrimartol A | (434) | Angelol H | (102) |
| Acetylpectachol | (426) | Angelol I | (99) |
| Acetylpectachol B | (428) | Angeloside A | (101) |
| Acetylumbelliferone | (68) | Angustifolin | (452) |
| Acrimarine A | (112) | Anhydronotoptol | (265) |
| Acrimarine B | (113) | Anhydrorutaretin | (367) |
| Acrimarine C | (114) | Anisocoumarin A | (455) |
| Adicardin | (65) | Anisocoumarin B | (246) |
| Aegelinol | (131) | Anisocoumarin C | (460) |
| Aegelinol benzoate | (133) | Anisocoumarin D | (461) |
| Albartin | (425) | Anisocoumarin E | (160) |
| Albartol | (423) | Anisocoumarin F | (156) |
| Aleuritin | (377) | Anisocoumarin G | (165) |
| Ammirin | (845) | Anisolactone | (266) |

| Name | Compound number | Name | Compound number |
|---|---|---|---|
| (+)-Anomalin | (868) | Capnolactone | (793) |
| Anpubesol | (103) | Casegravol | (184) |
| Apaensin | (403) | Casegravol isovalerate | (185) |
| Apetalolide | (531) | Cauferidin | (36) |
| Aphyllodenticulide | (641) | Cauferin | (32) |
| Apigravin | (354) | Cauferinin | (54) |
| Apiumetin | (359) | Cauferoside | (33) |
| Apiumoside | (356) | Cauloside | (34) |
| Apterin | (860) | Celereoin | (253) |
| Aquillochin | (448) | Celereoside | (255) |
| Arnocoumarin | (129) | Celerin | (353) |
| Arnottiacoumarin | (369) | Ceylantin | (407) |
| Arnottinin | (854) | Chalepin | (910) |
| Artanin | (384) | Chinchircumine | (599) |
| Artelin | (908) | Chloticol | (171) |
| Asacoumarin A | (17) | Citrubuntin | (94) |
| Asacoumarin B | (27) | Citrusal | (855) |
| Assafoetidin | (21) | Clausarin | (471) |
| Aurantiumal | (871) | Clausmarin A | (467) |
| Auraptenal | (790) | Clausmarin B | (468) |
| Auraptenol | (853) | Cleomiscosin A | (447) |
| Aureol | (737) | Cleomiscosin B | (444) |
|  |  | Cleomiscosin C | (448) |
| Badrakemin | (808) | Cleomiscosin D | (445) |
| Badrakemin acetate | (809) | Cleosandrin | (447) |
| Badrakemone | (810) | CM-c$_2$ | (159) |
| Balsamiferone | (456) | Cniforin A | (861) |
| Benahorin | (896) | Cniforin B | (202) |
| Bethancorin | (331) | Colladin | (807) |
| Bethancorol | (332) | Colladocin | (828) |
| Bhubaneswin | (753) | Colladonin | (806) |
| Bicoumol | (751) | Colladonin isovalerate | (31) |
| Bocconin | (221) | Conferdione | (821) |
| Bothrioclinin | (637) | Conferin | (820) |
| Brachycoumarin | (601) | Conferol | (812) |
| Bruceol | (875) | Conferol acetate | (813) |
| Bungediol | (324) | Conferone | (814) |
| Buntansin | (74) | Conferoside | (48) |
| (−)-Byakangelicin | (393) | Cordatolide A | (487) |
| (±)-Byakangelicol | (391) | Cordatolide B | (486) |
|  |  | Corylidin | (734) |
| Calustralin | (523) | Coumarsabin | (679) |
| Calophyllolide | (526) | Coumestrin | (728) |
| Campestrinol | (233) | Coumestrol | (714) |
| Campestrinoside | (217) | Coumurrayin | (877) |
| Campestrol | (223) | Coumurrin | (168) |
| Candicanin | (783) | Cyclobisuberodiene | (779) |
| Capensin | (412) | Cyclobrachycoumarin | (614) |

| Name | Compound number | Name | Compound number |
|---|---|---|---|
| Cycloethuliacoumarin | (628) | trans-Diversin | (788) |
| Cycloisobrachycoumarin | (612) | Dracunculin | (905) |
| Cyclolongipesin | (652) | Drimachone | (441) |
| Cyclolycoserone | (645) | Drimanthone | (438) |
| | | Drimartol A | (433) |
| Dalbergin | (541) | Drimartol B | (436) |
| Daphneticin | (352) | | |
| Daphnoretin | (768) | Edgeworin | (764) |
| Daphnorin | (769) | Edgeworoside A | (785) |
| Dauroside A | (66) | Edgeworthin | (767) |
| Dauroside B | (67) | Edulisin I | (204) |
| Dauroside C | (315) | Edulisin II | (203) |
| Dauroside D | (252) | Ekersenin | (940) |
| Deacetylkellerin | (59) | Elemiferone | (462) |
| Deacetyltadshikorin | (15) | 2'-Epicycloisobrachycoumarin | (613) |
| Decuroside I | (116) | 1'-Epi-6',7'-dehydrocyclolycoserone | (648) |
| Decuroside II | (117) | 1'-Epilycoserone | (606) |
| Decuroside III | (118) | 6',7'-Epoxyaurapten | (786) |
| Decuroside IV | (115) | 3',6'-Epoxycycloaurapten | (787) |
| Decuroside V | (125) | Epoxyfarnochrol | (417) |
| (−)-3'R-Decursinol | (131) | Eriobrucinol | (872) |
| exo-Dehydrochalepin | (459) | Eriocephaloside | (776) |
| 10',11'-Dehydrocyclolycoserone | (646) | Erioside | (408) |
| Dehydrogeijerin | (839) | Erlangeafusciol | (620) |
| Dehydroindicolactone | (365) | Erosmin | (735) |
| cis-Dehydroosthol | (172) | Ethuliacoumarin | (624) |
| trans-Dehydroosthol | (857) | Ethylsuberenol | (87) |
| Dehydropectanone | (430) | Euphorbetin | (756) |
| Demethyldaphnoretin | (767) | Exostemin | (547) |
| Demethylluvangetin | (371) | | |
| Demethylsuberosin | (836) | Farnesiferol C | (797) |
| Dentatin | (884) | Farnochrol | (416) |
| Deoxybruceol | (874) | Fatagarin | (766) |
| Deparnol | (419) | Fecarpin | (62) |
| Derrusnin | (701) | Fekolin | (22) |
| Desertorin A | (758) | Fekolone | (23) |
| Desertorin B | (759) | Fekrol | (24) |
| Desertorin C | (760) | Fekrynol | (28) |
| Desmethylkotanin | (762) | Fekrynol acetate | (29) |
| Desoxylacarol | (341) | Fepaldin | (55) |
| Dicoumarol | (777) | Ferilin | (37) |
| Dihydromelin A | (108) | Fernolin | (401) |
| Dihydromelin B | (109) | Ferocaulicin | (45) |
| Dihydrosuberenol | (79) | Ferocaulidin | (44) |
| Dihydroxanthyletin | (130) | Ferocaulin | (46) |
| Diospyroside | (309) | Ferocaulinin | (47) |
| Dioxyanomalin | (870) | Feropolidin | (815) |
| Diversin | (789) | Feropolol | (800) |

| Name | Compound number | Name | Compound number |
|------|-----------------|------|-----------------|
| Feropolone | (802) | Heliclactone | (687) |
| Ferprenin | (561) | Heraclenol acetonide | (362) |
| Ferruol | (919) | Heraclenol 2'-O-isovalerate | (363) |
| Ferucrinone | (60) | Heraclenol 2'-O-senecioate | (364) |
| Ferujol | (339) | Heraclesol | (380) |
| Ferukrin | (56) | Heratomin | (890) |
| Ferukrin acetate | (57) | Heratomol | (888) |
| Ferukrin isobutyrate | (58) | Hermandiol | (205) |
| Feselol | (811) | Honydusin | (302) |
| Feselol angelate | (41) | Hopeyhopin | (107) |
| Feshurin | (50) | Hortinone | (288) |
| Feterin | (35) | Hortiolone | (881) |
| Flemichapparin C | (730) | 10'-Hydroxycyclolycoserone | (647) |
| Foetidin | (555) | Hydroxyeriobrucinol | (873) |
| Foliferin | (26) | (E)-ω-Hydroxyferprenin | (562) |
| Funadonin | (77) | (Z)-ω-Hydroxyferprenin | (563) |
| Furopinnarin | (880) | 12'-Hydroxyferulenol | (556) |
| | | ω-Hydroxyferulenol | (557) |
| Galbanic acid | (804) | (E)-ω-Hydroxyferulenol | (556) |
| Galipein | (174) | (Z)-ω-Hydroxyferulenol | (557) |
| Gancaonin F | (744) | 8-Hydroxypereflorin | (680) |
| Geijerin | (838) | Hymexelsin | (310) |
| Geiparvarin | (792) | | |
| Gerberinol | (778) | Imperatorin | (895) |
| Glabrescin | (702) | Indicolactone | (365) |
| Gleinadiene | (280) | Indicolactonediol | (366) |
| Gleinene | (271) | Inophyllolide | (538) |
| Glycycoumarin | (477) | cis-(+)-Inophyllolide | (539) |
| Glycyrin | (478) | trans-(+)-Inophyllolide | (537) |
| Glycyrol | (741) | Inophyllum A | (536) |
| Gnidiacoumarin | (775) | Inophyllum B | (534) |
| Grandivittin | (132) | Inophyllum C | (537) |
| Grandivittinol | (93) | Inophyllum D | (535) |
| cis-Grandmarin | (289) | Inophyllum E | (539) |
| trans-Grandmarin isovalerate | (291) | Ipomopsin | (755) |
| Gravelliferone | (909) | Ismailin | (784) |
| Gynuron | (141) | Isoangelol | (105) |
| Gypothamniol | (642) | Isobergaptol | (286) |
| | | Isobyakangelicol | (392) |
| Hainanmurpanin | (189) | Isodalbergin | (544) |
| Haploperoside A | (312) | Isodrimachone | (442) |
| Haploperoside B | (314) | Isodrimartol A | (432) |
| Haploperoside C | (313) | Isoerlangeafusciol | (611) |
| Haploperoside D | (311) | Isoethuliacoumarin A | (625) |
| Haploperoside E | (316) | Isoethuliacoumarin B | (626) |
| Haplopinol | (318) | Isoethuliacoumarin C | (639) |
| Haptusinol | (411) | Isoeuphorbetin | (757) |
| Hassanon | (145) | Isofraxetin | (897) |

*References, pp. 283–316*

| Name | Compound number | Name | Compound number |
|------|-----------------|------|-----------------|
| Isofraxoside | (907) | Latilobinol | (19) |
| Isogerberacoumarin | (941) | Lehmferidin | (37) |
| Isoglycyrol | (743) | Lehmferin | (20) |
| Isolycoserone | (609) | Leptodactylone | (382) |
| Isomammeigin | (525) | Leptophyllidin | (359) |
| Isomexoticin | (281) | Leptophyllin | (891) |
| Isomurralonginol acetate | (148) | Leptophylloside | (892) |
| Isomurralonginol nicotinate | (149) | Libanotin A | (866) |
| Isomurranganon senecioate | (190) | Licopyranocoumarin | (479) |
| Isopeucenidin | (862) | Lomatin | (866) |
| Isophlojodicarpin | (187) | Lonchocarpenin | (711) |
| Isosabandin | (451) | Lonchocarpic acid | (710) |
| Isosamarcandin | (823) | Lucernol | (745) |
| Isosamarcandin angelate | (824) | Lycoserone | (607) |
| Isosibiricin | (878) | | |
| Isosojagol | (716) | MAB 1 | (520) |
| Isothamnosin A | (780) | Mammea A/AA | (521) |
| Isotriptiliocoumarin | (617) | Mammea A/AB | (520) |
| Isotriptospinocoumarin | (636) | Mammea A/AD | (518) |
| | | Mammea A/BA | (517) |
| Jatrophin | (443) | Mammea A/BB | (516) |
| Jayantinin | (754) | Mammea A/BB cyclo D | (524) |
| Jumutinol | (254) | Mammea B/AA | (925) |
| Junosmarin | (222) | Mammea B/AA cyclo F | (931) |
| | | Mammea B/AB | (924) |
| Kamolonol | (64) | Mammea B/AB cyclo F | (930) |
| Karatavic acid | (803) | Mammea B/AC | (923) |
| Karatavicin | (18) | Mammea B/AC cyclo F | (929) |
| Karatavicinol | (795) | Mammea B/BA | (922) |
| Kellerin | (831) | Mammea B/BA cyclo E | (935) |
| Kinocoumarin | (300) | Mammea B/BA cyclo F | (928) |
| Kiyomal | (144) | Mammea B/BB | (918) |
| Kokanidin | (51) | Mammea B/BB cyclo E | (933) |
| Kopeolin | (798) | Mammea B/BB cyclo F | (927) |
| Kopeolone | (25) | Mammea B/BC | (916) |
| Kopeoside | (799) | Mammea B/BC cyclo E | (932) |
| Kotanin | (763) | Mammea B/BC cyclo F | (926) |
| Kuhlmannin | (551) | Mammea B/BD | (917) |
| | | Mammea C/AB cyclo D | (936) |
| Lacarol | (381) | Mammea C/BB | (920) |
| Lacinartin | (348) | Mammea D/BB | (919) |
| Lacinartin epoxide | (349) | Mammea E/BA | (939) |
| Lacinartindiol | (350) | Mammea E/BB | (937) |
| Lanatin | (287) | Mammeigin | (529) |
| Lariside | (308) | Mammeisin | (521) |
| Laserpitin | (867) | Maoyancaosu | (333) |
| Lasiocephalin | (765) | Marmaricin | (30) |
| Lasioerin | (774) | Marmelide | (895) |

| Name | Compound number | Name | Compound number |
|------|----------------|------|----------------|
| Marmesin acetate | (119) | Murragleinin | (379) |
| Marmesin isovalerate | (120) | Murralongin | (858) |
| Marmin | (791) | Murranganon | (188) |
| Matsukazelactone | (752) | Murrangatin | (859) |
| Medicagol | (721) | Murrangatin acetate | (181) |
| Melanettin | (543) | Murrangatin palmitate | (182) |
| Melannein | (545) | Murraol | (159) |
| Meranzin hydrate | (856) | Murraxocin | (179) |
| Merillin | (162) | Murrayacarpin A | (142) |
| Mesuagin | (527) | Murrayacarpin B | (269) |
| Mesuarin | (528) | Murrayanone | (378) |
| Mesuol | (518) | Murrayatin | (169) |
| 8-Methoxycoumarsabin | (699) | Mutisicoumarin | (616) |
| (−)-8-Methoxyobliquin | (415) | Mutisifurocoumarin | (619) |
| O-Methylcedrelopsin | (336) | Myrsellin | (153) |
| Methyldalbergin | (546) | Myrsellinol | (154) |
| (−)-Methyldecursidinol | (134) | | |
| 5-Methylethuliacoumarin | (654) | Nachsmyrin | (128) |
| Methyl galbanate | (805) | Naphthoherniarin | (111) |
| trans-O-Methylgrandmarin | (290) | Nassauvirevolutin A | (594) |
| Methyllacarol | (383) | Nassauvirevolutin B | (596) |
| 9-Methyllongipesin | (656) | Nassauvirevolutin C | (597) |
| Methyl robustate | (705) | 2-epi-Nassauvirevolutin C | (598) |
| (E)-Methylsuberenol | (85) | Necatorin | (667) |
| (Z)-Methylsuberenol | (86) | Neoartanin | (386) |
| Mexolide | (782) | Neoartanindiol | (389) |
| (−)-Mexoticin | (281) | Neoartaninepoxide | (388) |
| Microlobiden | (61) | Neofolin | (480) |
| Microlobin | (63) | Nevskin | (822) |
| Micromelin | (843) | Nevskone | (53) |
| Microminutin | (150) | Nieshoutin | (887) |
| Minumicrolin | (177) | Nivegin | (492) |
| Mirificoumestan | (722) | 9′,13′-Norcyclolycoserone | (640) |
| Mirificoumestan glycol | (724) | Nordalbergin | (540) |
| Mirificoumestan hydrate | (723) | Nordentatin | (882) |
| Mogoltacin | (816) | Norwedelolactone | (738) |
| Mogoltavicin | (830) | Notopterol | (263) |
| Mogoltavidin | (829) | Notoptol | (264) |
| Mogoltavin | (818) | | |
| Mogoltavinin | (819) | Obliquetin | (886) |
| Mogoltin | (817) | Obliquetol | (885) |
| Moluccanin | (334) | Obliquin hydrate | (328) |
| Mulberroside B | (252) | Oblongulide | (485) |
| Murpanicin | (179) | Obtusicin | (413) |
| Murpanicol | (188) | Obtusidin | (473) |
| Murpanidin | (177) | Obtusifol | (906) |
| (−)-Murracarpin | (178) | Obtusifolin | (470) |
| Murraculatin | (279) | Obtusin | (414) |

References, pp. 283–316

| Name | Compound number | Name | Compound number |
|---|---|---|---|
| Obtusinin | (320) | Phlojodicarpin | (186) |
| Obtusinol | (319) | Piloselloidan | (602) |
| Obtusiprenin | (449) | Piloselloidol isovalerate | (651) |
| Obtusiprenol | (450) | Ponfolin | (299) |
| Obtusoside | (321) | Ponnalide | (524) |
| Officinalin | (833) | Praeroside I | (355) |
| Officinalin acetate | (834) | Praeroside II | (217) |
| Officinalin isobutyrate | (835) | Praeroside III | (216) |
| Omphalocarpin | (283) | Praeroside IV | (208) |
| Omphamurin | (276) | Praeroside V | (210) |
| Onognaphalin | (630) | (+)-Praeruptorin A | (225) |
| Oreojasmin | (773) | (+)-Praeruptorin B | (868) |
| Oroselol | (864) | Praeruptorin E | (227) |
| Oroselone | (865) | Prandiol | (846) |
| Osthenon | (146) | Prandiol senecioate | (127) |
| cis-Osthenon | (147) | Prangosin | (850) |
| Ostruthin | (844) | Prealtin A | (6) |
| Oxofarnochrol | (418) | Prealtin B | (7) |
| (E)-ω-Oxoferprenin | (566) | Prealtin C | (8) |
| (E)-ω-Oxoferulenol | (560) | Prealtin D | (13) |
| Oxyanomalin | (869) | Preethuliacoumarin | (621) |
| Oxycapnolactone | (794) | Prenyllacarol | (387) |
| | | 8-Prenylnodakenetin | (239) |
| Pachyrrhizin | (476) | Propacin | (446) |
| Panial | (193) | Psoralen | (849) |
| Paniculal | (143) | Psoralidin | (731) |
| Paniculin | (151) | Psoralidin oxide | (732) |
| Paniculonol isovalerate | (191) | Pterybinthinone | (123) |
| Pd–Ib | (235) | Pubesinol | (841) |
| Pd–III | (227) | Puerarol | (733) |
| Pd–C–I | (135) | Puerarostan | (748) |
| Pd–C–II | (139) | Pygmaeoherin | (665) |
| Pd–C–III | (136) | | |
| Pd–C–IV | (140) | | |
| Pd–C–V | (137) | Racemol | (891) |
| Pd–C–V | (138) | Racemosin | (407) |
| Pectachol | (424) | Ramosin | (852) |
| Pectachol B | (427) | Ramosinin | (469) |
| Pectanone | (429) | Reoselin | (796) |
| Pereflorin | (940) | Repensol | (725) |
| Pereflorin B | (700) | Robustic acid | (703) |
| Peroxyauraptenol | (174) | Robustin | (704) |
| Peroxymurraol | (176) | Rutalpinin | (489) |
| Peucedanol | (840) | Rutamarin | (911) |
| Peuruthenicin | (832) | Rutarensin | (770) |
| Phaseol | (736) | Rutaretin | (891) |
| Phebalin | (781) | Rutaretin methyl ether | (894) |
| (−)-Phebalosin | (173) | Rutarin | (893) |

| Name | Compound number | Name | Compound number |
|------|------|------|------|
| Samarcandin | (825) | Swietenocoumarin I | (457) |
| Samarcandin acetate | (826) | Swietenol | (78) |
| Samarcandone | (827) | | |
| (+)-Samidin | (231) | | |
| Sativol | (750) | Tadshiferin | (14) |
| Saxicolon | (206) | Tadshikorin | (16) |
| Scandenin | (709) | Tamarin | (84) |
| Scopodrimol A | (327) | Tavimolidin | (49) |
| Scopofarnol | (326) | Tenudiol | (346) |
| Secodrial | (422) | Tenuidin | (76) |
| Secodriol | (421) | Tephrosol | (747) |
| Secrolin | (200) | Thonningine A | (713) |
| Seravschanin | (218) | Thonningine B | (712) |
| Serratin | (491) | Toddalenone | (270) |
| Sesebrin | (274) | Toddanol | (249) |
| Sesebrinol | (282) | Toddanone | (250) |
| Seselinal | (278) | Toddasin | (782) |
| Seseloside | (370) | Tomenin | (901) |
| Seshadrin | (505) | Tomentin | (372) |
| Sesibiricol | (277) | Tomentin | (898) |
| Shijiaocaolactone A | (466) | Tomentolide A | (532) |
| (−)-Sibiricin | (273) | Tortuoside | (166) |
| Sibiricol | (272) | Tortuosidin | (157) |
| Sibirinol | (276) | Tortuosin | (262) |
| Sisafolin | (514) | Tortuosinin | (122) |
| Smirnioridin | (848) | Tortuosinol | (127) |
| Smirniorin | (847) | Trachyphyllin | (883) |
| Sojagol | (717) | Trichoclin | (361) |
| Sophoracoumestan A | (718) | Trifoliol | (726) |
| Sophoracoumestan B | (749) | Trigocoumarin | (690) |
| Soulattrolide | (533) | Trigoforin | (552) |
| Sphondin | (889) | Tripartol | (440) |
| (−)-Sprengelianin | (121) | Triphasiol | (161) |
| Stevenin | (542) | Triptiliocoumarin | (631) |
| Stylosin | (409) | Triptispinocoumarin | (635) |
| (Z)-Suberenol | (82) | Troupin | (490) |
| Suberosin | (837) | Tuberostan | (729) |
| Surangin A | (921) | Turgeniifolin | (237) |
| Surangin B | (938) | Turgeniifolin B | (224) |
| Surangin C | (484) | Turgeniifolin C | (230) |
| Swietenocoumarin A | (240) | | |
| Swietenocoumarin B | (295) | Ulismoncadin | (238) |
| Swietenocoumarin C | (241) | Ulismoncadin A | (465) |
| Swietenocoumarin D | (296) | Ulopterol | (841) |
| Swietenocoumarin E | (243) | Umckalin | (899) |
| Swietenocoumarin F | (298) | | |
| Swietenocoumarin G | (297) | | |
| Swietenocoumarin H | (242) | Ventilatone A | (662) |

*References, pp. 283–316*

| Name | Compound number | Name | Compound number |
|------|------|------|------|
| Ventilatone B | (678) | Wampetin | (365) |
| Versicolin | (195) | Wedelolactone | (740) |
| Villosin | (345) | | |
| Voludal | (515) | Xanthoxyletin | (876) |
| | | Xanthyletin | (851) |
| Wairol | (727) | Xerobioside | (310) |

# References

1. ABYSHEV, A.Z.: Latilobinol, a New Terpenoid Coumarin from *Prangos latiloba*. Khim. prirod. Soedinenii **1979**, 90; Chem. Abs. **91**, 52700 (1979).

2. —: Jumutinol, a New Dihydrofurocoumarin from *Seseli jomuticum*. Khim. prirod. Soedinenii **1980**, 250; Chem. Abs. **93**, 110565 (1980).

3. —: $^{13}$C NMR Spectra and Structure of Bungeidiol and its Transformation Products. Khim. prirod. Soedinenii **1982**, 294; Chem. Abs. **97**, 163267 (1982).

4. —: Stereochemistry of Latilobinol. Khim. prirod. Soedinenii **1984**, 712; Chem. Abs. **103**, 6523 (1985).

5. ABYSHEV, A.Z., and D.Z. ABYSHEV: The Coumarin Composition of *Seseli tortuosum*. Khim. prirod. Soedinenii **1983**, 704; Chem. Abs. **100**, 171544 (1984).

6. ABYSHEV, A.Z., B. AKHDAROV, and N.F. GASHIMOV: Oxypeucedanin hydrate acetonide, a New Component from *Peucedanum turcomanicum*. Khim. prirod. Soedinenii **1979**, 847; Chem. Abs. **93**, 41499 (1980)

7. ABYSHEV, A.Z., P.P. DENISENKO, D.Z. ABYSHEV, and Y.B. KERIMOV: Coumarin Composition of *Seseli grandivittatum*. Khim. prirod. Soedinenii **1977**, 640; Chem. Abs. **88**, 85987 (1978).

8. ABYSHEV, A.Z., P.P. DENISENKO, N.Y. ISAEV, and Y.B. KERIMOV: Coumarins from *Haplophyllum schelkovnikovii*. Khim. prirod. Soedinenii **1978**, 654; Chem. Abs. **90**, 83603 (1979).

9. ABYSHEV, A.Z., and N.F. GASHIMOV: Structure of Obtusifol. Khim. prirod. Soedinenii **1979**, 401; Chem. Abs. **92**, 41853 (1980).

10. — —: Obtusin, a New Coumarin from *Haplophyllum obtusifolium*. Khim. prirod. Soedinenii **1979**, 403; Chem. Abs. **92**, 41854 (1980).

11. — —: Haptusinol, a New Coumarin from *Haplophyllum obtusifolium*. Khim. prirod. Soedinenii **1979**, 485; Chem. Abs. **93**, 66024 (1980).

12. — —: 6-Geranyloxy-7-methoxycoumarin, a New Component from *Haplophyllum pedicellatum*. Khim. prirod. Soedinenii **1979**, 846; Chem. Abs. **93**, 41498 (1980).

13. — —: Coumarin Composition of *Haplophyllum bungei*. Khim prirod. Soedinenii **1982**, 648; Chem. Abs. **98**, 86225 (1983).

14. ABYSHEV, A.Z., N.Y. ISAEV, and Y.B. KERIMOV: Coumarins of Three Species of the Genus *Haplophyllum*. Khim. prirod. Soedinenii **1980**, 800; Chem. Abs. **94**, 171045 (1981).

15. ABYSHEV, A.Z., I.P. SIDOROVA, D.Z. ABYSHEV, V.N. FLORYA, V.P. ZMEIKOV, and Y.B. KERIMOV: Comparative Characterisation of the Coumarin Composition of *Seseli campestre* Growing in Moldavia and in the Caucasus. Khim. prirod. Soedinenii **1982**, 434; Chem. Abs. **99**, 155096 (1983).

16. ADINARAYANA, D., and T.R. SESHADRI: Chemical Components of the Indian Seeds of *Calophyllum inophyllum*; The Structure of a New 4-Phenylcoumarin, Ponnalide. Bull. Nat. Inst. Sci. India **31**, 91 (1965).

17. ADITYACHAUDHURY, N., and P.K. GUPTA: A New Pterocarpan and Coumestan in the Roots of *Flemingia chappar*. Phytochemistry **12**, 425 (1973).

18. AGRAWAL, A., I.R. SIDDIQUI and J. SINGH: Coumarins from the Roots of *Feronia limonia*. Phytochemistry **28**, 1229 (1989).

19. AHLUWALIA, V.K., D.R. BOYD, A.K. JAIN, C.H. KHANDURI, and N.D. SHARMA: Furanocoumarin Glucosides from the Seeds of *Apium graveolens*. Phytochemistry **27**, 1181 (1988).

20. AHLUWALIA, V.K., M. KHANNA, and R.P. SINGH: Synthesis of Obtusinin and 7-(3'-Hydroxymethylbut-2'-enyloxy)-6-methoxy-2H-1-benzopyran-2-one. Monatsh. Chem. **113**, 197 (1982).

21. AHLUWALIA, V.K., and D. KUMAR: Constitution and Synthesis of a Novel 4-Methoxycoumarin from *Platymiscium praecox*. Indian J. Chem. **15B**, 18 (1977).

22. AHLUWALIA, V.K., A.C. MEHTA, and T.R. SESHADRI: Constitution of Dalbergin. III. Synthesis of Dalbergin, Isodalbergin and their Ethers. Proc. Indian Acad. Sci. **45A**, 15 (1957).

23. AHLUWALIA, V.K., V.D. MEHTA, and C.H. KHANDURI: Syntheses of 7-Ethoxy-3,4-dimethylcoumarin and 5-Methoxy-4-methylcoumarin. Indian J. Chem. **27B**, 68 (1988).

24. AHLUWALIA, V.K., and I. MUKHERJEE: Synthesis of Coumarsabin and 9-Methoxy-coumarsabin. Indian J. Chem. **23B**, 272 (1984).

25. AHLUWALIA, V.K., I. MUKHERJEE, and K. MUKHERJEE: Synthesis of Racemosin. Indian J. Chem. **23B**, 270 (1984).

26. AHLUWALIA, V.K., and C. PRAKASH: Constitution of Peuruthenicin, Oficinalin, Its Acetate and Isobutyrate and Convenient Synthesis of 5-Methoxycoumarin. Indian J. Chem. **15B**, 423 (1977).

27. — —: Constitution and Synthesis of 5,7,8-Trimethoxycoumarin Isolated from *Ruta Sp Tene 29662*. Indian J. Chem. **15B**, 808 (1977).

28. AHLUWALIA, V.K., C. PRAKASH, and M.C. GUPTA: Structures of Isofraxetin, Umckalin, Arscotin and 7-Geranyloxy-5,6-dimethoxycoumarin. Indian J. Chem. **16B**, 286 (1978).

29. AHLUWALIA, V.K., C. PRAKASH, M.C. GUPTA, and S. MEHTA: Structure of Tomenin and 7-(3'-Methyl-2',3'-epoxybutyloxy)-5,6-methylenedioxycoumarin. Indian J. Chem. **16B**, 591 (1978).

30. AHLUWALIA, V.K., C. PRAKASH, and N. RANI: Synthesis of Furanocoumarins, Heratomol and Heratomin. Indian J. Chem. **15B**, 1137 (1977).

31. — — —: Synthesis of 4,6,7-Trimethoxy-5-Methylcoumarin and 4-Methoxy-5-methylcoumarin. Indian J. Chem. **16B**, 436 (1978).

32. AHLUWALIA, V.K., C. PRAKASH, and R.P. SINGH: Synthesis of some Novel 3,4-Dimethoxycoumarins: 3-Methoxypereflorin and 3,7-Dimethoxypereflorin. Chem. and Ind. **1980**, 464.

33. AHLUWALIA, V.K., and N. RANI: Partial Methylation of 7-Hydroxy- and 6,7-Dihydroxy-4-(4'-hydroxyphenyl)coumarins: Synthesis of Melanettin. Indian J. Chem. **15B**, 1000 (1977).

34. AHLUWALIA, V.K., and T.R. SESHADRI: Constitution of Dalbergin Part II. J. Chem. Soc. **1957**, 970.

35. AHLUWALIA, V.K., and SUNITA: Synthesis of Stevin. Indian J. Chem. **11**, 1241 (1973).

*36.* — —: A Convenient Method for the Synthesis of Coumarins: Constitution of Artelin. Indian J. Chem. **15B**, 936 (1977).

*37.* AHMAD, J., K.M. SHAMSUDDIN, and A. ZAMAN: A Pyranocoumarin from *Atalantia ceylanica*. Phytochemistry **23**, 2098 (1984).

*38.* AL-HAZIMI, H.M.G.: Terpenoids and a Coumarin from *Ferula sinica*. Phytochemistry **25**, 2417 (1986).

*39.* ANAND, N.K., N.D. SHARMA, and S.R. GUPTA: Coumarins from *Apium petroselinum* Seeds. Nat. Acad. Sci. Letters (India) **4**, 249 (1981).

*40.* ANGELES, L.R., O. LOCK DE U., I.C. SALKELD, and P. JOSEPH-NATHAN: A Coumarin from *Perezia coerulescens*. Phytochemistry **23**, 2094 (1984).

*41.* ANWER, F., A. SHOEB, R.S. KAPIL, and S.P. POPLI: Clausarin, a Novel Coumarin from *Clausena pentaphylla* (Roxb.) DC. Experientia **33**, 412 (1977).

*42.* APPENDINO, G., S. TAGLIAPIETRA, P. GARIBOLDI, G.M. NANO, and V. PICCI: ω-oxygenated Prenylated Coumarins from *Ferula communis*. Phytochemistry **27**, 3619 (1988).

*43.* APPENDINO, G., S. TAGLIAPIETRA, G.M. NANO, and V. PICCI: Ferprenin, a Prenylated Coumarin from *Ferula communis*. Phytochemistry **27**, 944 (1988).

*44.* AQUINO, R., M. D'AGOSTINO, F. DE SIMONE, and C. PIZZA: 4-Arylcoumarin Glycosides from *Coutarea hexandra*. Phytochemistry **27**, 1827 (1988).

*45.* ARISAWA, M., S.S. HANDA, D.D. MCPHERSON, D.C. LANKIN, G.A. CORDELL, H.H.S. FONG, and N.R. FARNSWORTH: Plant Anticancer Agents XXIX. Cleomiscosin A from *Simaba multiflora*, *Soulamea soulameoides*, and *Matayba arborescens*. J. Nat. Prod. **47**, 300 (1984).

*46.* ARISAWA, M., A.D. KINGHORN, G.A. CORDELL, and N.R. FARNSWORTH: Ipomopsin, a New Biscoumarin from *Ipomopsis aggregata*. J. Nat. Prod. **47**, 106 (1984).

*47.* ARNOLDI, A., A. ARNONE, and L. MERLINI: Synthesis of the Natural Coumarino-lignoids Propacin and Cleomiscosin A and B. An Empirical Spectroscopic Method to Distinguish Regioisomers of Natural Benzodioxane Lignoids. Heterocycles **22**, 1537 (1984).

*48.* ASAHARA, T., I. SAKAKIBARA, T. OKUYAMA, and S. SHIBATA: Studies on Coumarins of a Chinese Drug "Qian-Hu" V. Coumarin-glycosides from "Zi-Hua Qian-Hu". Planta Medica **50**, 488 (1984).

*49.* ASHEERVADAM, Y., P.S. RAO, and R.D.H. MURRAY: Adicardin, a New Apiose-containing Coumarin Glycoside from the Root Bark of *Adina cordifolia*. Fitoterapia **57**, 231 (1986).

*50.* AZIZ, M., and F. ROUESSAC: Syntheses de la (+)-Marmine, des (+)-Epoxy-6',7'-et (−)-Epoxy-3',6'-auraptenes et des Racemiques Correspondants. Tetrahedron Letters **28**, 2579 (1987).

*51.* — —: Syntheses en Série Racemique et en Série Optiquement Active d'une Famille de Derives Oxygenes Naturel de l'Ombelliferone. Structure Spatiale du (−)-Epoxy-3',6'-auraptene. Tetrahedron **44**, 101 (1988).

*52.* BABA, K., F. HAMASAKI, Y. TABATA, M. KOZAWA, G. HONDO, and M. TABATA: Chemical Studies on Chinese Crude Drug "She Chuang Zi". Shoyakugaku Zasshi **39**, 282 (1985); Chem. Abs. **105**, 85018 (1986).

*53.* BABA, K., Y. MATSUYAMA, T. ISHIDA, M. INOUE, and M. KOZAWA: Studies on Coumarins from the Root of *Angelica pubescens* Maxim. V. Stereochemistry of Angelols A-H. Chem. and Pharm. Bull. **30**, 2036 (1982).

*54.* BABA, K., Y. MATSUYAMA, and M. KOZAWA: Studies on Coumarins from the Root of *Angelica pubescens* Maxim. IV. Structure of Angelol-type Prenylcoumarins. Chem. and Pharm. Bull. **30**, 2025 (1982).

55. BABA, K., Y. TABATA, M. TANIGUTI, and M. KOZAWA: Coumarins from *Edgeworthia chrysantha*. Phytochemistry **28**, 221 (1989).
56. BAGIROV, V.Y.: A Coumarin from the Roots of *Ferula nevskii*. Khim. prirod. Soedinenii **1978**, 652; Chem. Abs. **90**, 118061 (1979).
57. BAGIROV, V.Y., and M.B. BELYI: An Investigation of *Seseli peucedanoides*. Khim. prirod. Soedinenii **1981**, 796; Chem. Abs. **96**, 177940 (1982).
58. BAGIROV, V.Y., V.I. SHEICHENKO, N.V. VESELOVSKAYA, Y.E. SKLYAR, A.A. SAVINA, and I.A. KIR'YANOVA: Structure and Stereochemistry of Galbanic Acid. Khim. prirod. Soedinenii **1980**, 620. Chem. Abs. **95**, 25279 (1981).
59. BALA, K.R., and T.R. SESHADRI: Isolation and Synthesis of Some Coumarin Components of *Mesua ferrea* Seed Oil. Phytochemistry **10**, 1131 (1971).
60. BALBAA, S.I., A.F. HALIM, F.T. HALAWEISH, and F. BOHLMANN: New 5-Methylcoumarins from *Ethulia conyzoides*. Phytochemistry **19**, 1519 (1980).
61. BAL-TEMBE, S., D.N. BEDI, N.J. DE SOUZA, and R.H. RUPP: Synthesis of ($\pm$)-Praeruptorin A and Related Khellactone Derivatives. Heterocycles **26**, 1239 (1987).
62. BANDARANAYAKE, W.M., S.S. SELLIAH, M.U.S. SULTANBAWA, and D.E. GAMES: Xanthones and 4-Phenylcoumarins of *Mesua thwaitesii*. Phytochemistry **14**, 265 (1975).
63. BANDOPADHYAY, M., S.B. MALIK, and T.R. SESHADRI: Candicanin, a Novel Bicoumarinyl Derivative from the Roots of *Heracleum candicans*. Tetrahedron Letters **1971**, 4221.
64. — — —: Coumarins from the Roots and Seeds of *Heracleum candicans*. Indian J. Chem. **11**, 410 (1973).
65. — — —: Synthesis of New Coumarin Components of *Heracleum candicans*. Indian J. Chem. **11**, 530 (1973).
66. BANERJEE, S.K., B.D. GUPTA, and C.K. ATAL: Coumarins of *Heracleum thomsoni* and Claisen Rearrangement of Lanatin. Phytochemistry **19**, 1256 (1980).
67. BANERJEE, S.K., B.D. GUPTA, R. KUMAR, and C.K. ATAL: New Coumarins from the Umbels of *Seseli sibiricum*. Phytochemistry **19**, 281 (1980).
68. BANERJI, A., D.L. LUTHRIA, and B.R. PRABHU: Prenylated Compounds from *Atalantia racemosa*: Isolation and Synthesis of Two Pyranoflavones. Phytochemistry **27**, 3637 (1988).
69. BANERJI, A., B. MALLICK, A. CHATTERJEE, H. BUDZIKIEWICZ, and M. BREUER: Assafoetidin and Ferocolicin, Two Sesquiterpenoid Coumarins from *Ferula assafoetida* Regel. Tetrahedron Letters **29**, 1557 (1988).
70. BANERJI, J., A.K. DAS, and B. DAS: Studies on Rutaceae 5. Synthesis of Arnottinin: a Lactone Constituent of *Xanthoxylum arnottiamum* Maxim. Chem. and Ind. **1987**, 395.
71. BARALDI, P.G., A. BARCO, S. BENETTI, A. CASOLARI, S. MANFREDINI, G.P. POLLINI, and D. SIMONI: Isoxazole-mediated Synthesis of Geiparvarin and Dihydrogeiparvarin. Tetrahedron **44**, 1267 (1988).
72. BARIK, B.R., A.K. DEY, and A. CHATTERJEE: Murrayatin, a Coumarin from *Murraya exotica*. Phytochemistry **22**, 2273 (1983).
73. BARIK, B.R., A.K. DEY, P.C. DAS, A. CHATTERJEE, and J.N. SHOOLERY: Coumarins of *Murraya exotica* – Absolute Configuration of Auraptenol. Phytochemistry **22**, 792 (1983).
74. BARIK, B.R., and A.B. KUNDU: A Cinnamic Acid Derivative and a Coumarin from *Murraya exotica*. Phytochemistry **26**, 3319 (1987).
75. BARUA, N.C., R.P. SHARMA, K.P. MADHUSUDANAN, G. THYAGARAJAN, and W. HERZ: Coumarins in *Artemisia caruifolia*. Phytochemistry **19**, 2217 (1980).

76. BASA, S.C.: Natural Bicoumarins. Phytochemistry **27**, 1933 (1988).
77. BASA, S.C., D.P. DAS, R.N. TRIPATHY, V. ELANGO, and M. SHAMMA: Bhubaneswin: a New Bicoumarin. Heterocycles **22**, 333 (1984).
78. BATIROV, E.K., D. BATSUREN, and V.M. MALIKOV: Components of *Haplophyllum dauricum*. Khim. prirod. Soedinenii **1984**, 244; Chem. Abs. **101**, 87464 (1984).
79. BATIROV, E.K., A.D. MATKARIMOV, V.M. MALIKOV, and E. SEITMURATOV: Coumarins of *Haplophyllum obtusifolium*. Structures of Two New Coumarin Glycosides. Khim. prirod. Soedinenii **1982**, 691; Chem. Abs. **98**, 122817 (1983).
80. BATIROV, E.K., A.D. MATKARIMOV, V.M. MALIKOV, M.R. YAGUDAEV, and E. SEITMURATOV: New Coumarins of *Haplophyllum obtusifolium*. Khim. prirod. Soedinenii **1980**, 785; Chem. Abs. **94**, 171043 (1981).
81. BATSURÉN, D., E.K. BATIROV, and V.M. MALIKOV: Coumarins of *Haplophyllum dauricum*. Khim. prirod. Soedinenii **1982**, 650; Chem. Abs. **98**, 86226 (1983).
82. BATSURÉN, D., E.K. BATIROV, V.M. MALIKOV, and M.R. YAGUDAEV: Structures of Daurosides A and B – New Acylated Coumarin Glycosides from *Haplophyllum dauricum*. Khim. prirod. Soedinenii **1983**, 142; Chem. Abs. **99**, 35939 (1983).
83. BEGLEY, M.J., L. CROMBIE, R.C.F. JONES, and C.J. PALMER: Synthesis of the Mammea Coumarins. Part 4. Stereochemical and Regiochemical Studies and Synthesis of ( – )-Mammea B/BB. J. Chem. Soc. Perkin 1 **1987**, 353.
84. BEGLEY, M.J., L. CROMBIE, D.A. SLACK, and D.A. WHITING: Rearrangement and Orientation in Citran Synthesis. X-Ray Crystal Structures of ( – )-Bruceol and a ( ± )-Deoxybruceol Derivative. J. Chem. Soc. Perkin 1 **1977**, 2402.
85. BELLINO, A., P. VENTURELLA, M.L. MARINO, O. SERVETTAZ, and G. VENTURELLA: Coumarins from *Seseli bocconi*. Phytochemistry **25**, 1195 (1986).
86. BESELOVSKAYA, N.V., and Y.E. SKLYAR: Deacetyltadzhikorin from *Ferula tadshikorum*. Khim. prirod. Soedinenii **1984**, 386.
87. BESSONOVA, I.A., E.K. BATIROV, and M.R. YAGUDAEV: Obtusifolin – a New Coumarin from Roots of *Haplophyllum obtusifolium*. Khim. prirod. Soedinenii **1988**, 187; Chem. Abs. **109**, 125887 (1988).
88. BEVALOT, F., J.A. ARMSTRONG, A.I. GRAY, and P.G. WATERMAN: Coumarins from Three *Phebalium* Species. Biochem. Syst. Ecol. **16**, 631 (1988).
89. — — — —: Coumarins from the Leaves of *Phebalium squameum*. Phytochemistry **27**, 1546 (1988).
90. BHANDARI, P., P. PANT, and R.P. RASTOGI: Aquillochin, a Coumarinolignan from *Aquilaria agallocha*. Phytochemistry **21**, 2147 (1982).
91. BHANDARI, P., and R.P. RASTOGI: A Novel Type of Bicoumarin Rhamnoside from *Lasiosiphon eriocephalus*. Phytochemistry **20**, 2044 (1981).
92. BHANDARI, P., S. TANDON, and R.P. RASTOGI: Erioside, a New Coumarin Glucoside from *Lasiosiphon eriocephalus*. Phytochemistry **19**, 1554 (1980).
93. BHARAT BHUSHAN, S. RANGASWAMI, and T.R. SESHADRI: Calaustralin, a New 4-Phenylcoumarin from the Seed Oil of *Calophyllum inophyllum* Linn. Indian J. Chem. **13**, 746 (1975).
94. BHARDWAJ, D.K., R. MURARI, T.R. SESHADRI, and R. SINGH: Isolation of 7-Acetoxy-4-methylcoumarin from *Trigonella foenum-graecum*. Indian J. Chem. **15B**, 94 (1977).
95. BHARGAVA, K.K., N.R. KRISHNASWAMY, and T.R. SESHADRI: Isolation of Desmethylwedelolactone and Its Glucoside from *Eclipta alba*. Indian J. Chem. **8**, 644 (1970).
96. BHATIA, S.K., and R.S. KAPIL: Synthesis of Clausarin. Indian J. Chem. **22B**, 1090 (1983).
97. BHATTACHARYA, A.K., and S.C. DAS: Lasiocephalin: a New Coumarin from *Lasiosiphon eriocephalus*. Chem. and Ind. **1971**, 885.

98. BHATTACHARYYA, P., P. CHAKRABARTY and B.K. CHOWDHUTY: Mesuarin: a New 4-Phenyl-coumarin from *Mesua ferrea*. Chem. and Ind. **1988**, 239.

99. BHATTACHARYYA, P., D. CHATTERJEE, A CHAKRABARTI, and D.P. CHAKRABORTY: Synthesis of Mesuagin, a Plant Antibiotic from *Mesua ferrea* Linn. Indian J. Chem. **17B**, 111 (1979).

100. BHIDE, K.S., R.B. MUJUMDAR, and A.V. RAMA RAO: Phenolics from the Bark of *Chloroxylon swietenia* DC. Indian J. Chem. **15B**, 440 (1977).

101. BICKOFF, E.M., A.N. BOOTH, R.L. LYMAN, A.L. LIVINGSTON, C.R. THOMPSON, and F. DE EDS: Coumestrol, a New Estrogen Isolated from Forage Crops. Science **126**, 969 (1957).

102. BICKOFF, E.M., A.L. LIVINGSTON, S.C. WITT, R.E. LUNDIN, and R.R. SPENCER: Isolation of 4'-O-Methylcoumestrol from Alfalfa. J. Agric. Food Chem. **13**, 597 (1965).

103. BICKOFF, E.M., R.L. LYMAN, A.L. LIVINGSTON, and A.N. BOOTH: Characterization of Coumestrol, a Naturally Occurring Plant Estrogen. J. Amer. Chem. Soc. **80**, 3969 (1958).

104. BICKOFF, E.M., R.R. SPENCER, B.E. KNUCKLES, and R.E. LUNDIN: 3'-Methoxycoumestrol from Alfalfa: Isolation and Characterization, J. Agric. Food Chem. **14**, 444 (1966).

105. BIGGS, D.R., and G.J. SHAW: Wairol, a New Coumestan from *Medicago sativa*. Phytochemistry **19**, 2801 (1980).

106. BITTNER, M., J. JAKUPOVIC, F. BOHLMANN, M. GRENZ, and M. SILVA: 5-Methylcoumarins and Chromones from *Triptilion* Species. Phytochemistry **27**, 3263 (1988).

107. BITTNER, M., J. JAKUPOVIC, F. BOHLMANN, and M. SILVA: 5-Methylcoumarins from *Nassauvia* Species. Phytochemistry **27**, 3845 (1988).

108. — — — —: Coumarins and Guaianolides from Further Chilean Representatives of the Subtribe Nassauriinae. Phytochemistry **28**, 2867 (1989).

109. BOHLMANN, F., S. BALBAA, A. HALIM, and F. HALAWEISH: A Terpene-coumarin Derivative from *Ethulia conyzoides*. Phytochemistry **20**, 177 (1981).

110. BOHLMANN, F., M. GRENZ, C. ZDERO, J. JAKUPOVIC, R.M. KING, and H. ROBINSON: Onognaphalin, a Further 5-Methylcoumarin from *Onoseris gnaphalioides*. Phytochemistry **24**, 1392 (1985).

111. BOHLMANN, F., and J. JAKUPOVIC: 8-Oxo-α-selinen und Neue Scopoletin-Derivate aus *Conyza*-Arten. Phytochemistry **18**, 1367 (1979).

112. BOHLMANN, F., J. JAKUPOVIC, L.N. MISRA, and V. CASTRO: 5-Methylcumarin-Derivate aus *Lycoseris latifolia*. Annalen **1985**, 1367.

113. BOHLMANN, F., and A. STEINMEYER: Synthesis of Brachycoumarin and Cyclobrachycoumarin. Tetrahedron Letters **27**, 5359 (1986).

114. BOHLMANN, F., and C. WIENHOLD: Natürlich vorkommende Cumarin-Derivate, XVII. Synthese natürlich vorkommender 5-Methylcumarine und -chromone. Chem. Ber. **112**, 2394 (1979).

115. BOHLMANN, F., and C. ZDERO: Natürlich vorkommende Cumarin-Derivate, XIV. Neue 5-Methyl-cumarine und -chromone aus *Erlangea rogersii* S. Moore. Chem. Ber. **110**, 1755 (1977).

116. — —: Über Inhaltstoffe der Tribus Mutisieae. Phytochemistry **16**, 239 (1977).

117. — —: Gynuron, ein neues Terpen-Cumarin-Derivat aus *Gynura crepioides*. Phytochemistry **16**, 494 (1977).

118. — —: Neuartige Cumarin-Derivate aus *Ethulia conyzoides*. Phytochemistry **16**, 1092 (1977).

119. — —: Neue 5-Alkylcumarine und Chromone aus *Bothriocline laxa*. Phytochemistry **16**, 1261 (1977).

120. — —: Neue Obliquin-Derivate aus *Helichrysum serpyllifolium*. Phytochemistry **19**, 331 (1980).

121. — —: Glaucolides and Other Constituents from South African *Vernonia* Species. Phytochemistry **21**, 2263 (1982).

122. BOHLMANN, F., C. ZDERO, R.M. KING and H. ROBINSON: New Heliangolides from *Conocliniopsis prasiifolia*. Phytochemistry **19**, 1547 (1980).

123. BOHLMANN, F., C. ZDERO and N. LE VAN: Neue Geranyl-Cumarin-Derivate und weitere Inhaltsstoffe aus der Tribus Mutisieae. Phytochemistry **18**, 99 (1979).

124. BORGES DEL CASTILLO, J., A.I. MARTINEZ-MARTIR, F. RODRIGUEZ-LUIS, J.C. RODRIGUEZ-UBIS, and P. VAZQUEZ-BUENO: Isolation and Synthesis of Two Coumarins from *Melampodium divaricatum*. Phytochemistry **23**, 859 (1984).

125. BRIGGS, L.H., and R.C. CAMBIE: The Constituents of *Phebalium nudum* Hook – 1. The Bark. Tetrahedron **2**, 256 (1958).

126. BRINK, C. VAN DER M., W. NEL, G.J.H. RALL, J.C. WEITZ, and K.G.R. PACHLER: *Neorautenia* isoflavonoids (II). Neofolin and Ficinin, Two Furoisoflavonoids from *N. ficifolia*. J. S. Afr. Chem. Inst. **19**, 24 (1966).

127. BROWN, K.L., A.I.R. BURFITT, R.C. CAMBIE, D. HALL, and K.P. MATHAI: The Constituents of *Phebalium nudum*. III. The Structures of Phebalin and Phebalarin. Austral. J. Chem. **28**, 1327 (1975).

128. BROWN, K.L., R.C. CAMBIE, and D. HALL: Structure of Phebalin: a Biscoumarin Derivative. Chem. and Ind. **1971**, 1020.

129. BUCHI, G.M., D.H. KLAUBERT, R.C. SHANK, S.M. WEINREB, and G.N. WOGAN: Structure and Synthesis of Kotanin and Desmethyl-kotanin. J. Org. Chem. **36**, 1143 (1971).

130. BUDDRUS, J., H. BAUER, E. ABU-MUSTAFA, A. KHATTAB, S. MISHAAL, E.A.M. EL-KHRISY, and M. LINSCHEID: Foetidin, a Sesquiterpenoid Coumarin from *Ferula assafoetida*. Phytochemistry **24**, 869 (1985).

131. BUKREEVA, T.V.: Structure of the Coumarin Glycoside Reoselin from the Roots of *Ferula kirialovii*. Khim. prirod. Soedinenii **1987**, 62; Chem. Abs. **107**, 93490 (1987).

132. BURKE, B.A., and H. PARKINS: Coumarins from *Amyris balsamifera*. Phytochemistry **18**, 1073 (1979).

133. BURKE, B.A., and S. PHILIP: Amyris of Jamaica. Coumarins of *Amyris elemifera* D.C., (Rutaceae). Heterocycles **16**, 897 (1981).

134. CAIRNS, N., L.M. HARWOOD, and D.P. ASTLES: Application of the Lewis-acid-catalysed Claisen Rearrangement of 4'-(1,1-Dimethylallyloxy)coumarates to the Synthesis of Demethylsuberosin. J. Chem. Soc., Chem. Commun. **1986**, 750.

135. — — —: Synthesis of Linear Coumarins via *para*-Claisen Rearrangement of Coumarate Ester Derivatives: Total Synthesis of Suberosin, Demethylsuberosin, and Ostruthin. J. Chem. Soc., Chem. Commun. **1986**, 1264.

136. — — —: An Iterative Procedure for the Synthesis of the Diprenylated Coumarins Balsamiferone and Gravelliferone from Umbelliferone via Multiple [3:3] Sigmatropic Rearrangements. J. Chem. Soc., Chem. Commun. **1987**, 400.

137. — — —: Efficient, Highly Regioselective Fries Rearrangement of Methyl 3-(2-acyloxy-4-methoxyphenyl)propanoates: the First Total Synthesis of the Linear Acylated Coumarins Geijerin and Dehydrogeijerin. Tetrahedron Letters **29**, 1311 (1988).

138. CAMPBELL, H.A., and K.P. LINK: Studies on the Hemorrhagic Sweet Clover Disease IV. The Isolation and Crystallization of the Hemorrhagic Agent. J. Biol. Chem. **138**, 21 (1941).

139. CAMPBELL, W.E., and G.M.L. CRAGG: A New Coumarin from *Phyllosma capensis*. Phytochemistry **18**, 688 (1979).

140. CARPENTER, I., E.J. MCGARRY, and F. SCHEINMANN: The Neoflavonoids and 4-Alkylcoumarins from *Mammea africana* G. Don. Tetrahedron Letters **1970**, 3983.

141. —————: Extractives from Guttiferae. Part XXI. The Isolation and Structure of Nine Coumarins from the Bark of *Mammea africana* G. Don. J. Chem. Soc. (C) **1971**, 3783.

142. CASTILLO, P., J.C. RODRIGUEZ-UBIS, and F. RODRIGUEZ: A High-yield Method for the Methylenation of *o*-Dihydroxyaromatic Compounds. Synthesis of Methylenedioxycoumarins. Synthesis **1986**, 839.

143. CECCHERELLI, P., M. CURINI, M.C. MARCOTULLIO, G. MADRUZZA, and A. MENGHINI: Tortuoside, a New Natural Coumarin Glucoside from *Seseli tortuosum*. J. Nat. Prod. **52**, 888 (1989).

144. CESKA, O., S.K. CHAUDHARY, P.J. WARRINGTON, and M.J. ASHWOOD-SMITH: Furocoumarins in the Cultivated Carrot, *Daucus carota*. Phytochemistry **25**, 81 (1986).

145. CHAKRABARTI, R., B. DAS, and J. BANERJI: *Bis*-Coumarins from *Edgeworthia gardneri*. Phytochemistry **25**, 557 (1986).

146. CHAKRABORTY, D.P., P. BHATTACHARYYA, and K.K. CHATTOPADHYAY: Marmelide, a Tyrinase Accelerating and Tryptophan Pyrrolase Inhibitory Furanocoumarin from *Aegle marmelos* Corr. Chem. and Ind. **1978**, 848.

147. CHAKRABORTY, D.P., and D. CHATTERJI: Structure of Mesuagin. A New 4-Phenylcoumarin. J. Org. Chem. **34**, 3784 (1969).

148. CHAKRABORTY, D.P., and B.C. DAS: The Structure of Mesuol. Tetrahedron Letters **1966**, 5727.

149. CHAKRABORTY, D.P., S. ROY, A. CHAKRABORTY, A.K. MANDAL, and B.K. CHOWDHURY: Structure and Synthesis of Mexolide: a New Antibiotic Dicoumarin from *Murraya exotica* Linn. [syn. *M. paniculata* (L) Jack.]. Tetrahedron **36**, 3563 (1980).

150. CHAKRAVARTI, K.K., A.K. BOSE, and S. SIDDIQUI: Chemical Examination of the Seeds of *Psoralea corylifolia*. J. Sci. Ind. Res. **7B**, 24, (1948).

151. CHATTERJEA, J.N., and N. PRASAD: Synthese von Furanverbindungen, XXVII. Synthese von Tri-*O*-methyl-wedelolacton und Dihydroerosnin. Chem. Ber. **97**, 1252 (1964).

152. CHATTERJEA, J.N., K.R.R.P. SINGH, I.S. JHA, Y. PRASAD, and S.C. SHAW: Synthesis of Gerberinol-I. Indian J. Chem. **25B**, 796 (1986).

153. CHATTERJEE, A., R. CHAKRABARTI, B. DAS, and J. BANERJI: New Coumarins from *Edgeworthia gardneri* Meissn. Indian J. Chem. **26B**, 81 (1987).

154. CHATTERJEE, A., D. GANGULY, and R. SEN: New Synthesis of 4-Phenyl Coumarins: Dalbergin and Nordalbergin. Tetrahedron **32**, 2407 (1976).

155. CHATTERJEE, A., S. SARKAR, and J.N. SHOOLERY: 7-Phenylacetoxy-coumarin from *Limonia crenulata*. Phytochemistry **19**, 2219 (1980).

156. CHATTERJEE, A., R. SEN, and D. GANGULY: Aegelinol, a Minor Lactonic Constituent of *Aegle marmelos*. Phytochemistry **17**, 328 (1978).

157. CHAWLA, H.M., and S.S. CHIBBER: Occurrence of 7-Hydroxy-4-methylcoumarin in *Dalbergia volubilis*. Indian J. Chem. **15B**, 492 (1977).

158. CHAWLA, H.M., and R.S. MITTAL: 7,3'-Dihydroxy-5,4'-dimethoxy-4-phenylcoumarin (Seshadrin), a New Neoflavonoid from *Dalbergia volubilis*. Indian J. Chem. **23B**, 375 (1984).

159. CHAWLA, H.M., R.S. MITTAL, and D.K. RASTOGI: Voludal, New Aldehydic Compound from *Dalbergia volubilis*. Indian J. Chem. **23B**, 175 (1984).

160. CHEN, K.-M., J.E. SEMPLE, and M.M. JOULLIE: Total Syntheses of Fungal Metabolites and Functionalized Furanones. J. Org. Chem. **50**, 3997 (1985).

161. CHEN, M., L. HOU, and G. ZHANG: The Diterpenoids from *Jatropha curcas* L. Zhiwu Xuebao **30**, 308 (1988); Chem. Abs. **110**, 21068 (1989).

162. CHEN, Y.S.: Lactone Constituents from *Boenninghausenia sessilicarpara* levl. Yaoxue Xuebao **24**, 260 (1989); Chem. Abs. **111**, 130761 (1989).
163. CHÊNEVERT, R., J. PAGÉ, and N. VOYER: Synthesis of ($\pm$)-Dehydroxyglaupalol and Analogues. Synth. Commun. **14**, 737 (1984).
164. CHIANG, M.T., M. BITTNER, M. SILVA, A. MONDACA, R. ZEMELMAN, and P.G. SAMMES: A Prenylated Coumarin with Antimicrobial Activity from *Haplopappus multifolius*. Phytochemistry **21**, 2753 (1982).
165. CLARK, E.P.: Scandenin – a Constituent of the Roots of *Derris scandens*. J. Org. Chem. **8**, 489 (1943).
166. CROMBIE, L., D.E. GAMES, N.J. HASKINS, and G.F. REED: Extractives of *Mammea americana* L. Part IV. Identification of New 7,8-Annulated Relatives of the Coumarins Mammea A/AA, A/AB, B/AA, and B/AB, and New Members of the 6-Acyl Family B/AA, B/AB, and B/AC. J. Chem. Soc. Perkin 1 **1972**, 2248.
167. CROMBIE, L., D.E. GAMES, N.J. HASKINS, G.F. REED, R.A. FINNEGAN, and K.E. MERKEL: Identification of New *Mammea* Coumarins: Four 7,8-Annulated Relatives of the *Mammea* Coumarins A/AA- A/AB, B/AA, B/AB, and Two of the 6-Acyl Family, B/AA (Isomammein) and B/AB. Tetrahedron Letters **1970**, 3979.
168. CROMBIE, L., D.E. GAMES, and A. McCORMICK: Isolation and Structure of Mammea A/BA, A/AB and A/BB: a Group of 4-Aryl-coumarin Extractives of *Mammea americana* L. Tetrahedron Letters **1966**, 415.
169. — — —: Extractives of *Mammea americana* L. Part II. The 4-Phenylcoumarins. Isolation and Structure of Mammea A/AA, A/A cyclo D, A/BA, A/AB and A/BB. J. Chem. Soc. (C) **1967**, 2553.
170. CROMBIE, L., R.C.F. JONES, and C.J. PALMER: Synthesis of the *Mammea* Coumarins. Part 1. The Coumarins of the Mammea A, B and C Series. J. Chem. Soc. Perkins 1 **1987**, 317.
171. — — —: Synthesis of the *Mammea* Coumarins. Part 3. The Insecticidal Coumarins of the Mammea E Series, Mammea D/BB, and a Dihydrocoumarin of the Mammea C Series. J. Chem. Soc. Perkin 1 **1987**, 345.
172. CROMBIE, L., S.D. REDSHAW, D.A. SLACK, and D.A. WHITING: Synthesis of ($\pm$)-Eriobrucinol and Regioisomeric Monoterpenoid Coumarins, Using Intramolecular Cycloadditions. J. Chem. Soc. Perkin 1 **1983**, 1411.
173. CUTLER, H.G., F.G. CRUMLEY, R.H. COX, O. HERNANDEZ, R.J. COLE, and J.W. DORNER: Orlandin: A Nontoxic Fungal Metabolite with Plant Growth Inhibiting Properties. J. Agric. Food Chem. **27**, 592 (1979).
174. D'AGOSTINO, M., V. DE FEO, F. DE SIMONE, and C. PIZZA: A 4-Arylcoumarin from *Coutarea hexandra*. Phytochemistry **28**, 1773 (1989).
175. D'AGOSTINO, M., V. DE FEO, F. DE SIMONE, F.F. VINCIERI, and C. PIZZA: Isolation of 8-Hydroxy-5,7,3′,4′-tetramethoxy-4-phenylcoumarin from *Coutarea hexandra*. Planta Medica **55**, 578 (1989).
176. DAILY, A., O. SELIGMANN, G. NONNENMACHER, B. FESSLER, S. WONG, and H. WAGNER: New Chromone, Coumarin and Coumestan Derivatives from *Mutisia acuminata* var. *hirsuta*. Planta Medica **54**, 50 (1988).
177. DAS, S., R.H. BARUAH, R.P. SHARMA, J.N. BARUA, P. KULANTHAIVEL, and W. HERZ: 7-Methoxycoumarins from *Micromelum minutum*. Phytochemistry **23**, 2317 (1984).
178. DAS, S.C., S. SENGUPTA, and W. HERZ: Revised Structure of Lasiocephalin: a New Coumarin from *Lasiosiphon eriocephalus* Decne. Chem. and Ind. **1973**, 792.
179. DAS GRAÇAS, M., B. ZOGHBI, N.F. ROQUE, and O.R. GOTTLIEB: Propacin, a Coumarinolignoid from *Protium opacum*. Phytochemistry **20**, 180 (1981).

180. DAWIDAR, A.-A., E. ABU-MUSTAFA, G. RÜCKER, E. ELKHRISY, and R. ABBAS: Marmaricin, a New Sequiterpenoid Coumarin from Ferula marmarica L. Chem. and Pharm. Bull. 27, 3153 (1979).

181. DEAN, F.M., A.M.B.S.R.C.S. COSTA, J.B. HARBORNE, and D.M. SMITH: Leptodactylone, a Yellow Coumarin from Leptodactylon and Linanthus Species. Phytochemistry 17, 505 (1978).

182. DEAN, F.M., B. PARTON, N. SOMVICHIEN, and D.A.H. TAYLOR: The Coumarins of Ptaeroxylon obliquum. Tetrahedron Letters 1967, 2147.

183. DEL CASTILLO, J.B., J.C. RODRIGUEZ-UBIS, and F.R. LUIS: Sitesis de Varias Dihidrofurocumarinas y Dihydropiranocumarinas Naturales. Anales de Quím. 81C, 106 (1985).

184. DEL CASTILLO, J.B., F.R. LUIS, and M. SECUNDINA: Angustifolin, a Coumarin from Ruta angustifolia. Phytochemistry 23, 2095 (1984).

185. DELGADO, G., and J. GARDUÑO: Pyranocoumarins from Arracacia nelsonii. Phytochemistry 26, 1139 (1987).

186. DELLE MONACHE, F., G. CAIRO VALERA, G.B. MARINI-BETTOLO, J.F. DE MELLO, and O. GONCALVES DE LIMA: Coumarins of Hortia arborea. II. Hortiolone and Hortinone. Gazzetta 107, 399 (1977).

187. DELLE MONACHE, F.G. CAIRO VALERA, D. SIALER DE ZAPATA, and G.B. MARINI-BETTOLO: 3-Aryl-4-methoxycoumarins and Isoflavones from Derris glabrescens. Gazzetta 107, 403 (1977).

188. DELLE MONACHE, G., B. BOTTA, and R. ALVES DE LIMA: A 4-Arylcoumarin from Coutarea hexandra. Phytochemistry 23, 1813 (1984).

189. DELLE MONACHE, G., B. BOTTA, F. DELLE MONACHE, and M. BOTTA: Synthesis of 4-Arylcoumarins from Coutarea hexandra. Phytochemistry 24, 1355 (1985).

190. DELLE MONACHE, G., B. BOTTA, F. MENICHINI, and R.M. PINHEIRO: Isolation from Coutarea hexandra and Synthesis of Exostemin and Its Methyl Ether. Bull. Chem. Soc. Ethiop. 1, 65 (1987); Chem. Abs. 110, 94760 (1989).

191. DELLE MONACHE, G., B. BOTTA, A. SERAFIM NETO, and R. ALVES DE LIMA: 4-Arylcoumarins from Coutarea hexandra. Phytochemistry 22, 1657 (1983).

192. DELLE MONACHE, G., B. BOTTA, V. VINCIGUERRA, and E. GACS-BAITZ: A New Neoflavanoid from Coutarea hexandra. Heterocycles 29, 355 (1989).

193. DE PASCUAL, J., A. SAN FELICIANO, J.M. MIGUEL DEL CORRAL, A.F. BARRERO, M. RUBIO, and L. MURIEL: 2,5-Dimethylcoumarins from leaves of Juniperus sabina. Phytochemistry 20, 2778 (1981).

194. DE PASCUAL TERESA, J., B. JIMENEZ, B. CORRALES, and M. GRANDE: Coumarinas de Ferulago GRUPO granatensis. Isovaleril marmesina. Anales de Quím. 75, 175 (1979).

195. DE SILVA, L.B., W.H.M.W. HERATH, R.C. JENNINGS, M. MAHENDRAN, and G.P. WANNIGAMA: Coumarins from Triphasia trifoliata. Phytochemistry 20, 2776 (1981).

196. DHARMARATNE, H.R.W., S. SOTHEESWARAN, S. BALASUBRAMANIAM, and E.S. WAIGHT: Triterpenoids and Coumarins from Leaves of Calophyllum cordato-oblongum. Phytochemistry 24, 1553 (1985).

197. DOGANCA, S., A. ULUBELEN, T. ISHIKAWA, and H. ISHII: (+)-Peucedanol Methyl Ether from Hippomarathrum cristatum, Umbelliferae. Chem. and Pharm. Bull. 27, 1049 (1979).

198. DOMINGUEZ, X.A., G. CANO, I. LUNA, and A. DIECK: Two Coumarins from the Aerial Parts of Amyris madrensis. Phytochemistry 16, 1096 (1977).

199. DOMINGUEZ, X.A., R. FRANCO, S. GARCÍA, A. MERIJANIAN, G. ESPINOZA, S. TAMEZ R., and A.B. ZILLI: Coumarins and Alkaloids of the Rutaceae Helietta parvifolia (Gray) Benth. ("Barreta"). Rev. Latinoam. Quim. 17, 60 (1986).

*200.* DOMINGUEZ, X.A., R. FRANCO, J. VERDE S., A. ZAMUDIO, and E.Y. GUEVARA Z.: Mexican Medicinal Plants LI. Coumarins from the Desert Rue *Thamnosma texana*, Gray (Torr). Rev. Latinoam. Quim. **17**, 60 (1986).

*201.* DONNELLY, B.J., D.M.X. DONNELLY, and A.M. O'SULLIVAN: Dalbergia Species: Constitution of Melannein. Chem. and Ind. **1966**, 1498.

*202.* — — —: *Dalbergia* Species – VI. The Occurrence of Melannein in the Genus Dalbergia. Tetrahedron **24**, 2617 (1968).

*203.* DONNELLY, D.M.X., and P.J. KAVANAGH: Isoflavanoids of *Dalbergia oliveri*. Phytochemistry **13**, 2587 (1974).

*204.* DONNELLY, D.M.X., J. O'REILLY, and W.B. WHALLEY: Neoflavonoids of *Dalbergia melanoxylon*. Phytochemistry **14**, 2287 (1975).

*205.* DONNELLY, D.M.X., J.C. THOMPSON, W.B. WHALLEY, and S. AHMAD: *Dalbergia* Species. Part IX. Phytochemical Examination of *Dalbergia stevensonii* Standl. J. Chem. Soc. Perkin 1 **1973**, 1739.

*206.* DREYER, D.L.: Constituents of *Thamnosma montana* Torr and Frem. Tetrahedron **22**, 2923 (1966).

*207.* DUFFLEY, R.P., and R. STEVENSON: Synthesis of Oroselol. Synth. Commun. **8**, 175 (1978).

*208.* DUKHOVLINOVA, L.I., Y.E. SKLYAR, and M.G. PIMENOV: Coumarins of *Seseli seravschanicum* Roots. Khim. prirod. Soedinenii **1980**, 832; Chem. Abs. **94**, 171050 (1981).

*209.* DUKHOVLINOVA, L.I., Y.E. SKLYAR, L.I. SDOBNINA, and M.G. PIMENOV: Secrolin, a New Dihydrofurocoumarin from *Seseli mucronatum*. Khim. prirod. Soedinenii **1979**, 721.

*210.* DUTT, P., N.C. DEB, and P.K. BOSE: A Preliminary Note on Mesuol, the Bitter Principle of *Mesua ferrea*. J. Indian Chem. Soc. **17**, 277 (1940).

*211.* DUTTA, P.K., D. BANERJEE, and N.L. DUTTA: Euphorbetin: A New Bicoumarin from *Euphorbia lathyris* L. Tetrahedron Letters **1972**, 601.

*212.* — — —: Isoeuphorbetin, a Novel Bicoumarin from *Euphorbia lathyris* Linn. Indian J. Chem. **11**, 831 (1973).

*213.* DUTTA, P.K., P.C. MAJUMDER, and N.L. DUTTA: Synthetic Approaches Towards Bicoumarins. Synthesis of Euphorbetin and Isoeuphorbetin. Tetrahedron **31**, 1167 (1975).

*214.* DZHAFOROV, Z.R., A.A. KULIEV, Z.A. KULIEV, V.M. MALIKOV, and N.M. ISMAILOV: A New Dihydrofurocoumarin from *Smyrniopsis aucheri*. Khim. prirod. Soedinenii **1988**, 754; Chem. Abs. **110**, 132183 (1989).

*215.* EAST, A.J., W.D. OLLIS, and R.E. WHEELER: Natural Occurrence of 3-Aryl-4-hydroxycoumarins. Part 1. Phytochemical Examination of *Derris robusta* (Roxb.) Benth. J. Chem. Soc. (C) **1969**, 365.

*216.* EISENBEISS, J., and H. SCHMID: Struktur des Erosnin (Norton and Hansberry's "Compound I"). Helv. Chim. Acta **42**, 61 (1959).

*217.* EL-BASYOUNI, S., and G.H.N. TOWERS: The Phenolic Acids in Wheat II. Natural Occurrence of Orthoferulic Acid (2-Hydroxy-3-methoxycinnamic Acid). Canad. J. Biochem. **42**, 493 (1964).

*218.* EMERSON, O.H., and E.M. BICKOFF: Synthesis of Coumestrol, 3,9-Dihydroxy-6*H*-benzofuro[3,2-c][1]benzopyran-6-one. J. Amer. Chem. Soc. **80**, 4381 (1958).

*219.* ENRÍQUEZ, R.G., M.L. ROMERO, L.I. ESCOBAR, P. JOSEPH-NATHAN, and W.F. REYNOLDS: High-performance Liquid Chromatography Study of *Casimiroa edulis*. II. Determination of Furocoumarins. J. Chromatogr. **287**, 209 (1984).

*220.* ERDELMEIER, C.A.J., B. MEIER, and O. STICHER: Reversed-phase High-performance Liquid Chromatographic Separation of Closely Related Furocoumarins. J. Chromatogr. **346**, 456 (1985).

221. ERDELMEIER, C.A.J., and O. STICHER: Coumarin Derivatives from *Eryngium campestre*. Planta Medica **1985**, 407.

222. ESPINOZA, S., R. GAONA, A. URZUA, and B.K. CASSELS: Two New Coumarins from *Gomortega keule*. Bol. Soc. Chil. Quim. **27**, 283 (1982); Chem. Abs. **96**, 214310 (1982).

223. FALSHAW, C.P., R.A. HARMER, W.D. OLLIS, R.E. WHEELER, V.R. LALITH, and N.V. SUBBA RAO: Natural Occurrence of 3-Aryl-4-hydroxycoumarins. Part II. Phytochemical Examination of *Derris scandens* (Roxb.) Benth. J. Chem. Soc. (C) **1969**, 374.

224. FAYEZ, M.B.E., F.K.A. EL-BEIH, B.A.H. TAWIL, and M.M. KHALIL: New Coumarin from Fruits of *Ammi majus* L. Pharmazie **37**, 53 (1982).

225. FINNEGAN, R.A., and C. DJERASSI: Naturally Occurring Oxygen Heterocycles. 4-Phenyl-5,7-dihydroxy-6-isovaleryl-8-isopentenylcoumarin. Tetrahedron Letters No. **13**, 11 (1959).

226. FINNEGAN, R.A., and K.E. MERKEL: Constituents of *Mammea americana*. IX. Oxidation of Mammein and Mammeisin. J. Pharm. Sci. **61**, 1603 (1972).

227. FINNEGAN, R.A., M.P. MORRIS, and C. DJERASSI: Naturally Occurring Oxygen Heterocycles. X. 4-Phenyl-5,7-dihydroxy-6-isovaleryl-8-isopentenylcoumarin. J. Org. Chem. **26**, 1180 (1961).

228. FINNEGAN, R.A., and W.H. MUELLER: Mammeigin: A New 4-Phenylcoumarin. Chem. and Ind. **1964**, 1065.

229. ――: Constituents of *Mammea americana* L. IV. The Structure of Mammeigin. J. Org. Chem. **30**, 2342 (1965).

230. FISCHER, H.A., RÖMER, B. ULBRICH, and H. ARENS: A New Biscoumarin Glucoside Ester from *Ruta chalepensis* Cell Cultures. Planta Medica **1988**, 398.

231. FISHER, J.F., and L.A. TRAMA: HPLC of Some Coumarins and Psoralens in Citrus Peel Oils. J. Agric. Food Chem. **27**, 1334 (1979).

232. FOMUM, Z.T., J.F. AYAFOR, J. WANDJI, W.G. FOMBAN, and A.E. NKENGFACK: Erthrinasinate, an Ester from Three *Erythrina* Species. Phytochemistry **25**, 757 (1986).

233. FORGACS, P., J.-F. DESCONCLOIS, J.-L. POUSSET, and A. RABARRON: Structure d'un Nouvel Héteroside Coumarinique: le Diospyroside. Tetrahedron Letters **1978**, 4783.

234. FOZDAR, B.I., S.A. KHAN, T. SHAMSUDDIN, K.M. SHAMSUDDIN, and J.P. KINTZINGER: Aleuritin, a Coumarinolignoid, and a Coumarin from *Aleurites fordii*. Phytochemistry **28**, 2459 (1989).

235. FUJIWARA, H., T. YOKOI, S. TANI, Y. SAIKI, and A. KATO: Studies on Constituents of *Angelica dahuricae* Radix. I. On a New Furocoumarin Derivative. Yakugaku Zasshi **100**, 1258 (1980); Chem. Abs. **94**, 136088 (1981).

236. FUKAI, T., Q.-H. WANG, T. KITAGAWA, K. KUSANO, T. NOMURA, and Y. IITAKA: Structures of Six Isoprenoid-substituted Flavonoids, Gancaoins F, G, H, I, Glycyrol, and Isoglycyrol from Xibei Licorice (*Glycyrrhiza* Sp.). Heterocycles **29**, 1761 (1989).

237. FUKUI, K., and M. NAKAYAMA: Total Synthesis of Erosnin. Tetrahedron Letters **1965**, 2559.

238. FURUKAWA, H., M. JU-ICHI, I. KAJIURA, and M. HIRAI: Ponfolin: a New Coumarin from Trifoliate Orange. Chem. and Pharm. Bull. **34**, 3922 (1986).

239. GALÁN, R.H., G.M. MASSANET, E. PANDO, F.R. LUIS, and J. SALVÁ: Synthesis of (±)-3-(1′,1′-Dimethylallyl)decursinol and 3-(1′,1′-Dimethylallyl)xanthyletin. Heterocycles **27**, 775 (1988).

240. ――――: Sigmatropic Rearrangements in Synthesis of C-8 Prenylated 3-(1′,1′-Dimethylallyl)coumarins: 3-(1′,1′-Dimethylallyl)seselin. Heterocycles **29**, 297 (1989).

241. GAMES, D.E., and N.J. HASKINS: Synthesis of some Dimethylpyrano- and 3-Methylbut-2-enyl-4-phenyl- and 4-n-propyl-coumarins. J. Chem. Soc. (D) **1971**, 1005.

242. GANGULY, A.K., B.S. JOSHI, V.N. KAMAT, and A.H. MANMADE: Synthesis of Claus-

enin, Xanthoxyletin, Alloxanthoxyletin, Xanthyletin and Nordalbergin. Tetrahedron **23**, 4777 (1967).

243. GANTIMUR, D., and A.A. SEMENOV: Coumarins from *Phlojodicarpus sibiricus*. Khim. prirod. Soedinenii **1981**, 47; Chem. Abs, **95**, 3372 (1981).

244. GANTIMUR, D., A.I. SYRCHINA, and A.A. SEMENOV: A Peucedanol Glucoside from *Phlojodicarpus turczaninovii*. Khim. prirod. Soedinenii **1985**, 190; Chem. Abs. **103**, 85043 (1985).

245. ———: New Glycosides from Plants of the Genus *Phlojodicarpus*. Khim. prirod. Soedinenii **1986**, 36; Chem. Abs. **105**, 94462 (1986).

246. ———: Khellactone Derivatives from *Phlojodicarpus sibiricus*. Khim. prirod. Soedinenii **1986**, 108; Chem. Abs. **105**, 75878 (1986).

247. GARCIA, M., M.H.C. KANO, D.M. VIEIRA, M.C. DO NASCIMENTO, and W.B. MORS: Isoflavonoids from *Derris spruceana*. Phytochemistry **25**, 2425 (1986).

248. GARCIA BILBAO, J.L., and B. RODRÍGUEZ: Flavanonas del *Rumex conglomeratus*. Anales de Quím. **74**, 1570 (1978).

249. GARG, S.K., S.R. GUPTA, and N.D. SHARMA: Apiumetin, a New Furanocoumarin from the Seeds of *Apium graveolens*. Phytochemistry **17**, 2135 (1978).

250. ———: Coumarins from *Apium graveolens* Seeds. Phytochemistry **18**, 1580 (1979).

251. ———: Apiumoside, a New Furocoumarin Glucoside from the Seeds of *Apium graveolens*. Phytochemistry **18**, 1764 (1979).

252. ———: Celerin, a New Coumarin from *Apium graveolens*. Planta Medica **38**, 186 (1980).

253. ———: Glucosides of *Apium graveolens*. Planta Medica **38**, 363 (1980).

254. GARG, S.K., N.D. SHARMA, and S.R. GUPTA: A New Dihydrofurocoumarin from *Apium graveolens*. Planta Medica **43**, 306 (1981).

255. GASHIMOV, N.F., A.Z. ABYSHEV, A.A. KAGRAMANOV, and L.I. ROZHKOVA: Coumarins of *Haplophyllum ramosissimum*. Khim. prirod. Soedinenii **1979**, 15; Chem. Abs. **91**, 105161 (1979).

256. ————: Versicolon, a New Coumarin from *Haplophyllum versicolor*. Khim. prirod. Soedinenii **1979**, 87; Chem. Abs. **91**, 52698 (1979).

257. GASHIMOV, N.F., and N.O. ORAZMUKHAMEDOVA: Coumarins of *Haplophyllum bungei*. Khim. prirod. Soedinenii **1978**, 653; Chem. Abs. **90**, 83602 (1979).

258. GHEN, Z.-X., B.-S. HUANG, Q.-L. SHE, and G.F. ZENG: Study on the Chemical Constituents of the Chinese Medicinal Plant, *Peucedanum praeruptorum* Dunn. Structures of Four New Coumarins. Yao Hsueh Hsueh Pao **14**, 486 (1979); Chem. Abs. **92**, 124903 (1980).

259. GHISALBERTI, E.L., P.R. JEFFERIES, C.L. RASTON, B.W. SKELTON, A.D. STUART, and A.H. WHITE: Structural Studies in the Bruceol System. J. Chem. Soc. Perkin 2 **1981**, 583.

260. GHISALBERTI, E.L., P.R. JEFFERIES, C.L. RASTON, B.W. SKELTON, A.H. WHITE, and G.K. WORTH: Structural Studies of Some Hydroxyeriobrucinol Derivatives. J. Chem. Soc. Perkin 2 **1981**, 576.

261. GHOSH, P., P. SIL, S.G. MAJUMDAR, and S. THAKUR: A Coumarin from *Limonia acidissima*. Phytochemistry **21**, 240 (1982).

262. GOLOVINA, L.A., T.K. KHASANOV, A.I. SAIDKHODZHAEV, V.M. MALIKOV, and U. RAKHMANKULOV: Coumarins and Esters of *Ferula microcarpa*. Khim. prirod. Soedinenii **1978**, 566; Chem. Abs. **90**, 83958 (1979).

263. GONZÁLEZ, A.G., J.T. BARROSO, R.J. CARDONA, J.M. MEDINA, and F.R. LUIS: Componentes de Umbeliferas. XVI. Cumarinas del *Peucedanum hispanicum* (Boiss) Endl. in Walpers. Anales de Quím. **73**, 1188 (1977).

264. GONZÁLEZ. A.G., J.T. BARROSO, E.D. CHICO, J.R. LUIS, and F.R. LUIS: Components de Umbeliferas. XIV. Estudio del *Heracleum granatense* Boiss. Elench. Anales de Quím. **73**, 858 (1977).

265. GONZÁLEZ. A.G., J.T. BARROSO, H.L. DORTA, J.R. LUIS, and F.R. LUIS: Components de Umbeliferas XVII. Cumarinas Sencillas del *Seseli tortuosum* L.B.S. Eur. Anales de Quím. **74**, 979 (1978).

266. ———————: Pyranocoumarin Derivatives from *Seseli tortuosum*. Phytochemistry **18**, 1021 (1979).

267. GONZÁLEZ, A.G., R.J. CARDONA, E.D. CHICO, H.L. DORTA, and F.R. LUIS: Nuevas Fuentes de Cumarinas Naturales. XXIII. Bicumarinas de la *Ruta* Sp. *Tene 29662*. Anales de Quím. **73**, 1510 (1977).

268. GONZÁLEZ, A.G., E.D. CHICO, H.L. DORTA, J.R. LUIS, and F.R. LUIS: Nuevas Fuentes de Cumarinas Naturales. XXXI. Dos Nuevas Cumarinas de la *Ruta* Sp. *Tene 29662*. Anales de Quím. **73**, 607 (1977).

269. GONZÁLEZ, A.G., H.L. DORTA, J.R. LUIS, and F.R. LUIS: Componentes de Umbeliferas. XXII. Otros Derivados Cumarinicos del *Seseli tortuosum* L.B.S. Eur. Anales de Quím. **78C**, 184 (1982).

270. GONZÁLEZ, A.G., B.M. FRAGA, M.G. HERNÁNDEZ, O. PINO, and A.G. RAVELO: Nuevas Cumarinas del *Cneorum tricoccum*. Rev. Latinoam. Quim. **9**, 205 (1978).

271. GONZÁLEZ, A.G., R.E. REYES, and J.B. ARENCIBIA: Nuevas Fuentes de Cumarinas Naturales. XXVIII. Dos Nuevas Bicumarinas de la *Ruta oreojasme* Weeb. Anales de Quím. **71**, 842 (1975).

272. GONZÁLEZ, A.G., R.E. REYES, and M.R. ESPINO: Two New Coumarins from *Ruta pinnata*. Phytochemistry **16**, 2033 (1977).

273. GÖREN, N., A. ULUBELEN, and S. ÖKSÜZ: A Sesquiterpene-coumarin Ether and an Acetylenic Compound from *Tanacetum heterotumum*. Phytochemistry **27**, 1527 (1988).

274. GOVINDACHARI, T.R., K. NAGARAJAN, and B.R. PAI: Chemical Examination of *Wedelia calendulacea*. Part I. Structure of Wedelolactone. J. Chem. Soc. (C) **1956**, 629.

275. GOVINDACHARI, T.R., K. NAGARAJAN, B.R. PAI, and P.C. PARTHASARATHY: Chemical Investigation of *Wedelia calendulacea*. Part II. The Position of the Methoxyl Group in Wedelolactone. J. Chem. Soc. **1957**, 545.

276. GRAF, E., and M. ALEXA: Über 5 neue Umbelliferonether aus Galbanumharz. Planta Medica **1985**, 428.

277. GRANDE, M., M.T. AGUADO, B. MANCHEÑO, and F. PIERA: Coumarins and Ferulol Esters from *Cachrys sicula*. Phytochemistry **25**, 505 (1986).

278. GRAY, A.I.: New Coumarins from *Coleonema album*. Phytochemistry **20**, 1711 (1981).

279. GRAY, A.I., C.J. MEEGAN, and N.B. O'CALLAGHAN: Coumarins from Two *Coleonema* Species. Phytochemistry **26**, 257 (1987).

280. GREGER, H., E. HASLINGER, and O. HOFER: Albartin, a New Sesquiterpene-coumarin Ether from *Artemisia alba*. Monatsh. Chem. **113**, 375 (1982).

281. GREGER, H., and O. HOFER: Sesquiterpene-coumarin Ethers and Polyacetylenes from *Brocchia cinerea*. Phytochemistry **24**, 85 (1985).

282. GREGER, H., O. HOFER, and A. NIKIFOROV: New Sesquiterpene-coumarin Ethers from *Achillea* and *Artemisia* Species. J. Nat. Prod. **45**, 455 (1982).

283. GREGER, H., O. HOFER, and W. ROBIEN: New Sesquiterpene Coumarin Esters from *Achillea ochroleuca*. $^{13}$C NMR of Isofraxidin-derived Open-chain, and bicyclic Sesquiterpene Ethers. J. Nat. Prod. **46**, 510 (1983).

284. ———: Types of Sesquiterpene-coumarin Ethers from *Achillea ochroleuca* and *Artemisia tripartata*. Phytochemistry **22**, 1997 (1983).

285. GU, L.-H., S.-X. WANG, X. LI, and T.-R. ZHU: Antibacterial Constituents from *Gerbera anandria* (L.) Sch. Bip. Acta Pharm. Sinica **22**, 272 (1987).

286. GUISE, G.B., E. RITCHIE, R.G. SENIOR, and W.C. TAYLOR: The Chemical Constituents of Australian *Zanthoxylum* Species. IV. Two New Coumarins from *Z. suberosum* C.T. White (syn. *Z. Dominianum* Merr. and Perry; *Z. ovalifolium* Wight.). Austral. J. Chem. **20**, 2429 (1967).

287. GUNASEKERA, S.P., G.S. JAYATILAKE, S.S. SELLIAH, and M.U.S. SULTANBAWA: Chemical Investigation of Ceylonese Plants. Part 27. Extractives of *Calophyllum cuneifolium* Thw. and *Calophyllum soulattri* Burm. J. Chem. Soc. Perkin 1 **1977**, 1505.

288. GUO, X., and Y. ZHANG: Chemical Studies on the Chinese Drug Qin Pi, the Bark of *Fraxinus stylosa*. Acta Pharm. Sinica **18**, 434 (1983).

289. GUPTA, B.D., S.K. BANERJEE, K.L. HANDA, and C.K. ATAL: 5-Hydroxy-8-(1′,1′-dimethylallyl)psoralen and 5-Hydroxyangelicin Two New Coumarins from *Heracleum thomsonii*. Indian J. Chem. **16B**, 38 (1978).

290. GUPTA, B.K., G.K. GUPTA, K.L. DHAR, and C.K. ATAL: Psoralidin Oxide, a Coumestan from the Seeds of *Psoralea corylifolia*. Phytochemistry **19**, 2232 (1980).

291. GUPTA, G.K., K.L. DHAR, and C.K. ATAL: Isolation and Constitution of Corylidin: a New Coumestrol from the Fruits of *Psoralea corylifolia*. Phytochemistry **16**, 403 (1977).

292. HALIM, A.F., E.-S.M. MARMAN, and F. BOHLMANN: 6-Hydroxy-4-methoxy-5-methylcoumarin from *Gerbera jamesonii*. Phytochemistry **19**, 2496 (1980).

293. HAMBURGER, M., M. GUPTA, and K. HOSTETTMANN: Coumarins from *Polygala paniculata*. Planta Medica **1985**, 215.

294. HAMBURGER, M., H. STOECKLI-EVANS, and K. HOSTETTMANN: A New Pyranocoumarin Diester from *Polygala paniculata* L. Helv. Chim. Acta **67**, 1729 (1984).

295. HANUMAIAH, T., D.S. MARSHALL, B.K. RAO, J.U.M. RAO, K.V.J. RAO, and R.H. THOMSON: Naphthoquinone-lactones and Extended Quinones from *Ventilago calyculata*. Phytochemistry **24**, 2669 (1985).

296. HARKAR, S., T.K. RAZDAN, and E.S. WAIGHT: Steroids, Chromone and Coumarins from *Angelica officinalis*. Phytochemistry **23**, 419 (1984).

297. HARPER, S.H.: The Active Principles of Leguminous Fish-poison Plants. Part VI. Robustic Acid. J. Chem. Soc. **1942**, 181.

298. HATA, K., T. NISHINO, Y. HIRAI, Y. WADA, and W. KOZAWA: On Coumarins from the Fruits of *Angelica pubescens* Maxim. Yakugaku Zasshi **101**, 67 (1981); Chem. Abs. **94**, 153448 (1981).

299. HATANO, T., T. YASUHARA, K. MIYAMOTO and T. OKUDA: Anti-human Immunodeficiency Virus Phenolics from Licorice. Chem. and Pharm. Bull. **36**, 2286 (1988).

300. HATTORI, M., K. MIYACHI, Y.-Z. SHU, N. KAKIUCHI, and T. NAMBA: Studies on Dental Caries Prevention by Traditional Medicines (IX). Plant Antibacterial Action of Coumarin Derivatives from Licorice Roots Against *Streptococcus mutans*. Shoyakugaku Zasshi **40**, 406 (1986); Chem. Abs. **107**, 46132 (1987).

301. HAYAKAWA, K., M. YODO, S. OHSUKI, and K. KANEMATSU: Novel Bicycloannulation via Tandem Vinylation and Intramolecular Diels-Alder Reaction of Five membered Heterocycles: a New Approach to Construction of Psoralen and Azapsoralen. J. Amer. Chem. Soc. **106**, 6735 (1984).

302. HERZ, W., S.V. BHAT, and P.S. SANTHANAM: Coumarins of *Artemisia dracunculoides* and 3′,6-Dimethoxy-4′,5,7-trihydroxyflavone in *A. arctica*. Phytochemistry **9**, 891 (1970).

303. HERZ, W., and M. BRUNO: Heliangolides, Kauranes and Other Constituents of *Helianthus heterophyllus*. Phytochemistry **25**, 1913 (1986).

298     R. D. H. MURRAY

*304.* HERZ, W., S.V. GOVINDAN, and N. KUMAR: Sesquiterpene Lactone and Other Constituents of *Eupatorium lancifolium* and *E. semiserratum*. Phytochemistry **20**, 1343 (1981).
*305.* HIFNAWY, M.S., J. VAQUETTE, T. SEVENET, J.-L. POUSSET, and A. CAVÉ: Produits Neutres et Alcaloides de *Myrtopsis macrocapra, M. myrtoidea, M. Novae-caledoniae* and *M. sellingi*. Phytochemistry **16**, 1035 (1977).
*306.* HIGA, T., and P.J. SCHEUER: Hawaiian Plant Studies. Part XVI. Coumarins and Flavones from *Pelea barbigera* (Gray) Hillebrand (Rutaceae). J. Chem. Soc. Perkin 1 **1974**, 1350.
*307.* HIRAKURA, K., I. SAIDA, T. FUKAI, and T. NOMURA: Mulberroside B, a New C-Glucosylcoumarin from the Cultivated Mulberry Tree (*Morus lhou* Koidz.). Heterocycles **23**, 2239 (1985).
*308.* HOFER, O.: Lanthanide-induced Shifts of Sterically Hindered Aromatic *o*-Dimethoxy Compounds: Model Compounds and *o*-Dimethoxycoumarins. J. Chem. Soc. Perkin 2 **1986**, 715.
*309.* HOFER, O., and H. GREGER: Naturally Occurring Sesquiterpene-coumarin Ethers, VI. New Sesquiterpene-isofraxidin Ethers from *Achillea depressa*. Monatsh. Chem. **115**, 477 (1984).
*310.* ——: Scopoletin Sesquiterpene Ethers from *Artemisia persica*. Phytochemistry **23**, 181 (1984).
*311.* ——: New Sesquiterpene-coumarin Ethers from *Anthemis cretica*. Liebigs Ann. Chem. **1985**, 1136.
*312.* HOFER, O., G. SZABO, H. GREGER, and A. NIKIFOROV: Leaf Coumarins from the *Artemisia laciniata* Group. Liebigs Ann. Chem. **1986**, 2142.
*313.* HOFER, O., W. WEISSENSTEINER, and M. WIDHALM: Absolute Configurations and Circular Dichroisms of Sesquiterpene-coumarin Ethers. Monatsh. Chem. **114**, 1399 (1983).
*314.* HOFER, G., M. WIDHALM, and H. GREGER: Circular Dichroism of Sesquiterpene-umbelliferone Ethers and Structure Elucidation of a New Derivative Isolated from the Gum Resin *Asa foetida*. Monatsh. Chem. **115**, 1207 (1984).
*315.* HORI, K., T. SATAKE, Y. SAIKI, T. MURAKAMI, and C.M. CHEN: Chemical and Chemotaxonomical Studies of Filices. LXIX. The Novel Coumarins of *Macrothelypteris torresiana* Ching var. *calvata* Holtt. (= *M. oligophlebia* Ching). Yakugaku Zasshi **107**, 491 (1987); Chem. Abs. **107**, 233103 (1987).
*316.* HUANG, S.-C., M.-T. CHEN, and T.-S. WU: Alkaloids and Coumarins from Stem Bark of *Citrus grandis*. Phytochemistry **28**, 3574 (1989).
*317.* HUTCHINSON, D.W., and J.A. TOMLINSON: The Structure of Dicoumarol and Related Compounds. Tetrahedron **25**, 2531 (1969).
*318.* IINUMA, M., T. TANAKA, K. HAMADA, M. MIZUNO, and F. ASAI: Flavonoids Syntheses VI. Synthesis and Spectral Properties of 4-Arylcoumarins (Neoflavones). Chem. and Pharm. Bull. **35**, 3909 (1987).
*319.* IMAI, F., K. ITOH, N. KISHIBUCHI, T. KINOSHITA, and U. SANKAWA: Constituents of the Root Bark of *Murraya paniculata* Collected in Indonesia. Chem. and Pharm. Bull. **37**, 119 (1989).
*320.* IMAI, F., T. KINOSHITA, A. ITAI, and U. SANKAWA: Acid-catalysed Rearrangement of an Epoxy Coumarin Phebalosin. The Revised Structure of Murralongin. Chem. and Pharm. Bull. **34**, 3978 (1986).
*321.* IMAI, F., T. KINOSHITA, and U. SANKAWA: Constituents of the Leaves of *Murraya paniculata* Collected in Taiwan. Chem. and Pharm. Bull. **37**, 358 (1989).
*322.* ———: New Coumarin Derivatives from *Murraya paniculata* (Linn.) Jack Leaves. Shoyakugaku Zasshi **41**, 157 (1987); Chem. Abs. **108**, 52809 (1988).

*323.* INGHAM, J.L., S. TAHARA, and S.Z. DZIEDZIC: Coumestans from the Roots of *Pueraria mirifica.* Z. Naturforsch. **43C**, 5 (1988).

*324.* ISHII, H., K. HOSOYA, T. ISHIKAWA, E. UEDA, and J. HAGINIWA: Chemical Constituents of Rutaceous Plants. XXI. Chemical Constituents of *Xanthoxylum [Zanthoxylum] arnottianum.* 2. Isolation of the Chemical Constituents of the Xylem of Stems. Yakugaku Zasshi **94**, 322 (1974); Chem. Abs. **81**, 132753 (1974).

*325.* ISHII, H., and T. ISHIKAWA: Structural Establishment of Arnocoumarin and Arnottia-coumarin due to Chemical Modification of Marmesin and Rutaretin Methyl Ether. Chem. and Pharm. Bull. **26**, 2598 (1978).

*326.* ISHII, H., J.-I. KOBAYASHI, and T. ISHIKAWA: Toddalenone: a New Coumarin from *Toddalia asiatica* (*T. aculeata*). Structural Establishment Based on the Chemical Conversion of Limettin into Toddalenone. Chem. and Pharm. Bull. **31**, 3330 (1983).

*327.* ISHII, H., F. SEKIGUCHI, and T. ISHIKAWA: Studies on the Chemical Constituents of Rutaceous Plants. XLI. Absolute Configuration of Rutaretin Methyl Ether. Tetrahedron **37**, 285 (1986).

*328.* ITO, C., and H. FURUKAWA: Constituents of *Murraya exotica* L. Structure Elucidation of New Coumarins. Chem. and Pharm. Bull. **35**, 4277 (1987).

*329.* — —: Two New Coumarins from *Murraya* Plants. Chem. and Pharm. Bull. **37**, 819 (1989).

*330.* — —: Three New Coumarins from *Murraya exotica.* Heterocycles **26**, 1731 (1987).

*331.* — — —: Three New Coumarins from Leaves of *Murraya paniculata.* Heterocycles **26**, 2959 (1987).

*332.* ITO, C., M. MATSUOKA, T. MIZUNO, K. SATO, Y. KIMURA, M. JU-ICHI, M. INOUE, I. KAJIURA, M. OMURA, and H. FURUKAWA: New Coumarins from Some *Citrus* Plants. Chem. and Pharm. Bull. **36**, 3805 (1989).

*333.* ITO, C., S. TANAHASHI, M. OMURA, and H. FURUKAWA: New Coumarins from *Citrus* Plant. Chem. and Pharm. Bull. **37**, 2217 (1989).

*334.* IVIE, G.W.: Linear Furocoumarins (Psoralens) from the Seeds of Texas *Ammi majus* L. (Bishop's Weed). J. Agric. Food Chem. **26**, 1394 (1978).

*335.* IWAGAWA, T., and T. HASE: A Coumarin Acetylglucoside from *Viburnum suspensum.* Phytochemistry **23**, 467 (1984).

*336.* JACKSON, R.F.W., and R.A. RAPHAEL: Novel Routes to Furan-3(2*H*)-ones. New Syntheses of Bullatenone and Geiparvarin. J. Chem. Soc. Perkin 1 **1984**, 535.

*337.* JAIN, A.C., and S.M. JAIN: Synthesis of Robustic Acid and Related 4-Hydroxy-3-phenylcoumarins. Tetrahedron **29**, 2803 (1973).

*338.* JAIN, A.C., S.M. JAIN, and J. SINGH: Synthesis of Robustin and Related 4-Hydroxy-3-phenylcoumarins and Isoflavones. Tetrahedron **30**, 2485 (1974).

*339.* JAIN, A.K., N.D. SHARMA, S.R. GUPTA, and D.R. BOYD: Coumarins from *Apium graveolens* Seeds. Planta Medica **1986**, 246.

*340.* JAKUPOVIC, J., R. BOEKER, A. SCHUSTER, F. BOHLMANN, and S.B. JONES: Further Guaianolides and 5-Alkylcoumarins from *Gutenbergia* and *Bothriocline* Species. Phytochemistry **26**, 1069 (1987).

*341.* JAKUPOVIC, J., V.P. PATHAK, F. BOHLMANN, R.M. KING, and H. ROBINSON: Obliquin Derivatives and Other Constituents from Australian *Helichrysum* Species. Phytochemistry **26**, 803 (1987).

*342.* JEFFREYS, J.A.D., M. bin ZAKARIA, P.G. WATERMAN, and S. ZHONG: A New Class of Natural Product: Homologues of Juglone Bearing 4-Hydroxy-5-methylcoumarin-3-yl Units from *Diospyros* Species. Tetrahedron Letters **24**, 1085 (1983).

*343.* JERRIS, P.J., and A.B. SMITH III: Synthesis and Configurational Assignment of Geiparvarin: a Novel Antitumor Agent. J. Org. Chem. **46**, 577 (1981).

344. JOHNSON, A.P., and A. PELTER: The Structure of Robustic Acid, a New 4-Hydroxy-3-phenylcoumarin. J. Chem. Soc (C) 1966, 606.

345. JOHNSON, A.P., A PELTER, and P. STAINTON: Extractives from *Derris scandens*. Part I. The Structures of Scandenin and Lonchocarpic Acid. J. Chem. Soc. (C) 1966, 192.

346. JONES, H.A.: Lonchocarpic Acid, a New Compound from a Species of *Lonchocarpus*. J. Amer. Chem. Soc. 56, 1247 (1934).

347. JOSEPH-NATHAN, P., J.D. HERNÁNDEZ, L.U. ROMÁN, E. GARCÍA G., V. MENDOZA, and S. MENDOZA: Coumarins and Terpenoids from *Perezia alamani* var. *oolepsis*. Phytochemistry 21, 1129 (1982).

348. JOSEPH-NATHAN, P., J. HIDALGO, and D. ABRAMO-BRUNO: A New Coumarin from *Perezia multiflora*. Phytochemistry 17, 583 (1978).

349. JOSHI, B.S., D.H. GAWAD, and K.R. RAVINDRANATH: Chemical Constituents of *Atlantia racemosa* Wt. and Arne. Structure and Synthesis of Racemosin, a Novel Pyranocoumarin. Proc. Indian Acad. Sci. 87A, 173 (1978).

350. JOSHI, P.C., S. MANDAL, and P.C. DAS: Jayantinin, a Dimeric Coumarin from *Boenninghausenia albiflora*. Phytochemistry 28, 1281 (1989).

351. JU-ICHI, M., M. INOUE, M. IKEGAMI, I. KAJIURA, M. OMURA, and H. FURUKAWA: Constituents of Domestic Citrus Plants. Part VII. New Coumarins from *Citrus funadoko*. Heterocycles 27, 1451 (1988).

352. JU-ICHI, M., M. INOUE, I. KAJIURA, M. OMURA, C. ITO, and H. FURUKAWA: The Structure of Acrimarines, the First Naturally Occurring Acridone-coumarin Dimers. Chem. and Pharm. Bull. 36, 3202 (1988).

353. JU-ICHI, M., M. INOUE, R. TSUDA, N. SHIBUKAWA, and H. FURUKAWA: Junosmarin, a New Khellactone Ester from *Citrus junos* Tanaka. Heterocycles 24, 2777 (1986).

354. JU-ICHI, M., H. KAGA, M. MURAGUCHI, M. INOUE, I. KAJIURA, M. OMURA, and H. FURUKAWA: New Acridone Alkaloid and Coumarin from Citrus Plants. Heterocycles 27, 2197 (1988).

355. JURD, L.: The Synthesis of Coumestrol from a Flavylium Salt. Tetrahedron Letters 1963, 1151.

356. KADYROV, A.S., A.I. SAIDKHODZHAEV, and V.M. MALIKOV: The Structure of Foliferin. Khim. prirod. Soedinenii 1978, 518; Chem. Abs. 90, 69072 (1979).

357. — — —: Structure of the New Coumarin Feshurin. Khim. prirod. Soedinenii 1979, 228; Chem. Abs. 92, 6724 (1980).

358. KAGRAMANOV, A.A., N.F. GASHIMOV, A.Z. ABYSHEV and L.I. ROZHKOVA: 6-Methoxymarmin Acetonide, a New Component of *Haplophyllum pedicellatum*. Khim. prirod. Soedinenii 1979, 88; Chem. Abs. 91, 52699 (1979).

359. KAJIMOTO, T., K. YAHIRO, and T. NOHARA: Sesquiterpenoid and Disulphide Derivatives from *Ferula assafoetida*. Phytochemistry 28, 1761 (1989).

360. KALRA, V.K., A.S. KUKLA, and T.R. SESHADRI: Synthesis of Lucernol and Sativol Dimethyl Ether. Tetrahedron Letters 1967, 2153.

361. KANG, S.H., and C.Y. HONG: Simple Synthetic Routes to Geiparvarin. Tetrahedron Letters 28, 675 (1987).

362. KATHPALIA, Y.P., and S. DUTT: Chemical Examination of a Crystalline Lactone Derived from the Wood of *Dalbergia sissoo*. Indian Soap J. 18, 213 (1953).

363. KAWASAKI, C., T. OKUYAMA, S. SHIBATA, and Y. IITAKA: Studies on Coumarins of a Chinese Drug "Qian-Hu"; VI. Coumarins of *Angelica edulis*. Planta Medica 50, 492 (1984).

364. KAWAZU, K., H. OHIGASHI, and T. MITSUI: The Piscidal Constituents of *Calophyllum inophyllum* Linn. Tetrahedron Letters 1968, 2383.

365. KAWAZU, K., H. OHIGASHI, N. TAKAHISHI, and T. MITSUI: Piscidal Constituents of

*Calophyllum inophyllum.* Bull. Inst. Chem. Res., Kyoto Univ. **50**, 160 (1972); Chem. Abs. **78**, 13744 (1973).

366. KHALID, S.A., and P.G. WATERMAN: Thonningine-A and Thonningine-B: Two 3-Phenylcoumarins from the Seeds of *Millettia thonningii.* Phytochemistry **22**, 1001 (1983).

367. KHAN, N.U., S.W.I. NAGRI, and K. ISHRATULLAH: Wampetin, a Furocoumarin from *Clausena wampi.* Phytochemistry **22**, 2624 (1983).

368. KHASANOV, T.K., V.M. MALIKOV, and S. MELIBAEV: Structure and Configuration of Tavimolidin. Khim. prirod. Soedinenii **1979**, 480; Chem. Abs. **92**, 111176 (1980).

369. KHASANOV, T.K., V.M. MALIKOV, and U. RAKHMANKULOV: Coumarins of *Ferula malacophyla.* Khim. prirod. Soedinenii **1979**, 226; Chem. Abs. **91**, 189775 (1979).

370. KHASTGIR, H.N., P.C. DUTTAGUPTA, and P. SENGUPTA: The Structure of Psoralidin. Tetrahedron **14**, 275 (1961).

371. KHURANA, S.K., V. KRISHNAMOORTHY, V.S. PARMAR, R. SANDUJA, and H.L. CHAWLA: 3,4,7-Trimethylcoumarin from *Trigonella foenumgraecum* Stems. Phytochemistry **21**, 2145 (1982).

372. KINOSHITA, T., T. SAITOH, and S. SHIBATA: A New 3-Arylcoumarin from Licorice Root. Chem. and Pharm. Bull. **26**, 135 (1978).

373. KIRKIACHARIAN, B., and C. MENTZER: Sur une Nouvelle Synthese de l'Ether Methylique de la Daphnoretine. Compt. rend. **260**, 197 (1965).

374. ——: Confirmation, par Synthese Totale, de la Structure de la Daphnoretine. Bull. Soc. Chim. France **1966**, 770.

375. KIR'YANOVA, I.A., and Y.E. SKLYAR: Ferucrin Isobutyrate and Ferucrinone from *Ferula foetidissima.* Khim. prirod. Soedinenii **1984**, 652; Chem. Abs. **102**, 59345 (1985).

376. KOKWARO, J.O., I. MESSANA, C. GALEFFI, M. PATAMIA, and G.B. MARINI BETTOLO: Research on African Medicinal Plants. V. Coumarins from *Zanthoxylum usambarense.* Planta Medica **47**, 251 (1983).

377. KOMATSU, M., I. YOKOE, and Y. SHIRATAKI: Studies on the Constituents of *Sophora* Species. XIV. Constituents of the Root of *Sophora franchetiana* Dunn. (1). Chem. and Pharm. Bull **29**, 532 (1981).

378. ———: Studies on the Constituents of *Sophora* Species. XV. Constituents of the Root of *Sophora franchetiana* Dunn. (2). Chem. and Pharm. Bull. **29**, 2069 (1981).

379. KOMISSARENKO, N.F., A.I. DERKACH, I.P. KOVALEV, and I.F. SATYSPEROVA: Coumarins of the Root of *Heracleum leskovii.* Khim. prirod. Soedinenii **1978**, 184; Chem. Abs. **89**, 103714 (1978).

380. KONG, Y.C., K.H. LAU, Y.Y. TAM, K.F. CHEUNG, P.G. WATERMAN, and R.C. CAMBIE: Dehydroindicolactone, a New Coumarin from *Clausena lansium.* Fitoterapia **54**, 47 (1983).

381. KOUL, S.K., K.L. DHAR, and R.S. THAKUR: A New Coumarin Glucoside from *Prangos pabularia.* Phytochemistry **18**, 1762 (1979).

382. KOZAWA, M., K. BABA, and Y. MATSUYAMA: Comparison of the Coumarin Components of Commercial Crude Drug Tou-Dokkatsu and the Roots of the Related Plants. Shoyakugaku Zasshi **36**, 202 (1982); Chem. Abs. **98**, 77991 (1983).

383. KOZAWA, M., K. BABA, Y. MATSUYAMA, and K. HATA: Studies on Coumarins from the Root of *Angelica pubescens* Maxim. Structures of Various Coumarins Including Angelin, a New Prenylcoumarin. Chem. and Pharm. Bull. **28**, 1782 (1980).

384. KOZAWA, M., K. BABA, K. OKUDA, T. FUKUMOTO, and K. HATA: Studies on Chemical Components of "Bai Zhi" (Supplement 1) On Coumarins from "Japanese Bai Zhi". Shoyakugaku Zasshi **35**, 90 (1981); Chem. Abs. **95**, 209449 (1981).

385. KOZAWA, M., M. FUKUMOTO, Y. MATSUYAMA, and K. BABA: Chemical Studies on the

Constituents of the Chinese Crude Drug "Quiang Huo". Chem. and Pharm. Bull. **31**, 2712 (1983).

*386.* KOZAWA, M., Y. MATSUYAMA, M. FUKUMOTO, and K. BABA: Chemical Studies of *Coelopleurum gmelinni* (D.C.) Ledeb. I. Constituents of the Root. Chem. and Pharm. Bull. **31**, 64 (1983).

*387.* KOZAWA, M., N. MORITA, K. BABA, and K. HATA: Chemical Components of the Roots of *Angelica keiskei* Koidzumi III. The Structure of a New Dihydrofurocoumarin. Yakugaku Zasshi **98**, 636 (1978); Chem. Abs. **89**, 129431 (1978).

*388.* KOZOVSKA V., and A. ZHELEVA: New Furocoumarin, 8-Methoxypeucedanin from Roots of *Peucedanum ruthenicum* M. B. Planta Medica **1980** (Suppl.) 60.

*389.* KUHNKE, J., and F. BOHLMANN: Enantioselective Synthesis of (−)-Lycoserone and Related Compounds. Liebigs Ann. Chem. **1988**, 743.

*390.* KULIEV, Z.A., and T.K. KHASANOV: The Structure of Ferocaulin, Ferocaulinin, Ferocaulidin and Ferocaulicin. Khim. prirod. Soedinenii **1978**, 322; Chem. Abs. **89**, 147082 (1978).

*391.* — —: Structures of Cauferin and Cauferidin. Khim. prirod. Soedinenii **1978**, 327; Chem. Abs. **89**, 163768 (1978).

*392.* KULIEV, Z.A., T.K. KHASANOV, and V.M. MALIKOV: Structure and Configuration of Cauferinin. Khim. prirod. Soedinenii **1979**, 151; Chem. Abs. **92**, 160520 (1980).

*393.* — — —: Terpenoid Coumarin Glycosides of *Ferula conocaula.* Khim. prirod. Soedinenii **1979**, 477; Chem. Abs. **92**, 160520 (1980).

*394.* — — —: Coumarins of *Ferula conocaula.* Khim. prirod. Soedinenii **1982**, 120; Chem. Abs. **96**, 177967 (1982).

*395.* KUMAR, R., B.D. GUPTA, S.K. BANERJEE, and C.K. ATAL: New Coumarins from *Seseli sibiricum.* Phytochemistry **17**, 2111 (1978).

*396.* KUMAR, S., A.B. RAY, C. KONNO, Y. OSHIMA, and H. HIKINO: Cleomiscosin D, a Coumarino-lignan from Seeds of *Cleome viscosa.* Phytochemistry **27**, 636 (1988).

*397.* KUMAR, V: Coumarins from *Murraya* Species. Chem. Sri Lanka **2**, 22 (1985).

*398.* KUMAR, V., J. REISCH, D.B.M. WICKREMARATNE, R.A. HUSSAIN, K.S. ADESINA, and S. BALASUBRAMANIAM: Gleinene and Gleinadiene 5,7-Dimethoxycoumarins from *Murraya gleinei* Root. Phytochemistry **26**, 511 (1987).

*399.* KUPCHAN, S.M., J.G. SWEENY, T. MURAE, M.-S. SHEN, and R.F. BRYAN: Structure of Gnidiacoumarin, a Novel Pentacyclic Dicoumarin from *Gnidia lamprantha.* J. Chem. Soc., Chem. Commun. **1975**, 94.

*400.* KUROYANAGI, M., M. SHIOTSU, T. EBIHARA, H. KAWAI, A. UENO, and S. FUKUSHIMA: Chemical Studies on *Viburnum awabuki* K. KOCH. Chem. and Pharm. Bull. **34**, 4012 (1986).

*401.* KUTNEY, J.P., T. INABA, and D.L. DREYER: Further Studies on Constituents of *Thamnosma montana* Torr. and Frem. The Structure of Thamnosin. A Novel Dimeric Coumarin System. Tetrahedron **26**, 3171 (1970).

*402.* KUZNETSOVA, G.A., E.V. MARKELOVA, M.E. PEREL'SON, E. KLEIN, and S. PAVLOVIC: (−)-3′(R)-Hydroxy-4′(S)-methoxy-3′,4′-dihydroxanthyletin from the Roots of *Peucedanum arenarium.* Phytochemistry **17**, 1805 (1978).

*403.* LAKSHMI, V., D. PRAKASH, R. RAJ, R.S. KAPIL, and S.P. POPLI: Furanocoumarin Lactones from *Clausena anisata.* Phytochemistry **23**, 2629 (1984).

*404.* LAMBERTON, J.A., and T.C. MORTON: The Epoxy Lactone Ring of the Coumarin Micromelin: Formation of a New Epimer. Austral. J. Chem. **38**, 1025 (1985).

*405.* LAMNAOUER, D., B. BODO, M.-T. MARTIN, and D. MOLHO: Ferulenol and ω-Hydroxyferulenol, Toxic Coumarins from *Ferula communis* var. *genuina.* Phytochemistry **26**, 1613 (1987).

406. LEMMICH, E., and L. GYLLE: A Dihydrofuranocoumarin from *Peucedanum oreoselinum*. Phytochemistry **27**, 3688 (1988).
407. LEMMICH, J., and S. HAVELUND: Coumarin Glycosides of *Seseli montanum*. Phytochemistry **17**, 139 (1978).
408. LEMMICH, J., S. HAVELUND, and O. THASTRUP: Dihydrofurocoumarin Glucosides from *Angelica archangelica* and *Angelica silvestris*. Phytochemistry **22**, 553 (1983).
409. LEMMICH, J., and M. SHABANA: Coumarin Sulphates of *Seseli libanotis*. Phytochemistry **23**, 863 (1984).
410. LE-VAN, N.: Coumestrin, a Coumestan Derivative from Soybean Roots. Phytochemistry **23**, 1204 (1984).
411. LIN, L.J., and G.A. CORDELL: Synthesis of Coumarinolignans Through Chemical and Enzymic Oxidation. J. Chem. Soc., Chem. Commun. **1984**, 160.
412. — —: Application of the SINEPT Pulse Programme in the Structure Elucidation of Coumarinolignans. J. Chem. Soc., Chem. Commun. **1986**, 377.
413. — —: Synthesis and Structure Elucidation of Coumarinolignans. J. Chem. Res. Synop. **1988**, 396.
414. LIVINGSTON, A.L., E.M. BICKOFF, R.E. LUNDIN, and L. JURD: Trifoliol, a New Coumestan from Ladino Clover. Tetrahedron **20**, 1963 (1964).
415. LIVINGSTON, A.L., S.C. WITT, R.E. LUNDIN, and E.M. BICKOFF: Medicagol, a New Coumestan from Alfalfa. J. Org. Chem. **30**, 2353 (1965).
416. MAGALHÃES, A.F., and R.T.S. FRIGHETTO: Synthesis of 7-(2,3-Epoxy-3-methyl-butyloxy)-5,6-(methylenedioxy)coumarin and Derivatives. Quim. Nova **6**, 165 (1983).
417. MAGALHÃES, A.F., E.G. MAGALHÃES, H.F. LEITÃO FILHO, R.T.S. FRIGHETTO, and S.M.G. BARROS: Coumarins from *Pterocaulon balansae* and *P. lanatum*. Phytochemistry **20**, 1369 (1981).
418. MAHANDRU, M.M., and V.K. RAVINDRAN: Surangin C, a Coumarin from *Mammea longifolia*. Phytochemistry **25**, 555 (1986).
419. MAHMOUD, Z.F., and N.A. ABDEL SALAM: Coumarins of *Euphorbia terracina* and *Euphorbia paralias* Growing in Egypt. Pharmazie **34**, 446, (1979).
420. MAHMOUD, Z.F., T.M. SARG, M.E. AMER, S.M. KHAFAGY, and F. BOHLMANN: A 5-Methylcoumarin Glucoside from *Ethulia conyzoides*. Phytochemistry **19**, 2029 (1980).
421. MAJUMBER, P.L., G.C. SENGUPTA, B.N. DINDA, and A. CHATTERJEE: Edgeworthin, a New *bis*-Coumarin from *Edgeworthia gardneri*. Phytochemistry **13**, 1929 (1974).
422. MALI, R.S., S.G. TILVE, K.S. PATIL, and G. NAGARAJAM: A Useful Synthesis of 3-Ethyl- and 3,4-Dimethyl-coumarins: Synthesis of Trigoforin, a Coumarin from *Trigonella foenumgraecum*. Indian J. Chem. **24B**, 1271 (1985).
423. MALI, R.S., V.J. YADAEV, and R.N. ZAWARE: An Improved Synthesis of Naturally Occurring Coumarins. Synthesis of Herniarin, Scoparone and 7-Methoxy-6-methylcoumarin. Indian J. Chem. **21B**, 759 (1982).
424. MANANDHAR, M.D.: 8-Substituted 7-Methoxycoumarins from *Murraya exotica* Linn. Indian J. Chem. **19B**, 1006 (1980).
425. MASSANET, G.M., E. PANDO, F.R. LUIS, and J. SALVÁ: Synthesis of 3-(1',1'-Dimethylallyl)coumarins: Gravelliferone, Chalepin and Rutamarin. Heterocycles **26**, 1541 (1987).
426. MATA, R., F. CALZADA, and M. DEL ROSARIO GARCIA: Chemical Studies on Mexican Plants Used in Traditional Medicine, VI. Additional New 4-Phenylcoumarins from *Exostema caribaeum*. J. Nat. Prod. **51**, 851 (1988).
427. MATA, R., F. CALZADA, ROSARIO GARCIA, and TERESA REGUERO: Chemical Studies on Mexican Plants Used in Traditional Medicine, III: New 4-Phenylcoumarins from *Exostema caribaeum*. J. Nat. Prod. **50**, 866 (1987).

428. MATANO, Y., T. OKUYAMA, S. SHIBATA, M. HOSON, T. KAWADA, H. OSADA, and T. NOGUCHI: Studies on Coumarins of a Chinese Drug "Qian- Hu"; VII. Structures of New Coumarin-glycosides of Zi-Hua Qian-Hu and Effect of Coumarin-glycosides on Human Platelet Aggregation. Planta Medica **1986**, 135.

429. MATKARIMOV, A.D., E.K. BATIROV, V.M. MALIKOV, and E. SEITMURATOV: Obtusinin, a New Coumarin from *Haplophyllum obtusifolium*. Khim. prirod. Soedinenii **1980**, 328; Chem. Abs. **93**, 164316 (1980).

430. ————: Coumarins of *Haplophyllum obtusifolium*. Khim. prirod. Soedinenii **1980**, 565; Chem. Abs. **93**, 200982 (1980).

431. ————: Obtusoside, New Coumarin Glycoside from Aerial Parts of *Haplophyllum obtusifolium*. Khim. prirod. Soedinenii **1980**, 831; Chem. Abs. **94**, 171049 (1981).

432. ————: Structure of the New Coumarin Obtusiprenol. Khim. prirod. Soedinenii **1981**, 795; Chem. Abs. **96**, 214258 (1982).

433. ————: A Study of the Coumarins of *Haplophyllum obtusifolium*. The Structure of Obtusidin and Obtusiprenin. Khim. prirod. Soedinenii **1982**, 173; Chem. Abs. **97**, 123923 (1982).

434. MCHALE, D., P.P. KHOPKAR, and J.B. SHERIDAN: Coumarin Glycosides from *Citrus* Flavedo. Phytochemistry **26**, 2547 (1987).

435. MÉNDEZ, J., and J. CASTRO-PECEIRO: Furocoumarins from *Angelica pachycarpa*. Phytochemistry **22**, 2599 (1983).

436. MENG, Q., and W. CHEN: Pygmaeoherin: a Novel Coumarin from *Pygmaeopremna herbacea*. Planta Medica **54**, 48 (1988).

437. MENZEL, D., R. KAZLAUSKAS, and J. REICHELT: Coumarins in the Siphonalean Green Algal Family Dasycladaceae Kuetzing (Chlorophyceae). Botanica Marina **26**, 23 (1983).

438. MEYER, B.N., M.E. WALL, M.C. WANI, and H.L. TAYLOR: Plant Anti-tumor Agents, 21. Flavones, Coumarins, and an Alkaloid. J. Nat. Prod. **48**, 952 (1985).

439. MIYAKADO, M., N. OHNO, H. YOSHIOKA, and T.J. MABRY: Trichoclin, a New Furocoumarin from *Trichocline incana*. Phytochemistry **17**, 143 (1978).

440. MIYAZAKI, T., and S. MIHASHI: Studies on the Constituents of *Boenninghausenia albiflora* Meissner var. *japonica* S. Suzuki I. Structure of Matsukaze-lactone. (1). Chem. and Pharm. Bull. **12**, 1232 (1964).

441. MIYAZAKI, T., S. MIHASHI, and K. OKABAYASHI: Studies on the Constituents of *Boenninghausenia albiflora* Meissner var. *japonica* S. Suzuki. II. Structure of Matsukaze-lactone (2). Chem. and Pharm. Bull. **12**, 1236 (1964).

442. MUJUMDAR, R.B., S.S. RATHI, and A.V. RAMA RAO: Heartwood Constituents of *Chloroxylon swietenia* DC. Indian J. Chem. **15B**, 200 (1977).

443. MUKAIYAMA, T., and N. IWASAWA: The Asymmetric Diels-Alder Reactions of α,β-Unsaturated Carboxylic Amides Derived from (−)-Phenylglycinol and the Asymmetric Total Synthesis of (+)-Farnesiferol C. Chem. Letters **1981**, 29.

444. MUKERJEE, S.K., T. SAROJA, and T.R. SESHADRI: The Structure of Exostemin and Synthesis of Some Related 4-Phenyl Coumarins. Tetrahedron **24**, 6527 (1968).

445. ————: Partial Methylation of 4-Phenylcoumarins and a Synthesis of Melannein. Indian J. Chem. **7**, 844 (1969).

446. ————: Dalbergiachromene. A New Neoflavonoid from Stem-bark and Heartwood of *Dalbergia sissoo*. Tetrahedron **27**, 799 (1971).

447. MURAKAMI, T., N. TANAKA, T. SATAKE, Y. SAIKI, and C.-M. CHEN: Chemical and Chemotaxonomical Studies of Filices. 57. Constituents of *Colysis hemionitidea* (Wall) Presi. and *Microsorium fortunei* (Moore) Ching. Yakugaku Zasshi **105**, 655 (1985); Chem. Abs. **103**, 157319 (1985).

448. MURRAY, R.D.H.: Naturally Occurring Plant Coumarins. Fortschr. Chem. Org. Naturstoffe 35, 199 (1978).

449. MURRAY, R.D.H., M.M. BALLANTYNE, and K.P. MATHAI: Claisen Rearrangements-III. Convenient Syntheses of the Coumarins, Osthenol, Demethylsuberosin and Coumurrayin. Tetrahedron 27, 1247 (1971).

450. MURRAY, R.D.H., and D.A. HALL: Structure Revision of the Coumarin, Ceylantin. Phytochemistry 24, 2465 (1985).

451. MURRAY, R.D.H., and Z.D. JORGE: Claisen Rearrangements-XI. Synthesis of the Coumarins, Obliquetol, Obliquetin and Nieshoutin. Tetrahedron 39, 3163 (1983).

452. — —: Claisen Rearrangements-XII. Synthesis of the Coumarins, 5-Methoxyseselin, Trachyphyllin, Coumurrayin and Xanthoxyletin. Tetrahedron 40, 3129 (1984).

453. — —: Claisen Rearrangements-XIII. Synthesis of the Natural Coumarins, Nordentatin, Dentatin and Clausarin. Tetrahedron 40, 3133 (1984).

454. — —: Claisen Rearrangements-XV. Structure Revision of the Coumarin, Celerin, by Synthesis. Tetrahedron 40, 5229 (1984).

455. MURRAY, R.D.H., Z.D. JORGE, and D.M. BOAG: Claisen Rearrangements-XIV. Synthesis of the Coumarin, Benahorin and Revision of the Structure of Marmelide. Tetrahedron 40, 5225 (1984).

456. MURRAY, R.D.H., Z.D. JORGE, and K.W.M. LAWRIE: Claisen Rearrangements-X. Synthesis of the Coumarins, Hortiolone and Hortinone. Tetrahedron 39, 3159 (1983).

457. MURRAY, R.D.H., and K.W.M. LAWRIE: Claisen Rearrangements-IX. Synthesis of the Coumarin, Furopinnarin. Tetrahedron 35, 697 (1979).

458. MURRAY, R.D.H., J. MÉNDEZ, and S.A. BROWN: The Natural Coumarins: Occurrence, Chemistry and Biochemistry. Chichester: Wiley, 1982.

459. MURRAY, R.D.H., and M. STEFANOVÍC: 6-Methoxy-7,8-methylenedioxycoumarin from Artemisia dracunculoides and Artemisia vulgaris. J. Nat. Prod. 49, 550 (1986).

460. MURRAY, R.D.H., and S. ZEGHDI: Synthesis of the Natural Coumarins, Murraol (CM-$c_2$), trans-Dehydroosthol and Swietenocoumarin G. Phytochemistry 28, 227 (1989).

461. MURTI, V.V.S., P.S.S. KUMAR, and T.R. SESHADRI: Structure of Ponnalide. Indian J. Chem. 10, 255 (1972).

462. NABIEV, A.A., T.K. KHASANOV, and V.M. MALIKOV: New Terpenoid Coumarins of Ferula kopetdaghensis. Khim. prirod. Soedinenii 1978, 516; Chem. Abs. 89, 193845 (1978).

463. — — —: A Chemical Study of the Roots of Ferula kopetdaghensis. Khim. prirod. Soedinenii 1989, 17; Chem. Abs. 91, 105162 (1979).

464. — — —: A New Terpenoid Coumarin from Ferula kopetdaghensis. Khim. prirod. Soedinenii 1982, 48; Chem. Abs. 96, 214275 (1982).

465. — — —: Terpenoid Coumarins of Ferula kokanica. Khim. prirod. Soedinenii 1982, 578; Chem. Abs. 98, 86223 (1983).

466. NABIEV, A.A., T.K. KHASANOV, and S. MELIBAEV: Coumarins of Ferula diversivittata. Khim. prirod. Soedinenii 1978, 517; Chem. Abs. 89, 211931 (1978).

467. NABIEV, A.A., and V.M. MALIKOV: Microlobin, a New Coumarin from Ferula microloba. Khim. prirod. Soedinenii 1983, 700; Chem. Abs. 100, 171543 (1984).

468. — —: Microlobidene, a Terpenoid Coumarin from Ferula microloba. Khim. prirod. Soedinenii 1983, 781; Chem. Abs. 100, 171551 (1983).

469. NABIEV, A.A., V.M. MALIKOV, and T.K. KHASANOV: Karatavicin, a New Coumarin from Ferula karatavica. Khim. prirod. Soedinenii 1983, 526; Chem. Abs. 100, 3489 (1984).

470. NAGARAJAN, G.R., U. RANI, and V.S. PARMAR: Coumarins from Fraxinus floribunda Leaves. Phytochemistry 19, 2494 (1980).

471. NAGARAJAN, G.R., U. RANI, and V.S. PARMAR: Floribin, New Coumarin from *Fraxinus floribunda* Wall. Pharmazie **38**, 72 (1983).

472. NAGEM, T.J., and M. DE A. E SILVA: Xanthones and Phenyl-coumarins from *Kielmeyera pumila*. Phytochemistry **27**, 2961 (1988).

473. NAGUMO, S., K. IMAMURA, T. INOUE, and M. NAGAI: Cyanogenic Glycosides and 4-Hydroxycoumarin Glycosides from *Gerbera jamesonii hybrida*. Chem. and Pharm. Bull. **33**, 4803 (1985).

474. NAGUMO, S., T. TOYONAGA, T. INOUE, and M. NAGAI: New Glucosides of a 4-Hydroxy-5-methylcoumarin and a Dihydropyrone from *Gerbera jamesonii hybrida*. Chem. and Pharm. Bull. **37**, 2621 (1989).

475. NAIR, A.G.R.: Cleosandrin, a Novel 7-Phenoxycoumarin from Seeds of *Cleome icosandra*. Indian J. Chem. **17B**, 438 (1979).

476. NARANTUYAA, S., D. BATSURÉN, E.K. BATIROV, and V.M. MALIKOV: A Chemical Study of the Mongolian Flora. Lariside, a New Scopoletin Glycoside from *Salsola laricifolia*. Khim. prirod. Soedinenii **1986**, 288; Chem. Abs. **105**, 168910 (1986).

477. NASIROV, S.M., A.I. SAIDKHODZHAEV, T.K. KHASANOV, M.R. YAGUDAEV, and V.M. MALIKOV: Stereochemistry of Terpenoid Coumarins. Crystal and Molecular Structure of Samarcandin. Khim. prirod. Soedinenii **1985**, 184; Chem. Abs. **103**, 68269 (1985).

478. NGADJUI, B.T., J.F. AYAFOR, B.L. SONDENGAM, and J.D. CONNOLLY: Prenylated Coumarins from the leaves of *Clausena anisata*. J. Nat. Prod. **52**, 243 (1989).

479. —————: Coumarins from *Clausena anisata*. Phytochemistry **28**, 585 (1989).

480. NIGAM, S.K., C.R. MITRA, G. KUNESCH, B.C. DAS, and J. POLONSKY: Constituents of *Calophyllum tomentosum* and *Calophyllum apetalum* Nuts: Structure of a New 4-Alkyl- and Two New 4-Phenylcoumarins. Tetrahedron Letters **1967**, 2633.

481. NISHIZAWA, M., H. TAKENAKA, and Y. HAYASHI: Total Synthesis of (±)-Karatavic Acid: Structure Confirmation of the First Seco-drimane Type Sesquiterpenoid. Tetrahedron Letters **25**, 437 (1984).

482. NORTON, L.B., and R. HANSBERRY: Constituents of the Insecticidal Resin of the Yam Beam (*Pachyrrhizus erosus*). J. Amer. Chem. Soc. **67**, 1609 (1945).

483. NOZAWA, K., H. SEYEA, S. NAKAJIMA, S. UDAGAWA, and K. KAWAI: Studies on Fungal Products. Part 10. Isolation and Structures of Novel Bicoumarins, Desertorins A, B, and C, from *Emericella desertorum*. J. Chem. Soc. Perkin 1 **1987**, 1735.

484. NYIREDY, S., C.A.J. ERDELMEIER, K. DALLENBACK-TOELKE, K. NYIREDY-MIKITA, and O. STICHER: Preparative On-line Overpressure Layer Chromatography (OPLC): a New Separation Technique for Natural Products. J. Nat. Prod. **49**, 885 (1986).

485. OHSHIMA, Y., T. OKUYAMA, K. TAKAHASHI, T. TAKIZAWA, and S. SHIBATA: Isolation and High Performance Liquid Chromatography (HPLC) of Isoflavonoids from the *Pueraria* Root. Planta Medica **54**, 250 (1988).

486. OKOGUN, J.I., V.U. ENYENIHI, and D.E.U. EKONG: Spectral Studies on Coumarins and the Determination of the Constitution of Ekersenin by Total Synthesis. Tetrahedron **34**, 1221 (1978).

487. OKUYAMA, T., and S. SHIBATA: Studies on Coumarins of a Chinese Drug "Qian-Hu". Planta Medica **42**, 89 (1981).

488. OKUYAMA, T., M. TAKATA, and S. SHIBATA: Studies on Coumarins of a Chinese Drug Qian-Hu, IX: Structures of Linear Furano- and Simple-Coumarin Glycosides of Bai-Hua Qian-Hu. Planta Medica **55**, 64 (1989).

489. OLLIS, W.D., B.T. REDMAN, R.J. ROBERTS, I.O. SUTHERLAND, O.R. GOTTLIEB, and M.T. MAGALHÃES: Neoflavonoids and the Cinnamylphenol Kuhlmannistyrene from *Machaerium kuhlmannii* and *M. nictitans*. Phytochemistry **17**, 1383 (1978).

490. O'NEILL, M.J.: Aureol and Phaseol, Two New Coumestans from *Phaseolus aureus* Roxb. Z. Naturforsch. **38C**, 698 (1983).

491. O'NEILL, M.J., S.A. ADESANYA, and M. F. ROBERTS: Isosojagol, a Coumestan from *Phaseolus coccineus*. Phytochemistry **23**, 2704 (1984).

492. ORMANCEY-POTIER, A., A. BUZAS, and E. LEDERER: Sur le calophyllolide et l'acide calophyllique isolés des graines de *Calophyllum inophyllum*. Bull. Soc. Chim. France **1951**, 577.

493. PAKNIKAR, S.K., and J. VEERAVALLI: Structure of Karatavic Acid: The First Representative of the 3,4-Seco-drimane Skeleton. Chem. and Ind. **1978**, 431.

494. PAN, J.-X., K. LAMY, B. ARISON, J. SMITH, and G.-Q. HAN: Isolation and Identification of Isoangelol, Anpubesol and Other Coumarins from *Angelica pubescens* Maxim. Acta Pharm. Sinica **22**, 380 (1987).

495. PANETTA, J.A., and H. RAPOPORT: New Syntheses of Coumarins. J. Org. Chem. **47**, 946 (1982).

496. PARMAR, V.S., H.N. JHA, S.K. SANDUJA, and R. SANDUJA: Trigocoumarin, a New Coumarin from *Trigonella foenum-graecum*. Z. Naturforsch. **37B**, 521 (1982).

497. PARMAR, V.S., J.S. RATHORE, S. SINGH, A.K. JAIN, and S.R. GUPTA: Troupin, a 4-Methylcoumarin from *Tamarix troupii*. Phytochemistry **24**, 871 (1985).

498. PARMAR, V.S., S. SINGH, and J.S. RATHORE: A Structure Revision of Trigocoumarin. J. Chem. Res. Synop. **1984**, 378.

499. PARTHASARATHY, M.R., and K.P. SARADHI: Synthesis of Coumarinolignans. Indian J. Chem. **23B**, 1105 (1984).

500. — —: A Coumarino-lignan from *Jatropha glandulifera*. Phytochemistry **23**, 867 (1984).

501. PELTER, A., and A.P. JOHNSON: The Structures of Scandenin and Lonchocarpic Acid. Tetrahedron Letters **1964**, 2817.

502. PEREL'SON, M.E., V.I. SHEICHENKO, Y.E. SKLYAR, and V.B. ANDRIANOVA: Configuration of Kellerin and Samarcandin. Khim.-Farm. Zhur. **11**, 33 (1977); Chem. Abs. **88**, 23159 (1978).

503. PEREL'SON, M.E., Y.E. SKLYAR, N.V. VESELOVSKAYA, and M.G. PIMENOV: Ferukrin, a New Terpenoid from *Ferula krylovii*. Khim.-Farm. Zhur. **11**, 78 (1977); Chem. Abs. **87**, 18965 (1977).

504. PEREL'SON, M.E., A.I. SOKOLOVA, and Y.E. SKLYAR: The Structure of Feterin, a New Terpenoid Coumarin from *Ferula teterrima*. Khim. prirod. Soedinenii **1978**, 318; Chem. Abs. **89**, 147081 (1978).

505. PEREL'SON, M.E., V.V. VANDYSHEV, Y.E. SKLYAR, K. VEZHKHOVSKA-RENKE, N.N. VESELOVSKAYA, and M.G. PIMENOV: New Terpenoid Coumarins from *Ferula tadshikorum*. Khim. prirod. Soedinenii **1976**, 593; Chem. Abs. **86**, 86159 (1977).

506. PHADKE, C.P., S.L. KELKAR, and M.S. WADIA: A General Synthesis of Bis-(Coumarinyl) Ethers: Synthesis of Daphnoretin Methyl Ether. Synthesis **1986**, 413.

507. PHILIPS, S., and B.A. BURKE: *Amyris* of Jamaica. New Coumarins from *Amyris elemifera* L. (Rutaceae). Heterocycles **23**, 1921 (1985).

508. PINAR, M.: Cumarinas de la *Magydaris panacifolia, Cachrys sicula, Cachrys libanotis y Lafuenta rotundifolia*. Anales de Quím. **73**, 599 (1977).

509. PINAR, M., and M.P. GALAN: Coumarins from *Eryngium ilicifolium*. J. Nat. Prod. **48**, 853 (1985).

510. PINAR, M., and B. RODRÍGUEZ: A New Coumarin from *Ferula loscosii* and the Correct Structure of Coladonin. Phytochemistry **16**, 1987 (1977).

511. POLONSKY, J: Structure chimique du calophyllolide, de l'inophyllolide et de l'acide

calophyllique, constituants des noix de *Calophyllum inophyllum*. Bull Soc. Chim. France **1957**, 1079.

*512.* POLONSKY, J., and R. TOUBIANA: Sur les constituants des noix de *Calophyllum inophyllum*: isolement d'une nouvelle lactone, l'inophyllolide. Compt. rend. **242**, 2877 (1956).

*513.* PRAKASH, D., K. RAJ, R.S. KAPIL, and S.P. POPLI: Coumarins from *Clausena indica*. Phytochemistry **17**, 1194 (1978).

*514.* PRASAD, A.V.K., R.S. KAPIL, and S.P. POPLI: Studies of Pterocarponoids: Anhydrotuberosin, 3-*O*-Methylanhydrotuberosin and Tuberostan from *Pueraria tuberosa*. Indian J. Chem. **24B**, 236 (1985).

*515.* PURUSHOTHAMAN, K.K., A. SARADA, A. SARASWARTHY, K. VANAJA, and A. RAJENDRAN: Structure of Racemol, a Rare Coumarin from *Atalantia racemosa* W and A. Indian Drugs **23**, 487 (1986); Chem. Abs. **105**, 168851 (1986).

*516.* RAIZADA, M.B., S.K. GARG, and S.R. GUPTA: Synthesis of Apigravin. Indian J. Chem. **20B**, 918 (1981).

*517.* RAJAGOPALAN, P., and A.I. KOSAK: The Synthesis of Pachyrrhizin. Tetrahedron Letters No. **21**, 5 (1959).

*518.* RAJANI, P., and P.N. SHARMA: A Coumestone from the Roots of *Tephrosia hamiltonii*. Phytochemistry **27**, 648 (1988).

*519.* RAMAKRISHNA, K.V., R.A. KHAN, and R.S. KAPIL: New Isoflavone and Coumestan from *Pueraria tuberosa*. Indian J. Chem. **27B**, 285 (1988).

*520.* RAMA RAO, A.V., K.S. BHIDE, and R.B. MUJUMDAR: Phenolics from the Bark of *Chloroxylon swietenia* DC.: Part II – Isolation of Swietenocoumarins G, H and I. Indian J. Chem. **19B**, 1046 (1980).

*521.* RAO, P.P., and G. SRIMANNARAYANA: Tephrosol, a New Coumestone from the Roots of *Tephrosia villosa*. Phytochemistry **19**, 1272 (1980).

*522.* RAO, P.S., Y. ASHEERVADAM, M. KHALEELULLAH, N. SUBBA RAO, and R.D.H. MURRAY: Hymexelsin, an Apiose-containing Scopoletin Glycoside from the Stem Bark of *Hymenodictyon excelsum*. J. Nat. Prod. **51**, 959 (1988).

*523.* RAY, A.B., S.K. CHATTOPADHYAY, C. KONNO, and H. HIKINO: Structure of Cleomiscosin A, a Coumarino-lignoid of *Cleome viscosa* Seeds. Tetrahedron Letters **21**, 4477 (1980).

*524.* —————: Structure of Cleomiscosin B, a Coumarino-lignoid of *Cleome viscosa* Seeds. Heterocycles **19**, 19 (1982).

*525.* RAY, A.B., S.K. CHATTOPADHYAY, S. KUMAR, C. KONNO, Y. KISO, and H. HIKINO: Structures of Cleomiscosins, Coumarino-lignoids of *Cleome viscosa* Seeds. Tetrahedron **41**, 209 (1985).

*526.* REHER, G., and L. KRAUS: New Neoflavonoids from *Coutarea latifolia*. J. Nat. Prod. **47**, 172 (1984).

*527.* REHER, G., L. KRAUS, V. SINNWELL, and W.A. KÖNIG: A Neoflavonoid from *Coutarea hexandra* (Rubiaceae). Phytochemistry **22**, 1524 (1983).

*528.* REISCH, J., and A. BATHE: Naturstoffchemie, 118. Synthese der Cumarine 6- und 8-Naphthoherniarin, Dehydrogeijerin und Murraol. Liebigs Ann. Chem. **1988**, 543.

*529.* REISCH, J., and I. MESTER: Psoralen und Angelicin durch Umsetzung von 7-Hydroxycoumarin mit 4-Chlor-1,3-dioxolan-2-on. Chem. Ber. **112**, 1491 (1979).

*530.* REISCH, J., I. NOVÁK, K. SZENDREI, and E. MINKER: Über weitere $C_3$-substituierte Cumarin-Derivate aus *Ruta graveolens*: Daphnoretin und Daphnoretin-methylather. Planta Medica **16**, 372 (1968).

*531.* REISCH, J., and E. PODPETSCHNIG: Contribution to the Search for New Natural

Products by Means of Synthetic Authentic Samples: *exo*-Dehydrochalepin from the Roots of *Ruta graveolens*. Sci. Pharm. **56**, 171 (1988).

532. REISCH, J., and H. STROBEL: New Constituents from *Toddalia aculeata* Pers., Rutaceae. Pharmazie **37**, 862 (1982).

533. REISCH, J., K. SZENDREI, E. MINKER, and I. NOVÁK: Natural Substances. XXI. Coumarins from *Ruta graveolens* Roots. Xanthyletin and Byakangelicin. Planta Medica **17**, 116 (1969).

534. RIZZACASA, M.A., and M.V. SARGENT: Synthesis of Desertorin C, a Bicoumarin from the Fungus *Emericella desertorum*. J. Chem. Soc. Perkin 1 **1988**, 2425.

535. RODIGHIERO, P., A. GUIOTTO, G. PASTORINI, P. MANZINI, F. DALL'ACQUA, G. INNOCENTI, and G. CAPORALE: Synthesis of (±)-5-Hydroxymarmesin: Biogenetic Precursor of the Skin Photosensitising Agent Bergapten. Gazzetta **110**, 167 (1980).

536. RODIGHIERO, P., P. MANZINI, G. PASTORINI, and A. GUIOTTO: New Synthesis of (±)-5-Hydroxymarmesin, Biogenetic Precursor of the Skin Photosensitizing Agent Bergapten (5-MOP). J. Heterocyclic Chem. **18**, 447 (1981).

537. RÓZSA, Z., I. MESTER, J. REISCH, and K. SZENDREI: Naphthoherniarin, an Unusual Coumarin Derivative from *Ruta graveolens*. Planta Medica **39**, 219 (1980).

538. — — — —: Natural Product Chemistry. 101. Naphthoherniarin: an Unusual Coumarin Derivative from *Ruta graveolens*. Planta Medica **55**, 68 (1989).

539. RUSTAIYAN, A., L. NAZARIANS and F. BOHLMANN: Two New 5-Methylcoumarins from *Erlangea fusca*. Phytochemistry **19**, 1254 (1980).

540. SADAVONGVIVAD, C., and P. SUPAVILAI: Three Monohydroxycoumarins from *Alyxia lucida*. Phytochemistry **16**, 1451 (1977).

541. SAGITDINOVA, G.V., A.I. SAIDKHODZHAEV, and V.M. MALIKOV: Structure and Stereochemistry of the Coumarins of *Ferula lehmanni*. Khim. prirod. Soedinenii **1983**, 709; Chem. Abs. **100**, 171545 (1984).

542. SAIDKHODZHAEV, A.I., A.S. KADYROV, and V.M. MALIKOV: Stereochemistry of Feshurin, Nevskin and Colladocin. Khim. prirod. Soedinenii **1979**, 308; Chem. Abs. **92**, 76696 (1980).

543. SAIDKHODZHAEV, A.I., A.Y. KUSHMURADOV, A.S. KADYROV, and V.M. MALIKOV: Fepaldin, Terpenoid Coumarin from *Ferula pallida*. Khim. prirod. Soedinenii **1980**, 716; Chem. Abs. **94**, 117789 (1981).

544. SAIDKHODZHAEV, A.I., and V.M. MALIKOV: Stereochemistry of Terpenoid Coumarins. Khim. prirod. Soedinenii **1978**, 707; Chem. Abs. **19**, 57212 (1979).

545. — —: The Stereochemistry of Feropolol, Feropolin, Feropolone and Feropolidin. Khim. prirod. Soedinenii **1978**, 799; Chem. Abs. **90**, 200277 (1979).

546. SAIKI, Y., K. MORINAGA, O. OKEGAWA, S. SAKAI, Y. AMAYA, A. UENO, and S. FUKUSHIMA. Coumarins of the Roots of *Angelica dahurica*. Yakugaku Zasshi **91**, 1313 (1971).

547. SAIMOTO, H., T. HIYAMA, and H. NOZAKI: Regiocontrolled Formation of 4,5-Dihydro-3(2*H*)-furanones from 2-Butyne-1,4-diol Derivatives. Synthesis of Bullatenone and Geiparvarin. Bull. Chem. Soc. Japan **56**, 3078 (1983).

548. SAITOH, T., and S. SHIBATA: Chemical Studies on the Oriental Plant Drugs. XXII. Some New Constituents of Licorice Root. (2). Glycyrol, 5-*O*-Methylglycyrol and Isoglycyrol. Chem. and Pharm. Bull. **17**, 729 (1969).

549. SAKAI, T., H. ITO, A. YAMAWAKI, and A. TAKEDA: A Convenient Synthesis of Geiparvarin. Tetrahedron Letters **25**, 2987 (1984).

550. SAKAKIBARA, I., T. OKUYAMA, and S. SHIBATA: Studies on Coumarins of a Chinese Drug "Qian-Hu". III. Coumarins from "Zi-Hua Qian-Hu". Planta Medica **44**, 199 (1982).

*551.* SAKAKIBARA, I., T. OKUYAMA, and S. SHIBATA: Studies on Coumarins of a Chinese Drug "Qian-Hu", IV. Coumarins from "Zi-Hua Qian-Hu" (Supplement). Planta Medica **50**, 117 (1984).

*552.* SÁNCHEZ-VIESCA, F.: Structure of Exostemin, a New 4-Phenylcoumarin Isolated from *Exostemma caribaeum*. Phytochemistry **8**, 1821 (1969).

*553.* SÁNCHEZ-VIESCA, F., E. DÍAZ, and G. CHÁVEZ: Chemistry of a New Component from *Exostemma caribaeum*. Ciencia **25**, 135 (1967).

*554.* SASAKI, H., H. TAGUCHI, T. ENDO, and I. YOSIOKA: The Constituents of *Glehnia littoralis* Fr. Schmidt et Miq. Structure of a New Coumarin Glycoside, Osthenol-7-*O*-β-gentiobioside. Chem. and Pharm. Bull. **28**, 1847 (1980).

*555.* SATI, S.P., D.C. CHAUKIYAL, O.P. SATI, F. YAMADA, and M. ONO: 2D NMR Structure Elucidation of a New Coumarin Glycoside from *Xeromphis spinosa*. J. Nat. Prod. **52**, 376 (1989).

*556.* SATYANARAYANA, P., P. SUBRAHAMANYAM, R. KASAI, and O. TANAKA: An Apiose-containing Coumarin Glycoside from *Gmelina arborea* Root. Phytochemistry **24**, 1862 (1985).

*557.* SAXENA, V.K., K.P. TIWARI, and S.P. TANDON: A New 4-Phenylcoumarin, Sisafolin, from *Dalbergia latifolia*. Proc. Nat. Acad. Sci. India **40A**, 165 (1970).

*558.* SCHNEIDERS, G.E., and R. STEVENSON: Synthesis of the Natural Furanocoumarin, Prangosine. J. Chem. Res. Synop. **1982**, 182.

*559.* SCHREIBER, F.G., and R. STEVENSON: Synthesis of the Angular Furocoumarin Oroselone and the Linear Analogue 2'-Isopropenylpsoralene. J. Chem. Res. Synop. **1978**, 92.

*560.* SENGUPTA, P., M. SEN, P. KARURI, E. WENKERT, and T.D.J. HALLS: Structure of Gerberinol. Novel Dimethyldicoumarol from *Gerbera lanuginosa* Benth. J. Indian Chem. Soc. **62**, 916 (1985).

*561.* SENGUPTA, S., and S.C. DAS: Structure of Lasioerin: a Novel Coumarin from *Lasiosiphon eriocephalus* Decne (Thymelaceaceae). Chem. and Ind. **1978**, 954.

*562.* SERKEROV, S.V., and N.F. MIR-BABAEV: A New Terpenoid Coumarin trans-Diversin from *Ferula litwinowiana*. Khim. prirod. Soedinenii **1987**, 360; Chem. Abs. **107**, 233106 (1987).

*563.* SESHADRI, T.R., and VISHWAPAUL: Recent Advances in Naturally Occurring Coumarins. J. Sci. Ind. Res. **32**, 227 (1973).

*564.* SHAH, V.R., G.D. SHAH, and R.C. SHAH: New Synthesis of 4-Hydroxycoumarins. J. Org. Chem. **25**, 677 (1960).

*565.* SHAMSUDDIN, T., W. RAHMAN, S.A. KHAN, K.M. SHAMSUDDIN, and J.P. KINTZINGER: Moluccanin, a Coumarino-lignoid from *Aleurites moluccana*. Phytochemistry **27**, 1908 (1988).

*566.* SHANG, T.-M., L. WANG, H.-T. LIANG, X.-Y. LIN, and S.L. NIU: Studies on the Constituents of Mao-Yan-Cao (*Euphorbia lunulata* Bge.). Hua Hsüeh Hsüeh Pao **37**, 119 (1979).

*567.* SHARMA, B.R., R.K. RATTAN and P. SHARMA: Anhydrorutaretin, a New Furano-coumarin and Other Minor Constituents of *Apium leptophyllum* Seeds. Phytochemistry **19**, 1556 (1980).

*568.* SHARMA, B.R., and P. SHARMA: Structure of Leptophyllidin and Identity of Leptophyllin and Leptophylloside with Rutaretin and Isorutarin. Indian J. Chem. **17B**, 647 (1979).

*569.* — —: New 6-Methylcoumarins and Other Minor Components of *Trachyspermum roxburghianum* Seeds. Indian J. Chem. **19B**, 85 (1980).

*570.* SHARMA, D.K., and T.R. SESHADRI: Oxidative Coupling of Esculetin and Isoscopoletin by [Fe(DMF)$_3$Cl$_2$][FeCl$_4$], Potassium Hexacyanoferrate(III) and Manganese Tris (acetylacetonate). Indian J. Chem. **15B**, 939 (1977).

571. SHARMA, P., M. SHARMA, and S. RANGASWAMI: Structure of Leptophyllidin, Leptophyllin and Leptophylloside, New Dihydrofurocoumarins from the Seeds of *Apium leptophyllum*. Indian J. Chem. **16B**, 563 (1978).

572. SHARMA, P.N., A. SHOEB, R.S. KAPIL, and S.P. POPLI: Synthesis of Toddanol and Toddanone. Indian J. Chem. **19B**, 938 (1980).

573. — — — —: Toddasin, a New Dimeric Coumarin from *Toddalia asiatica*. Phytochemistry **19**, 1258 (1980).

574. — — — —: Toddanol and Toddanone, Two Coumarins from *Toddalia asiatica*. Phytochemistry **20**, 335 (1981).

575. SHARMA, R.B., and R.S. KAPIL: Synthesis of 3-(1,1-Dimethylallyl)xanthyletin, 3-(1,1-Dimethylallyl)-8-(3,3-dimethylallyl)xanthyletin and *dl*-Chalepin. Indian J. Chem. **22B**, 538 (1983).

576. SHARMA, R.B., D. SWAROOP, and R.S. KAPIL: Synthesis of 3-(1,1-Dimethylallyl)xanthyletin. Indian J. Chem. **20B**, 153 (1981).

577. SHAW, G.J., M.K. YATES, and D.R. BIGGS: Synthesis and Structural Proof of Wairol, a New Coumestan from *Medicago sativa*. Phytochemistry **21**, 249 (1982).

578. SHIBATA, S., and M. NOGUCHI: Two New Coumarins in *Boenninghausenia albiflora*. Phytochemistry **16**, 291 (1977).

579. SHIGEMATSU, N., I. KOUNO, and N. KAWANO: On the Isolation of ( + )-Samidin from the Roots of *Peucedanum japonicum*. Yakugaku Zasshi **102,** 392 (1982).

580. SHIMOMURA, H., Y. SASHIDA, H. NAKATA, J. KAWASAKI, and Y. ITO: Plant Growth Regulators from *Heracleum lanatum*. Phytochemistry **21**, 2213 (1982).

581. SHIMOMURA, H., Y. SASHIDA, and Y. OHSHIMA: Coumarins from *Artemisia apiacea*. Phytochemistry **18**, 1761 (1979).

582. — — —: The Chemical Components of *Artemisia apiaceae* Hance II. More Coumarins from the Flower Heads. Chem. and Pharm. Bull. **28**, 347 (1980).

583. SHIOZAWA, T., S. URATA, T. KINOSHITA, and T. SAITOH: Revised Structures of Glycyrol and Isoglycyrol, Constituents of the Root of *Glycyrrhiza uralensis*. Chem. and Pharm. Bull. **37**, 2239 (1989).

584. SHIPCHANDLER, M., and T.O. SOINE: A Revised Structure of Columbianin and Pertinent Mass Spectral Studies. J. Pharm. Sci. **57**, 747 (1968).

585. SHOEB, A., M.D. MANANDHAR, R.S. KAPIL, and S.P. POPLI: Clausmarins A and B: Two Novel Spasmolytic Terpenoid Coumarins from *Clausena pentaphylla* (Roxb.) DC. J. Chem. Soc., Chem. Commun. **1978**, 281.

586. SHUKLA, V.S., S.C. DUTTA, R.N. BARUAH, R.P. SHARMA, G. THYAGARAJAN, W. HERZ, N. KUMAR, K. WATANABE, and J.F. BLOUNT: New 5-Methylcoumarins from *Ethulia conyzoides*. Phytochemistry **21**, 1725 (1982).

587. SIBANDA, S., B. NDENGU, G. MULTARI, V. POMPI, and C. GALEFFI: A coumarin Glucoside from *Xeromphis obovata*. Phytochemistry **28**, 1550 (1989).

588. SIMONITSCH, E., H. FREI, and H. SCHMID: Die Konstitution des Pachyrrhizins. Monatsh. Chem. **88**, 541 (1957).

589. SINGH, H., A.S. CHAWLA, V.K. KAPOOR, N. KUMAR, D.M. PIATAK, and W. NOWICKI: Investigation of *Erythrina* Spp. IX. Chemical constituents of *Erythrina stricta* Bark. J. Nat. Prod. **44**, 526 (1981).

590. SINGH, M.M., D.N. GUPTA, V. WADHWA, G.K. JAIN, N.M. KHANNA, and V.P. KAMBOJ: Contraceptive Efficacy and Hormonal Profile of Ferujol: a New Coumarin from *Ferula jaeschkeana*. Planta Medica **1985**, 268.

591. SINGH, R.P., V.B. PANHEY, and S. SEPULVEDA: A New Neoflavonoid from *Echinops niveus*. Chem. and Ind. **1987**, 828.

592. SOKOLOVA, A.I., Y.E. SKLYAR, and M.G. PIMENOV: Coumarins of *Seseli saxicolum*. Khim. prirod. Soedinenii **1980**, 715; Chem. Abs. **94**, 117788 (1980).

593. SOTO R., B., F. DIAZ C., R. YANEZ O., O. COLLERA C., and F. GARCIA J.: Two 4-Phenylcoumarins from *Hintonia latifolia* (Rubiaceae). Spectroscopy (Ottawa) **6**, 123 (1988).

594. SPENCER, R.R., E.M. BICKOFF, R.E. LUNDIN, and B.E. KNUCKLES: Lucernol and Sativol, Two New Coumestans from Alfalfa (*Medicago sativa*). J. Agric. Food Chem. **14**, 162 (1966).

595. SPENCER, R.R., B.E. KNUCKLES, and E.M. BICKOFF: 7-Hydroxy-11,12-dimethoxy-coumestan. Characterization and Synthesis. J. Org. Chem. **31**, 988 (1966).

596. SPENCER, R.R., S.C. WITT, R.E. LUNDIN, and E.M. BICKOFF: Bicoumol, a New Bicoumarinyl, from Ladino Clover. J. Agric. Food Chem. **15**, 536 (1967).

597. SPRING, M.S., and J.R. STOKER: The Biosynthesis of 4-Hydroxycoumarin by *Penicillium jenseni*. Canad. J. Biochem. **47**, 301 (1969).

598. STAHMANN, M.A., C.F. HUEBNER, and K.P. LINK: Studies on the Hemorrhagic Sweet Clover Disease. V. Identification and Synthesis of the Hemorrhagic Agent. J. Biol. Chem. **138**, 513 (1941).

599. STAHMANN, M.A., I. WOLFF, and K.P. LINK: Studies on 4-Hydroxycoumarins. I. The Synthesis of 4-Hydroxycoumarins. J. Amer. Chem. Soc. **65**, 2285 (1943).

600. STANLEY, W.L., and S.H. VANNIER: Psoralens and Substituted Coumarins from Expressed Oil of Lime. Phytochemistry **6**, 585 (1967).

601. STECK, W., and M. MAZUREK: Indetification of Natural Coumarins by Nmr Spectroscopy. Lloydia **35**, 418 (1972).

602. STEFANOVIĆ, M., M. DERMANOVIĆ, and M. VERENČEVIĆ: Chemical Investigation of the Plant Species of *Artemisia vulgaris* L. (Compositae). Glas. Hem. Drus. Beograd. **47**, 7 (1982); Chem. Abs. **96**, 177968 (1982).

603. STEFANOVIĆ, M., S. MLADENOVIĆ, M. DERMANOVIĆ, S. MATIĆ, I. KRSTANOVIĆ, and L. KARANOVIĆ: Stereoisomeric Pyranocoumarins (Khellactone Esters), Pyrano- and Furanochromones from *Peucedanum austriaca* (Jacq). Glas. Hem. Drus. Beograd. **49**, 5 (1984); Chem. Abs. **101**, 107325 (1984).

604. SUN, H., and J. JAKUPOVIC: Further Heraclenol Derivatives from *Angelica archangelica*. Pharmazie **41**, 888 (1986).

605. SUN, H., Z. LIN, F. NIU, and J. DING: Studies on Chinese Herbs in Umbelliferae. II. New Coumarins in *Peucedanum turgeniifolium* Wolff. Acta Botanica Yunnanica **3**, 173 (1981); Chem. Abs. **96**, 40789 (1982).

606. —— —— —— ——: Studies on the Chinese Drugs of Umbelliferae. IV. Structure of Apaensin. Acta Botanica Yunnanica **3**, 279 (1981); Chem. Abs. **96**, 48961 (1982).

607. SUORTTI, T., and A. VON WRIGHT: Isolation of a Mutagenic Fraction from Aqueous Extracts of the Wild Edible Mushroom *Lactarius necator* (a Preliminary Note). J. Chromatogr. **255**, 529 (1983).

608. SUORTTI, T., A. VON WRIGHT, and A. KOSKINEN: Necatorin, a Highly Mutagenic Compound from *Lactarius necator*. Phytochemistry **22**, 2873 (1983).

609. SWAGER, T.M., and J.H. CARDELLINA II: Coumarins from *Musineon divaricatum*. Phytochemistry **24**, 805 (1985).

610. SWAROOP, D., R.B. SHARMA, and R.S. KAPIL: Synthesis of 3-(3,3-Dimethyl-allyl)xanthyletin. Indian J. Chem. **22B**, 105 (1983).

611. —— ——: Synthesis of Balsamiferone and *dl*-3-(3,3-Dimethylallyl)marmesin. Indian J. Chem. **22B**, 408 (1983).

612. SZABÓ, G., H. GREGER, and O. HOFER: Coumarin-hemiterpene Ethers from *Artemisia* species. Phytochemistry **24**, 537 (1985).

613. TAKATA, M., T. OKUYAMA, and S. SHIBATA: Studies on Coumarins of a Chinese Drug "Qian-Hu"; VIII. Structures of New Coumarin-glycosides of "Bai-Hua Qian-Hu". Planta Medica **54**, 323 (1988).

*614.* TALAPATRA, S.K., N.C. GANGULY, S. GOSWAMI, and B. TALAPATRA: Chemical Constituents of *Casearia graveolens*: Some Novel Reactions and the Preferred Molecular Conformation of the Major Coumarin, Micromelin. J. Nat. Prod. **46**, 401 (1983).

*615.* TALAPATRA, S.K., S. GOSWAMI, N.C. GANGULY, and B. TALAPATRA: Structure of Casegravol: a New Monomeric Coumarindiol from *Casearia graveolens* Dalz. Chem. and Ind. **1980**, 154.

*616.* TANAKA, H., M. ISHIHARA, K. ICHINO, and K. ITO: Total Synthesis of Coumarinolignans, Propacin and Its Regioisomer. Chem. and Pharm. Bull. **36**, 1738 (1988).

*617.* ———: Total Synthesis of Coumarinolignans, Aquillochin (Cleomiscosin C) and Cleomiscosin D. Chem. and Pharm. Bull. **36**, 3833 (1988).

*618.* ———: Syntheses of Natural Coumarinolignans: Oxidative Coupling of 7,8-Dihydroxycoumarins and Phenylpropenes in the Presence of Diphenyl Selenoxide. Heterocycles **27**, 2651 (1988).

*619.* TANAKA, H., I. KATO, and K. ITO: Total Synthesis of Cleomiscosin A, a Coumarinolignoid. Chem. and Pharm. Bull. **33**, 3218 (1985).

*620.* ———: Total Synthesis of Cleomiscosin B, a Coumarinolignoid. Heterocycles **23**, 1991 (1985).

*621.* ———: Total Synthesis of Daphneticin, a Coumarinolignoid. Chem. and Pharm. Bull. **34**, 628 (1986).

*622.* RUANGRUNGSI, N., V. VAISIRIROJ, D.C. LANKIN, N.S. BHACCA, R.P. BORRIS, G.A. CORDELL, and L.F. JOHNSON: Microminutin, a Novel Cytotoxic Coumarin from *Micromelum minutum* (Rutaceae). J. Org. Chem. **48**, 268 (1983).

*623.* TATUM, J.H., and R.E. BERRY: Coumarins and Psoralens in Grapefruit Peel Oil. Phytochemistry **18**, 500 (1979).

*624.* THASTRUP, O., and J. LEMMICH: Furocoumarin Glucosides of *Angelica archangelica* subsp. *litoralis*. Phytochemistry **22**, 2035 (1983).

*625.* THEBTARANONTH, C., S. IMRAPORN, and N. PADUNGKUL: Phenylcoumarins from *Ochrocarpus siamensis*. Phytochemistry **20**, 2305 (1981).

*626.* THOMPSON, E.B., G.H. AYNILIAN, R.H. DOBBERSTEIN, G.A. CORDELL, H.H.S. FONG, and N.R. FARNSWORTH: Biological and Phytochemical Investigation of Plants. XV. *Pteryxia terebinthina* var. *terebinthina* (Umbeliferae). J. Nat. Prod. **42**, 120 (1979).

*627.* THOMPSON, H.J., and S.A. BROWN: Separations of Some Coumarins of Higher Plants by Liquid Chromatography. J. Chromatogr. **314**, 323 (1984).

*628.* TSCHESCHE, R., U. SCHACHT, and G. LEGLER: Über Daphnoretin, ein natürlich vorkommendes Derivat des 3,7'-Dicumaryläthers. Liebigs Ann. Chem. **662**, 113 (1963).

*629.* ———: Über Daphnorin, ein neues Cumaringlucoside aus *Daphne mezereum*. Naturwissenschaften **50**, 521 (1963).

*630.* TSUGE, O., S. KANEMASA, and H. SUGA: An Entry to 5-(1-Alkenyl)-3(2H)-furanones Through Cycloaddition of Phosphorus-Functionalized Nitrile Oxide to Acetylene Alcohols. Chem. Letters **1987**, 323.

*631.* ULUBELEN, A., R.R. KERR, and T.J. MABRY: Two New Neoflavonoids and C-Glycosylflavones from *Passiflora serratodigitata*. Phytochemistry **21**, 1145 (1982).

*632.* ULUBELEN, A., S. ÖKSUZ, and E. TUZLACI: Flavonoids and Coumarins from *Achillea schischkinii*. Planta Medica **53**, 507 (1987).

*633.* ULUBELEN, A., and B. TEREM: Alkaloids and Coumarins from Roots of *Ruta chalepensis*. Phytochemistry **27**, 650 (1988).

*634.* ULUBELEN, A., B. TEREM, and E. TUZLACI: Coumarins and Flavonoids from *Daphne gnidioides*. J. Nat. Prod. **49**, 692 (1986).

635. VALLE, M.G., G. APPENDINO, G.M. NANO, and V. PICCI: Prenylated Coumarins and Sesquiterpenoids from *Ferula communis*. Phytochemistry 26, 253 (1987).

636. VAN WAGENEN, B.C., J. HUDDLESTON, and J.H. CARDELLINA II. Native American Food and Medicinal Plants, 8. Water-soluble Constituents of *Lomatium dissectum*. J. Nat. Prod. 51, 136 (1988).

637. VDOVIN, A.D., D. BATSURÉN, E.K. BATIROV, M.R. YAGUDAEV, and V.M. MALIKOV: $^1$H and $^{13}$C NMR Spectra and the Structure of a New Coumarin, *C*-Glycoside Dauroside D, from *Haplophyllum dauricum*. Khim. prirod. Soedinenii 1983, 441; Chem. Abs. 100, 82693 (1984).

638. VENTURELLA, P., A. BELLINO, and M.L. MARINO: Synthesis of Terpenoid Coumarins. An Approach to the Synthesis of Piloselliodan. Gazzetta 112, 433 (1982).

639. — — —: Synthesis of Coumarsabin and 4,6,7-Trimethoxy-5-methyl-2*H*-1-benzopyran-2-one. Gazzetta 113, 819 (1983).

640. VESELOVSKAYA, N.V., and Y.E. SKLYAR: Ferilin from *Ferula iliensis*. Khim. prirod. Soedinenii 1984, 387; Chem. Abs. 102, 92935 (1985).

641. VESELOVSKAYA, N.V., Y.E. SKLYAR, D.A. FESENKO and M.G. PIMENOV: Fekrol, a Terpenoid Coumarin from *Ferula krylovii*. Khim. prirod. Soedinenii 1979, 851; Chem. Abs. 93, 41500 (1980).

642. VESELOVSKAYA, N.V., Y.E. SKLYAR, M.E. PEREL'SON, and M.G. PIMENOV: Terpenoid Coumarins of *Ferula krylovii*. Khim. prirod. Soedinenii 1979, 227; Chem. Abs. 91, 171682 (1979).

643. VESELOVSKAYA, N.V., Y.E. SKLYAR, and M.G. PIMENOV: Terpenoid Coumarin from *Ferula aitchisonii*. Khim. prirod. Soedinenii 1982, 397.

644. VESELOVSKAYA, N.V., Y.E. SKLYAR, and A.A. SAVINA: Fekrynol and its Acetate from *Ferula krylovii*. Khim. prirod Soedinenii 1981, 798; Chem. Abs. 96, 199908 (1982).

645. VISWANATHAN, N., and V. BALAKRISHNAN: Structure and Synthesis of Lasiocephalin. Indian J. Chem. 12, 450 (1974).

646. VUORELA, H., C.A.J. ERDELMEIER, S. NYIREDY, K. DALLENBACH-TOELKE, C. ANKLIN, R. HILTUNEN, and O. STICHER: Isobyakangelicin Angelate; a Novel Furanocoumarin from *Peucedanum palustre*. Planta Medica 54, 538 (1988).

647. WAKHARKAR, R.D., V.H. DESHPANDE, A.B. LANDGE, and B.K. UPADHYE: Synthesis of Microminutin and 4-Methylmicrominutin. Org. Prep. Proced. Int. 20, 527 (1988).

648. WANG, C., and J. CHEN: Chemical Constituents of the Root of *Libanotis buchtorimensis* (Fisch) DC. Acta Bot. Sinica 28, 192 (1986).

649. WANG, M.-S., W.-C. LIU, J.-R. LI, F.-L. SHEN, X.-Y. LIN, and Q.-T. ZHENG: Structure Elucidation of Heliclactone. Acta Chim. Sinica 46, 768 (1988).

650. WANZLICK, H.-W., R. GRITZKY, and H. HEIDEPRIEM: Die Synthese des Wedelolactons. Chem. Ber. 96, 305 (1963).

651. WAYKOLE, P., S. SHAIKH, and R.N. USGAONKAR: Two New Syntheses of Xanthyletin. Indian J. Chem. 19B, 238 (1980).

652. WICKRAMARATNE, D.B.M., V. KUMAR, and S. BALASUBRAMANIAM: Murragleinin, a Coumarin from *Murraya gleinei* Leaves. Phytochemistry 23, 2964 (1984).

653. WILZER, K.A., F.R. FRONCZEK, L.E. URBATSCH, and N.H. FISCHER: Coumarins from *Aster prealtus*. Phytochemistry 28, 1729 (1989).

654. WIRASUTISNA, K.R., J. GLEYE, C. MOULIS, E. STANISLAS, and C. MORETTI: Galipein, a Coumarin from *Galipea trifoliata*. Phytochemistry 26, 3372 (1987).

655. WOLFRUM, C., and F. BOHLMANN: Synthese weiterer natürlich vorkommender 5-Methylcoumarine. Liebigs Ann. Chem. 1989, 295.

656. WONG, E., and G.C.M. LATCH: Coumestans in Diseased White Clover. Phytochemistry 10, 466 (1971).

657. WU, T.-S.: Omphamurin, a New Coumarin from *Murraya omphalocarpa*. Phytochemistry **20**, 178 (1981).

658. —: Coumarins from the Leaves of *Murraya paniculata*. Phytochemistry **27**, 2357 (1988).

659. —: Alkaloids and Coumarins of *Citrus grandis*. Phytochemistry **27**, 3717 (1988).

660. WU, T.-S., and H. FURUKAWA: Acridone Alkaloids. VII. Constituents of *Citrus sinensis* Osbeck var. *brasiliensis* Tanaka. Isolation and Characterization of Three New Acridone Alkaloids, and a New Coumarin. Chem. and Pharm. Bull. **31**, 901 (1983).

661. WU, T.-S., S.-C. HUANG, T.-T. JONG, J.-S. LAI, and C-S. KUOH: Coumarins, Acridone Alkaloids and a Flavone from *Citrus grandis*. Phytochemistry **27**, 585 (1988).

662. WU, T.-S., C.-S. KUOH, and H. FURUKAWA: Acridone Alkaloids and a Coumarin from *Citrus grandis*. Phytochemistry **22**, 1493 (1983).

663. WU, T.-S., M.-J. LIOU, and C.-S. KUOH: Coumarins of the Flowers of *Murraya paniculata*. Phytochemistry **28**, 293 (1989).

664. WULFF, W.D., J.S. MCCALLUM, and F.A. KUNNG: Two Regiocomplementary Approaches to Angular Furanocoumarins with Chromium Carbene Complexes: Synthesis of Sphondin, Thiosphondin, Heratomin and Angelicin. J. Amer. Chem. Soc. **110**, 7419 (1988).

665. YAMAHARA, J., M. KOZUKA, T. SWADA, H. FUJIMURA, F. NAKANO, T. TOMIMATSU, and T. NOHARA: Biologically Active Principles of Crude Drugs. Anti-allergic Principles in *Cnidii monnieri*. Chem. and Pharm. Bull. **33**, 1676 (1985).

666. YANG, J., and Y. SU: Studies on the Constituents of *Murraya paniculata* (L) Jack. Acta Pharm. Sinica **18**, 760 (1983).

667. YANG, J.-S., and M.-H. DU: Constituents of *Murraya paniculata* (L). Jack Grown in Hainan. Acta Chim. Sinica **42**, 1308 (1984).

668. YANG, P., and A.D. KINGHORN: Coumarin Constituents of the Chinese Tallow Tree (*Sapium sebiferum*). J. Nat. Prod. **48**, 486 (1985).

669. YE, J., H. ZHANG, and C. YUAN: Isolation and Identification of Coumarin Praeruptorin E from the Root of the Chinese Drug *Peucedanum praeruptorum* Dunn (Umbelliferae). Acta Pharm. Sinica **17**, 431 (1982).

670. YULDASHEV, M.P., E.K. BATIROV, and V.M. MALIKOV: Coumarin Glycosides of *Haplophyllum perforatum*. Khim. prirod. Soedinenii **1980**, 168; Chem. Abs. **93**, 110556 (1980).

671. — — —: Acylated Coumarin Glycoside from *Haplophyllum perforatum*. Khim. prirod. Soedinenii **1980**, 412; Chem. Abs. **93**, 164323 (1980).

672. YULDASHEV, M.P., E.K. BATIROV, V.M. MALIKOV, and M.E. PEREL'SON: The Structure of Haploperoside B, an Acylated Coumarin Glycoside from *Haplophyllum perforatum*. Khim. prirod. Soedinenii **1981**, 718; Chem. Abs. **96**, 181532 (1982).

673. YULDASHEV, M.P., E.K. BATIROV, A.D. VDOVIN, V.M. MALIKOV, and M.R. YAGUDAEV: Coumarin Glycosides of *Haplophyllum perforatum*. Structures of Haploperosides C, D, and E. Khim. prirod. Soedinenii **1985**, 27; Chem. Abs. **103**, 101990 (1985).

674. ZAKARIA, M.B., I. SAITO, and T. MATSUURA: Coumarins of *Merrillia caloxylon*. Phytochemistry **28**, 657 (1989).

675. ZDERO, C., F. BOHLMANN, R.M. KING, and H. ROBINSON: Further 5-Methylcoumarins and Other Constituents from the Subtribe Mutisiinae. Phytochemistry **25**, 509 (1986).

676. — — — —: Diterpene Glycosides and Other Constituents from Argentinian *Baccharis* Species. Phytochemistry **25**, 2841 (1986).

677. — — — —: α-Isocedrene Derivatives, 5-Methylcoumarins and Other Constituents from the Subtribe Nassauviinae of the Compositae. Phytochemistry **25**, 2873 (1986).

678. ZDERO, C., F. BOHLMANN, and H.M. NIEMEYER: 5-Methylcoumarin Derivatives from *Aphyllocladus denticulatus*. Phytochemistry **27**, 1821 (1988).

679. ———: Diterpenes and 5-Methylcoumarin Derivatives from *Gypothamnium pinifolium* and *Plaza daphnoides*. Phytochemistry **27**, 2953 (1988).

680. ZDERO, C., F. BOHLMANN, and J. SOLOMON: Further 5-Methylcoumarin Derivatives from *Mutisia orbignyana*. Phytochemistry **27**, 891 (1988).

681. ZHANG, L.G., O. SELIGMAN, and H. WAGNER: Daphneticin, a Coumarinolignoid from *Daphne tangutica*. Phytochemistry **22**, 617 (1983).

682. ZILG, H., and H. GRISEBACH: Biosynthesis of Isoflavones. XVII. Identification and Biosynthesis of Coumestanes in *Soja hispida*. Phytochemistry **7**, 1765 (1968).

(*Received December 17, 1990*)

# Author Index

ABBAS, R.  *292*
ABDEL SALAM, N.A.  *303*
ABRAMO-BRUNO, D.  *300*
ABU-MUSTAFA, E.  *289, 292*
ABYSHEV, A.Z.  *283, 295, 300*
ABYSHEV, D.Z.  *283*
ADESANYA, S.A.  *307*
ADESINA, K.S.  *302*
ADINARAYANA, D.  *284*
ADITYACHAUDHURY, N.  *284*
AGNOLO, G.D.  *77*
AGRAWAL, A.  *284*
AGUADO, M.T.  *296*
AHLUWALIA, V.K.  *284, 285*
AHMAD, J.  *285*
AHMAD, S.  *293*
AHMED, Z.H.  *75*
AIZAWA, M.  *74*
AJAZ, A.A.  *74, 77*
AKHDAROV, B.  *283*
ALBERS-SCHÖNBERG, G.  *75*
ALEXA, M.  *296*
AL-HAZIMI, H.M.G.  *285*
ALVES DE LIMA, R.  *292*
AMAYA, Y.  *309*
AMER, M.E.  *303*
AMY, C.M.  *79*
ANAND, N.K.  *285*
ANDRIANOVA, V.B.  *307*
ANGELES, L.R.  *285*
ANKLIN, C.  *314*
ANWER, F.  *285*
APPENDINO, G.  *285, 314*
AQUINO, R.  *285*
ARENCIBIA, J.B.  *296*
ARENS, H.  *294*
ARISAWA, M.  *285*
ARISON, B.  *307*
ARMSTRONG, J.A.  *287*
ARNOLDI, A.  *285*
ARNONE, A.  *285*
ASAHARA, T.  *285*
ASAI, F.  *298*

ASHEERVADAM, Y.  *285, 308*
ASHWOOD-SMITH, M.J.  *290*
ASHWORTH, D.M.  *73–75, 77*
ASTLES, D.P.  *289*
ATAL, C.K.  *286, 297, 302*
AWAYA, J.  *77*
AYAFOR, J.F.  *294, 306*
AYNILIAN, G.H.  *313*
AZIZ, M.  *285*

BABA, K.  *285, 286, 301, 302*
BAGIROV, V.Y.  *286*
BAILEY, C.R.  *80*
BALA, K.R.  *286*
BALAKRISHNAN, V.  *314*
BALASUBRAMANIAM, S.  *292, 302, 314*
BALBAA, S.I.  *286, 288*
BALDWIN, J.E.  *34, 76*
BALLANTYNE, M.M.  *305*
BAL-TEMBE, S.  *286*
BALTZ, R.H.  *76, 80*
BANDARANAYAKE, W.M.  *286*
BANDOPADHYAY, M.  *286*
BANERJEE, D.  *293*
BANERJEE, S.K.  *286, 297, 302*
BANERJI, A.  *286*
BANERJI, J.  *290*
BARALDI, P.G.  *286*
BARCO, A.  *286*
BARIK, B.R.  *286*
BARRERO, A.F.  *292*
BARROS, S.M.G.  *303*
BARROSO, J.T.  *295, 296*
BARUA, J.N.  *291*
BARUA, N.C.  *286*
BARUAH, R.H.  *291, 311*
BASA, S.C.  *287*
BATHE, A.  *308*
BATIROV, E.K.  *287, 304, 306, 314, 315*
BATSURÉN, D.  *287, 306, 314*
BAUER, H.  *289*
BECK, J.  *79, 80*
BECK, K.-F.  *80*

BECKMAN, R.J.   *80*
BEDI, D.N.   *286*
BEGLEY, M.J.   *287*
BELLINO, A.   *287, 314*
BELYI, M.B.   *286*
BENETTI, S.   *286*
BERRY, R.E.   *313*
BESELOVSKAYA, N.V.   *287*
BESSONOVA, I.A.   *287*
BEVALOT, F.   *287*
BEVITT, D.J.   *81*
BHACCA, N.S.   *313*
BHANDARI, P.   *287*
BHARAT BHUSHAN   *287*
BHARDWAJ, D.K.   *287*
BHARGAVA, K.K.   *287*
BHAT, S.V.   *297*
BHATIA, S.K.   *287*
BHATTACHARYA, A.K.   *287*
BHATTACHARYYA, P.   *288, 290*
BHIDE, K.S.   *288, 308*
BIBB, M.J.   *81*
BICKOFF, E.M.   *288, 293, 303, 312*
BIGGS, D.R.   *288, 311*
BIN ZAKARIA, M.   *299*
BIRCH, A.J.   *6, 7, 73*
BIRO, S.   *81*
BITTNER, M.   *288, 291*
BLOCH, K.   *77*
BLOCK, M.H.   *77*
BLOUNT, J.F.   *311*
BOAG, D.M.   *305*
BODDY, I.   *75*
BODO, B.   *302*
BOEKER, R.   *299*
BOHLMANN, F.   *286, 288, 289, 297, 299, 302, 303, 309, 314–316*
BOOTH, A.N.   *288*
BORDERS, D.B.   *75*
BORGES DEL CASTILLO, J.   *289*
BORRIS, R.P.   *313*
BOSE, A.K.   *290*
BOSE, P.K.   *293*
BOTTA, B.   *292*
BOTTA, M.   *292*
BOYD, D.R.   *284, 299*
BRENDELBERGER, G.   *74*
BREUER, M.   *286*
BRIGGS, L.H.   *289*
BROCKMANN, H.   *7, 73*
BROWN, K.L.   *289*
BROWN, S.A.   *305, 313*
BRUNO, M.   *297*
BRYAN, R.F.   *302*
BÜCHI, G.M.   *289*
BUDDRUS, J.   *289*

BUDESINSKY, M.   *75*
BUDZIKIEWICZ, H.   *286*
BUKREEVA, T.V.   *289*
BULLOCK, M.W.   *75*
BULSING, M.J.   *73*
BURFITT, A.I.R.   *289*
BURKE, B.A.   *289, 307*
BUTLER, M.J.   *80*
BUZAS, A.   *307*

CAIRNS, N.   *289*
CAIRO VALERA, G.   *292*
CALZADA, F.   *303*
CAMBIE, R.C.   *289, 301*
CAMPBELL, H.A.   *289*
CANE, D.E.   *20, 23, 43, 44, 46, 73–77*
CANO, G.   *292*
CAPORALE, G.   *309*
CARDELLINA II, J.H.   *312, 314*
CARDONA, R.J.   *295, 296*
CARPENTER, I.   *290*
CARTER, G.T.   *75*
CASOLARI, A.   *286*
CASSELS, B.K.   *294*
CASTILLO, P.   *290*
CASTRO, V.   *288*
CASTRO-PECEIRO, J.   *304*
CAVÉ, A.   *298*
CECCHERELLI, P.   *290*
CELMER, W.D.   *43, 44, 46, 50, 73, 77*
CESKA, O.   *290*
CHAKRABARTI, A.   *288*
CHAKRABARTI, R.   *290*
CHAKRABARTY, P.   *288*
CHAKRABORTY, A.   *290*
CHAKRABORTY, D.P.   *288, 290*
CHAKRAVARTI, K.K.   *290*
CHAMBERLIN, J.W.   *73*
CHANEY, M.O.   *76*
CHANG, C.   *77*
CHANG, C.-J.   *73*
CHANG, S.-I.K.   *79*
CHATER, K.F.   *81*
CHATTERJEA, J.N.   *290*
CHATTERJEE, A.   *286, 290, 303*
CHATTERJI, D.   *290*
CHATTOPADHYAY, K.K.   *290*
CHATTOPADHYAY, S.K.   *308*
CHAUDHARY, S.K.   *290*
CHAUKIYAL, D.C.   *310*
CHÁVEZ, G.   *310*
CHAWLA, A.S.   *311*
CHAWLA, H.L.   *301*
CHAWLA, H.M.   *290*
CHEN, C.-M.   *298, 304*
CHEN, J.   *314*

CHEN, K.-M. *290*
CHEN, M. *290*
CHEN, M.-T. *298*
CHEN, S. *73*
CHEN, W. *304*
CHEN, Y.S. *291*
CHÊNEVERT, R. *291*
CHEUNG, K.F. *301*
CHIANG, M.T. *291*
CHIBBER, S.S. *290*
CHICO, E.D. *296*
CHIRALA, S.S. *79*
CHOU, H.-N. *74*
CHOWDHURY, B.K. *288, 290*
CLARK, E.P. *291*
CLORE, G.M. *78*
COLE, R.J. *291*
COLLERA, C., O. *312*
COLLINS, J.F. *81*
CONNOLLY, J.D. *306*
CORCORAN, J.W. *73*
CORDELL, G.A. *285, 303, 313*
CORRALES, B. *292*
CORTEZ, J. *80, 81*
COSTA, A.M.B.S.R.C.S. *292*
COX, K. *80*
COX, R.H. *291*
CRAGG, G.M.L. *289*
CRAW, P.A. *75*
CROMBIE, L. *287, 291*
CRONAN, J.E. *78*
CROSSLEY, M.J. *76*
CRUMLEY, F.G. *291*
CURINI, M. *290*
CUTLER, H.G. *291*

D'AGOSTINO, M. *285, 291*
DAILY, A. *291*
DALL'ACQUA, F. *309*
DALLENBACH-TOELKE, K. *306, 314*
DAS, A.K. *286*
DAS, B. *286, 290*
DAS, B.C. *290, 306*
DAS, D.P. *287*
DAS, P.C. *286, 300*
DAS, S. *291*
DAS, S.C. *287, 310*
DAS GRAÇAS, M. *291*
DAVID, L. *73, 75*
DAVIES, A. *73, 75*
DAVIES, A.B. *73, 75*
DAVIES, D.H. *73*
DAVIS, D.H. *73*
DAVIS, N.K. *81*
DAWIDAR, A.-A. *292*

DAY, L.E. *73*
DE A. E SILVA, M. *306*
DEAN, F.M. *292*
DEB, N.C. *293*
DE EDS, F. *288*
DE FEO, V. *291*
DEL CASTILLO, J.B. *292*
DELGADO, G. *292*
DELLE MONACHE, F. *292*
DELLE MONACHE, G. *292*
DEL ROSARIO GARCIA, M. *303*
DE MELLO, J.F. *292*
DEMETRIADOU, A.K. *75*
DENISENKO, P.P. *283*
DE PASCUAL TERESA, J. *292*
DERKACH, A.I. *301*
DERMANOVIĆ, M. *312*
DESCONCLOIS, J.-F. *294*
DESHPANDE, V.H. *314*
DE SILVA, L.B. *292*
DE SIMONE, F. *285, 291*
DE SOUZA, N.J. *286*
DEWICK, P.M. *73*
DEY, A.K. *286*
DHAR, K.L. *297, 301*
DHARMARATNE, H.R.W. *292*
DHILLON, N. *80*
DÍAZ, E. *310*
DIAZ C., F. *312*
DIECK, A. *292*
DIMARE, M. *76*
DIMROTH, P. *79*
DINDA, B.N. *303*
DING, J. *312*
DJERASSI, C. *294*
DOBBERSTEIN, R.H. *313*
DOBLER, M. *73*
DODDRELL, D.M. *75*
DOGANCA, S. *292*
DOHERTY, A.M. *73*
DOLPHIN, D. *73*
DOMINGUEZ, X.A. *292, 293*
DONADIO, S. *80, 81*
DO NASCIMENTO, M.C. *295*
DONAUBAUER, J. *76*
DONNELLY, B.J. *293*
DONNELLY, D.M.X. *293*
DONOVAN, F.W. *73*
DORMAN, D.E. *73*
DORNER, J.W. *291*
DORTA, H.L. *296*
DOUGLAS, A.W. *75*
DREYER, D.L. *293, 302*
DU, M.-H. *315*
DUESLER, E.N. *72*
DUFFLEY, R.P. *293*

DUKHOVLINOVA, L.I. *293*
DUNCAN, J.S. *76*
DUNITZ, J.D. *73*
DUTT, P. *293*
DUTT, S. *300*
DUTTA, N.L. *293*
DUTTA, P.K. *293*
DUTTA, S.C. *311*
DUTTAGUPTA, P.C. *301*
DYER, U.C. *76*
DZHAFOROV, Z.R. *293*
DZIEDZIC, S.Z. *299*

EAST, A.J. *293*
EBIHARA, T. *302*
EDWARDS, P.D. *76*
EISENBEISS, J. *293*
EKONG, D.E.U. *306*
ELANGO, V. *287*
EL-BASYOUNI, S. *293*
EL-BEIH, F.K.A. *294*
EL-KHRISY, E.A.M. *289, 292*
EMADZADEH, S. *73*
EMERSON, O.H. *293*
ENDO, T. *310*
ENRÍQUEZ, R.G. *293*
ENYENIHI, V.U. *306*
EPP, J.K. *80*
ERDELMEIER, C.A.J. *293, 294, 306, 314*
ESCOBAR, L.I. *293*
ESPINO, M.R. *296*
ESPINOZA, G. *292*
ESPINOZA, S. *294*
EVANS, D.A. *76*

FALSHAW, C.P. *73, 294*
FARNSWORTH, N.R. *285, 313*
FAYEZ, M.B.E. *294*
FESENKO, D.A. *314*
FESSLER, B. *291*
FIALA, R.R. *75*
FINNEGAN, R.A. *291, 294*
FISCHER, H.A. *294*
FISCHER, N.H. *314*
FISHER, J.F. *294*
FLORYA, V.N. *283*
FLOSS, H.J. *74*
FOMBAN, W.G. *294*
FOMUM, Z.T. *294*
FONG, H.H.S. *285, 313*
FORD, L. *80*
FORGACS, P. *294*
FOZDAR, B.I. *294*
FRAGA, B.M. *296*
FRANCO, R. *292, 293*
FREI, H. *311*

FRIEND, E.J. *80*
FRIGHETTO, R.T.S. *303*
FRONCZEK, F.R. *314*
FUJIMURA, H. *315*
FUJIWARA, H. *294*
FUKAI, T. *294, 298*
FUKUI, K. *294*
FUKUMOTO, M. *302*
FUKUMOTO, T. *301*
FUKUSHIMA, S. *302, 309*
FULCO, A.J. *78*
FUNABASHI, H. *77*
FURUKAWA, H. *294, 299, 300, 315*

GACS-BAITZ, E. *292*
GALAN, M.P. *307*
GALÁN, R.H. *294*
GALEFFI, C. *301, 311*
GAMES, D.E. *286, 291, 294*
GANGULY, A.K. *294*
GANGULY, D. *290*
GANGULY, N.C. *313*
GANI, G. *74*
GANNETT, F. *76*
GANTIMUR, D. *295*
GAONA, R. *294*
GARCÍA, E.G. *300*
GARCIA, M. *295*
GARCÍA, S. *292*
GARCIA BILBAO, J.L. *295*
GARCIA, J., F. *312*
GARDUÑO, J. *292*
GARG, S.K. *295, 308*
GARIBOLDI, P. *285*
GARWIN, J.L. *78*
GASHIMOV, N.F. *283, 295, 300*
GAWAD, D.H. *300*
GHEN, Z.-X. *295*
GHISALBERTI, E.L. *295*
GHOSH, P. *295*
GIL, J.A. *81*
GLEYE, J. *314*
GODFREY, O. *80*
GOLGBERG, I. *77*
GOLOVINA, L.A. *295*
GONCALVES DE LIMA, O. *292*
GONZÁLEZ, A.G. *295, 296*
GOODMAN, J.J. *75*
GORDEE, E.Z. *73*
GÖREN, N. *296*
GORMAN, M. *73*
GORST-ALLMAN, C.P. *73*
GOSWAMI, S. *313*
GOTTLIEB, O.R. *291, 306*
GOVINDACHARI, T.R. *296*
GOVINDAN, S.V. *298*

GRAF, E. *296*
GRANDE, M. *292, 296*
GRAY, A.I. *287, 296*
GREGER, H. *296, 298, 312*
GRENZ, M. *288*
GRISEBACH, H. *316*
GRITZKY, R. *314*
GRONENBORN, A.M. *78*
GU, L.-H. *297*
GUEVARA Z., E.Y. *293*
GUIOTTO, A. *309*
GUISE, G.B. *297*
GUNASEKERA, S.P. *297*
GUO, X. *297*
GUPTA, B.D. *286, 297, 302*
GUPTA, B.K. *297*
GUPTA, D.N. *311*
GUPTA, G.K. *297*
GUPTA, M. *297*
GUPTA, M.C. *284*
GUPTA, P.K. *284*
GUPTA, S.R. *285, 295, 299, 307, 308*
GYLLE, L. *303*

HAGINIWA, J. *299*
HALAWEISH, F.T. *286, 288*
HALE, R.S. *80*
HALIM, A.F. *286, 288, 297*
HALL, D. *289*
HALL, D.A. *305*
HALLAM, S.E. *80, 81*
HALLS, T.D.J. *310*
HAMADA, K. *298*
HAMASAKI, F. *285*
HAMBURGER, M. *297*
HAMILL, R.L. *73, 76*
HAMMES, G.G. *79*
HAN, G.-Q. *307*
HANDA, K.L. *297*
HANDA, S.S. *285*
HANEY, M.E. *72*
HANSBERRY, R. *306*
HANUMAIAH, T. *297*
HARBORNE, J.B. *292*
HARKAR, S. *297*
HARMER, R.A. *294*
HARPER, S.H. *297*
HARWOOD, L.M. *289*
HASE, T. *299*
HASKINS, N.J. *291, 294*
HASLER, H. *74*
HASLINGER, E. *296*
HATA, K. *297, 301, 302*
HATAKEYAMA, S. *74*
HATANO, T. *297*

HATTORI, M. *297*
HAUCK, P.R. *72*
HAVELUND, S. *303*
HAYAKAWA, K. *297*
HAYASHI, M. *76*
HAYASHI, Y. *306*
HAYDOCK, S.F. *81*
HEIDEPRIEM, H. *314*
HENKEL, W. *7, 73*
HENSENS, O.D. *75*
HERATH, W.H.M.W. *292*
HERMSMEIER, M. *72*
HERNÁNDEZ, J.D. *300*
HERNÁNDEZ, M.G. *296*
HERNANDEZ, O. *291*
HERZ, W. *286, 291, 297, 298, 311*
HIDALGO, J. *300*
HIFNAWY, M.S. *298*
HIGA, T. *298*
HIKINO, H. *302, 308*
HILTUNEN, R. *314*
HIRAI, M. *294*
HIRAI, Y. *297*
HIRAKURA, K. *298*
HIYAMA, T. *309*
HOEHN, M.M. *72*
HOELLERER, E. *79*
HOELTKE, H.-J. *79*
HOFER, O. *296, 298, 312*
HOFFMAN, B. *79*
HOLAK, T.A. *78*
HOLMES, D.S. *75, 76*
HOLZER, K.P. *79*
HONDO, G. *285*
HONG, C.Y. *300*
HOPWOOD, D.A. *79–81*
HORI, K. *73, 298*
HOSON, M. *304*
HOSOYA, K. *299*
HOSTETTMANN, K. *297*
HOU, L. *290*
HOYE, T.R. *31, 75*
HUANG, B.-S. *295*
HUANG, S.-C. *298, 315*
HUANG, W.-Y. *79*
HUBBARD, B.R. *75*
HUBER, M.L. *80*
HUDDLESTON, J. *314*
HUEBNER, C.F. *312*
HULIN, B. *75*
HUMPHREYS, G.O. *80*
HUNAITI, A.R. *77*
HUNTER, I. *80*
HUSSAIN, R.A. *302*
HUTCHINSON, C.R. *52, 74, 76, 77, 80, 81*
HUTCHINSON, D.W. *298*

ICHINO, K. *313*
IINUMA, M. *298*
IITAKA, Y. *294, 300*
IKEGAMI, M. *300*
IMAI, F. *298*
IMAMURA, K. *306*
IMRAPORN, S. *313*
INABA, T. *302*
INGHAM, J.L. *299*
INNOCENTI, G. *309*
INOUE, M. *285, 299, 300*
INOUE, T. *306*
ISAEV, N.Y. *283*
ISHIDA, T. *285*
ISHIHARA, M. *313*
ISHII, H. *292, 299*
ISHIKAWA, T. *292, 299*
ISHRATULLAH, K. *301*
ISMAILOV, N.M. *293*
ITAI, A. *298*
ITO, C. *299, 300*
ITO, H. *309*
ITO, K. *313*
ITO, Y. *311*
ITOH, K. *298*
IVIE, G.W. *299*
IWAGAWA, T. *299*
IWASAKI, S. *77*
IWASAWA, N. *304*

JACKOWSKI, S. *78*
JACKSON, R.F.W. *299*
JAIN, A.C. *299*
JAIN, A.K. *284, 299, 307*
JAIN, G.K. *311*
JAIN, S.M. *299*
JAKUPOVIC, J. *288, 299, 312*
JAMES, J.C. *75*
JAYATILAKE, G.S. *297*
JEFFERIES, P.R. *295*
JEFFREYS, J.A.D. *299*
JENNINGS, R.C. *292*
JERRIS, P.J. *299*
JHA, H.N. *307*
JHA, I.S. *290*
JIMENEZ, B. *292*
JOHNSON, A.P. *300, 307*
JOHNSON, L.F. *313*
JONES, H.A. *300*
JONES, N.D. *76*
JONES, R.C.F. *287, 291*
JONES, S.B. *299*
JONG, T.-T. *315*
JORDAN, P.M. *79*
JORGE, Z.D. *305*
JOSEPH-NATHAN, P. *285, 293, 300*

JOSHI, B.S. *294, 300*
JOSHI, P.C. *300*
JOULLIE, M.M. *290*
JU-ICHI, M. *294, 299, 300*
JURD, L. *300, 303*

KACZMAREK, S. *80*
KADYROV, A.S. *300, 309*
KAGA, H. *300*
KAGRAMANOV, A.A. *295, 300*
KAJIMOTO, T. *300*
KAJIURA, I. *294, 299, 300*
KAKIUCHI, N. *297*
KALRA, V.K. *300*
KAMAT, V.N. *294*
KAMBOJ, V.P. *311*
KANEMASA, S. *313*
KANEMATSU, K. *297*
KANG, S.H. *300*
KANO, M.H.C. *295*
KAPIL, R.S. *248, 285, 287, 302, 308, 311, 312*
KAPLAN, L. *75*
KAPOOR, V.K. *311*
KARANOVIĆ, L. *312*
KARURI, P. *310*
KASAI, R. *310*
KASTURI, R. *79*
KATHPALIA, Y.P. *300*
KATO, A. *294*
KATO, I. *313*
KATZ, L. *70, 81*
KAUPPINEN, S. *78*
KAVANAGH, P.J. *293*
KAWADA, T. *304*
KAWAGUCHI, A. *77*
KAWAI, H. *302*
KAWAI, K. *306*
KAWANO, N. *311*
KAWASAKI, C. *300*
KAWASAKI, J. *311*
KAWAZU, K. *300*
KAZLAUSKAS, R. *304*
KEARSLEY, S.K. *78*
KELKAR, S.L. *307*
KELLER, P.J. *74*
KELLER-JUSLEN, C. *73*
KEMPF, D. *72*
KERIMOV, Y.B. *283*
KERR, R.R. *313*
KHAFAGY, S.M. *303*
KHALEELULLAH, M. *308*
KHALID, S.A. *301*
KHALIL, M.M. *294*
KHAN, N.U. *301*
KHAN, R.A. *308*

KHAN, S.A. *294, 310*
KHANDURI, C.H. *284*
KHANNA, M. *284*
KHANNA, N.M. *311*
KHASANOV, T.K. *295, 301, 302, 305, 306*
KHASTGIR, H.N. *301*
KHATTAB, A. *289*
KHOPKAR, P.P. *304*
KHURANA, S.K. *301*
KIESER, H.M. *80, 81*
KIM, Y. *78*
KIMURA, Y. *299*
KING, H.D. *73*
KING, R.M. *288, 289, 299, 315*
KING, T.J. *73*
KINGHORN, A.D. *285, 315*
KINOSHITA, K. *76*
KINOSHITA, T. *298, 301, 311*
KINTZINGER, J.P. *294, 310*
KIRKIACHARIAN, B. *301*
KIRST, H.A. *76*
KIR'YANOVA, I.A. *286, 301*
KISHI, K. *17, 74*
KISHIBUCHI, N. *298*
KISO, Y. *308*
KITAGAWA, T. *294*
KLAGES, A.L. *78*
KLAUBERT, D.H. *289*
KLEIN, E. *34, 76, 302*
KNUCKLES, B.E. *288, 312*
KOBAYASHI, J.-I. *299*
KOKWARO, J.O. *301*
KOLUTTUKUDY, P.E. *77*
KOMATSU, M. *301*
KOMISSARENKO, N.F. *301*
KONG, Y.C. *301*
KÖNIG, W.A. *308*
KONNO, C. *302, 308*
KOSAK, A.I. *308*
KOSKINEN, A. *312*
KOTTIG, H. *79*
KOUL, S.K. *301*
KOUNO, I. *311*
KOVALEV, I.P. *301*
KOWAL, C. *73*
KOZAWA, M. *285, 286, 301, 302*
KOZAWA, W. *297*
KOZOVSKA, V. *302*
KOZUKA, M. *315*
KRAUS, L. *308*
KRISHNAMOORTHY, V. *301*
KRISHNASWAMY, N.R. *287*
KRSTANOVIĆ, I. *312*
KUHN, M. *73*
KUHNKE, J. *302*
KUHSTOSS, S. *80*

KUKLA, A.S. *300*
KULANTHAIVEL, P. *291*
KULIEV, A.A. *293*
KULIEV, Z.A. *293, 302*
KUMAR, D. *284*
KUMAR, N. *298, 311*
KUMAR, P.S.S. *305*
KUMAR, R. *286, 302*
KUMAR, S. *302, 308*
KUMAR, V. *302, 314*
KUNDU, A.B. *286*
KUNESCH, G. *306*
KUNNG, F.A. *315*
KUOH, C.-S. *315*
KUPCHAN, S.M. *302*
KUROYANAGI, M. *302*
KUSANO, K. *294*
KUSHMURADOV, A.Y. *309*
KUTNEY, J.P. *302*
KUZIORA, M.A. *79*
KUZNETSOVA, G.A. *302*

LAI, J.-S. *315*
LAKSHMI, V. *302*
LALITH, V.R. *294*
LAMBERTON, J.A. *302*
LAMNAOUER, D. *302*
LAMY, K. *307*
LANDGE, A.B. *314*
LANKIN, D.C. *285, 313*
LATCH, G.C.M. *314*
LAU, K.H. *301*
LAUE, E.D. *73, 75*
LAWRIE, K.W.M. *305*
LEADLAY, P.F. *55, 69, 70, 78, 80, 81*
LEDERER, E. *307*
LEE, J.J. *73*
LEE, M.S. *74*
LEEPER, F.J. *73, 75*
LEGLER, G. *313*
LEHTONEN, E.-M.M. *76*
LEITÃO FILHO, H.F. *303*
LEMMICH, E. *303*
LEMMICH, J. *303, 313*
LEUNG, J.O. *81*
LE VAN, N. *289, 303*
LEWANDOWSKI-SKARBEK, M. *80*
LEWIS, M.D. *74*
LEY, S.V. *73*
LI, J.-R. *314*
LI, X. *297*
LI, Z. *77*
LIANG, H.-T. *310*
LIANG, T.-C. *74, 75, 77*
LIN, L.J. *303*
LIN, X.-Y. *310, 314*

LIN, Z. *312*
LINK, K.P. *289, 312*
LINSCHEID, M. *289*
LIOU, M.-J. *315*
LIU, W. *79*
LIU, W.-C. *314*
LIVINGSTON, A.L. *288, 303*
LOCK DE O., U. *285*
LOOSLI, H.R. *73*
LOSICK, R. *80*
LUIS, F.R. *292, 294–296, 303*
LUIS, J.R. *296*
LUKACS, G. *74*
LUNA, I. *292*
LUNDIN, R.E. *288, 303, 312*
LUTHRIA, D.L. *286*
LYMAN, R.L. *288*
LYNEN, F. *61, 78, 79*

MABELIS, R.P. *73*
MABRY, T.J. *304, 313*
MADHUSUDANAN, K.P. *286*
MADRUZZA, G. *290*
MAGALHÃES, A.F. *303*
MAGALHÃES, E.G. *303*
MAGALHÃES, M.T. *306*
MAHANDRU, M.M. *303*
MAHENDRAN, M. *292*
MAHMOUD, Z.F. *303*
MAJUMBER, P.L. *303*
MAJUMDAR, S.G. *295*
MAJUMDER, P.C. *293*
MALI, R.S. *303*
MALIK, S.B. *286*
MALIKOV, V.M. *287, 293, 295, 300–302, 304–306, 309, 314, 315*
MALLICK, B. *286*
MALPARTIDA, F. *80, 81*
MANANDHAR, M.D. *303, 311*
MANCHEÑO, B. *296*
MANDAL, A.K. *290*
MANDAL, S. *300*
MANFREDINI, S. *286*
MANMADE, A.H. *294*
MANZINI, P. *309*
MARCOTULLIO, M.C. *290*
MARINI-BETTOLO, G.B. *292, 301*
MARINO, M.L. *287, 314*
MARKELOVA, E.V. *302*
MARMAN, E.-S.M. *297*
MARSHALL, D.S. *297*
MARTIN, J.H. *75*
MARTIN, M.-T. *302*
MARTINEZ-MARTIR, A.I. *289*
MASIROV, S.M. *306*
MASON, I. *75*

MASSANET, G.M. *294, 303*
MATA, R. *303*
MATANO, Y. *304*
MATHAI, K.P. *289, 305*
MATIĆ, S. *312*
MATKARIMOV, A.D. *287, 304*
MATSUBARA, H. *77, 78*
MATSUOKA, M. *299*
MATSUSHIMA, P. *80*
MATSUURA, T. *315*
MATSUYAMA, Y. *285, 301, 302*
MATTICK, J.S. *78, 79*
MAYO, K.H. *78*
MAZUREK, M. *312*
MCCALLUM, J.S. *315*
MCCORMICK, A. *291*
MCGARRY, E.J. *290*
MCHALE, D. *304*
MCINNES, A.G. *74, 77*
MCKILLOP, C. *80*
MCPHERSON, D.D. *285*
MEDINA, J.M. *295*
MEEGAN, C.J. *296*
MEHTA, A.C. *284*
MEHTA, S. *284*
MEHTA, V.D. *284*
MEIER, B. *293*
MELIBAEV, S. *301, 305*
MÉNDEZ, J. *304, 305*
MENDOZA, S. *300*
MENDOZA, V. *300*
MENG, Q. *304*
MENGHINI, A. *290*
MENICHINI, F. *292*
MENTZER, C. *301*
MENZEL, D. *304*
MERIJANIAN, A. *292*
MERKEL, K.E. *291, 294*
MERLINI, L. *285*
MESSANA, I. *301*
MESTER, I. *308, 309*
MEYER, B.N. *304*
MIGUEL DEL CORRAL, J.M. *292*
MIHASHI, S. *304*
MINKER, E. *308, 309*
MIR-BABAEV, N.F. *310*
MISHAAL, S. *289*
MISRA, L.N. *288*
MITRA, C.R. *306*
MITSUHASHI, O. *77*
MITSUI, T. *300*
MITTAL, R.S. *290*
MITZUGAKI, M. *78*
MIYACHI, K. *297*
MIYAKADO, M. *304*
MIYAMOTO, K. *297*

MIYAZAKI, T. *304*
MIZUNO, M. *298*
MIZUNO, T. *299*
MLADENOVIĆ, S. *312*
MODY, N.H. *79*
MOHAMED, A.H. *79*
MOLHO, D. *302*
MONDACA, A. *291*
MORETTI, C. *314*
MORI, S. *73*
MORINAGA, K. *309*
MORITA, N. *302*
MORRIS, M.P. *294*
MORS, W.B. *295*
MORTON, G.O. *75*
MORTON, T.C. *302*
MOTAMEDI, H. *80, 81*
MOULIS, C. *314*
MOWERY, P.C. *75*
MUELLER, G. *80*
MUELLER, W.H. *294*
MUJUMDAR, R.B. *288, 304, 308*
MUKAIYAMA, T. *304*
MUKERJEE, S.K. *304*
MUKHERJEE, I. *284*
MUKHERJEE, K. *284*
MULLER, G. *79*
MULTARI, G. *311*
MURAE, T. *302*
MURAGUCHI, M. *300*
MURAKAMI, T. *298, 304*
MURARI, R. *287*
MURIEL, L. *292*
MURPHY, C.M. *78*
MURRAY, R.D.H. *83, 285, 305, 308*
MURTI, V.V.S. *305*

NABIEV, A.A. *305*
NAGAI, M. *306*
NAGARAJAM, G. *303*
NAGARAJAN, G.R. *305*
NAGARAJAN, K. *296*
NAGEM, T.J. *306*
NAGGERT, J. *79*
NAGRI, S.W.I. *301*
NAGUMO, S. *306*
NAIR, A.G.R. *306*
NAKAJIMA, S. *306*
NAKANISHI, K. *74*
NAKANO, F. *315*
NAKASHIMA, T.T. *74*
NAKATA, H. *311*
NAKAYAMA, M. *294*
NALLIN, M.K. *75*
NAMBA, T. *297*
NANO, G.M. *285, 314*

NARANTUYAA, S. *306*
NASIROV, S.M. *306*
NAZARIANS, L. *309*
NDENGU, B. *311*
NEL, W. *289*
NESS, T. *73*
NGADJUI, B.T. *306*
NICKLESS, J. *78*
NIEMEYER, H.M. *316*
NIGAM, S.K. *306*
NIKIFOROV, A. *296, 298*
NILGES, M. *78*
NISHINO, T. *297*
NISHIZAWA, M. *306*
NIU, F. *312*
NIU, S.L. *310*
NKENGFACK, A.E. *294*
NOGUCHI, M. *311*
NOGUCHI, T. *304*
NOHARA, T. *300, 315*
NOMURA, K. *73*
NOMURA, S. *77*
NOMURA, T. *294, 298*
NONNENMACHER, G. *291*
NORTON, L.B. *306*
NOVÁK, I. *308, 309*
NOWICKI, W. *311*
NOZAKI, H. *309*
NOZAWA, K. *306*
NOZOE, S. *77*
NYIREDY, S. *306, 314*
NYIREDY-MIKITA, K. *306*

O'CALLAGHAN, N.B. *296*
OGASAWARA, K. *76*
O'HAGAN, D. *46, 50, 73, 74*
OHIGASHI, H. *300*
OHNO, N. *304*
OHSHIMA, Y. *306, 311*
OHSUKI, S. *297*
OIKAWA, H. *75*
OKABAYASHI, K. *304*
OKEGAWA, O. *309*
OKOGUN, J.I. *306*
ÖKSÜZ, S. *296, 313*
OKUDA, K. *301*
OKUDA, S. *77*
OKUDA, T. *297*
OKUYAMA, T. *285, 300, 304, 306, 309, 310, 312*
OLLIS, W.D. *293, 294, 306*
OMURA, M. *299, 300*
OMURA, S. *55, 74, 77, 78*
O'NEILL, M.J. *307*
ONO, M. *310*
ORAZMUKHAMEDOVA, N.O. *295*

O'REILLY, J. *293*
ORMANCEY-POTIER, A. *307*
OSADA, H. *304*
OSHIMA, Y. *302*
O'SULLIVAN, A.M. *293*
OTT, W.R. *76*

PACHE, W. *73*
PACHLER, K.G.R. *289*
PADUNGKUL, N. *313*
PAGÉ, J. *291*
PAI, B.R. *296*
PAKNIKAR, S.K. *307*
PALMER, C.J. *287, 291*
PAN, J.-X. *307*
PANDO, E. *294, 303*
PANETTA, J.A. *307*
PANHEY, V.B. *311*
PANT, P. *287*
PARKINS, H. *289*
PARMAR, V.S. *301, 305, 307*
PARTHASARATHY, M.R. *307*
PARTHASARATHY, P.C. *296*
PARTON, B. *292*
PASCHAL, J.W. *73, 76*
PASTORINI, G. *309*
PATAMIA, M. *301*
PATEL, D.V. *76*
PATERSON, I. *31, 75*
PATHAK, V.P. *299*
PATIL, K.S. *303*
PAUL, I.C. *72*
PAVLOVIC, S. *302*
PAZIRANDEH, M. *79*
PELTER, A. *300, 307*
PEREL'SON, M.E. *302, 307, 314, 315*
PETCHER, T.J. *73*
PHADKE, C.P. *307*
PHILIP, S. *289, 307*
PIATAK, D.M. *311*
PICCI, V. *285, 314*
PIERA, F. *296*
PIMENOV, M.G. *293, 307, 311, 314*
PINAR, M. *307*
PINHEIRO, R.M. *292*
PINO, O. *296*
PIZZA, C. *285, 291*
PODPETSCHNIG, E. *309*
POLLINI, G.P. *286*
POLONSKY, J. *306–308*
POMPI, V. *311*
POPLI, S.P. *285, 302, 308, 311*
POSPISIL, S. *75*
POUSSET, J.-L. *294, 298*
PRABHU, B.R. *286*
PRAKASH, C. *284*

PRAKASH, D. *302, 308*
PRASAD, A.V.K. *75, 308*
PRASAD, N. *290*
PRASAD, Y. *290*
PRESTEGARD, J.H. *78*
PRUESS, D.L. *73*
PRUESS, R.G. *73*
PRZYBYLA, C.A. *76*
PURUSHOTHAMAN, K.K. *308*

QIN, G.-W. *74*

RABARRON, A. *294*
RAHMAN, S.K. *72*
RAHMAN, W. *310*
RAIZADA, M.B. *308*
RAJ, K. *308*
RAJ, R. *302*
RAJAGOPALAN, P. *308*
RAJAN, S. *75*
RAJANI, P. *308*
RAJENDRAN, A. *308*
RAKHMANKULOV, U. *295, 301*
RALL, G.J.H. *289*
RAMAKRISHNA, K.V. *308*
RAMA RAO, A.V. *288, 304, 308*
RANDHAWA, Z. *79*
RANGASWAMI, S. *287, 311*
RANI, N. *284*
RANI, U. *305*
RAO, B.K. *297*
RAO, J.U.M. *297*
RAO, K.V.J. *297*
RAO, P.P. *308*
RAO, P.S. *285, 308*
RAO, R.A. *80*
RAO, R.N. *80*
RAPHAEL, R.A. *299*
RAPOPORT, H. *307*
RASTOGI, D.K. *290*
RASTOGI, R.P. *287*
RASTON, C.L. *295*
RATHI, S.S. *304*
RATHORE, J.S. *307*
RATTAN, R.K. *310*
RAVELO, A.G. *296*
RAVINDRAN, V.K. *303*
RAVINDRANATH, K.R. *300*
RAY, A.B. *302, 308*
RAZDAN, T.K. *297*
REDMAN, B.T. *306*
REDSHAW, S.D. *291*
REED, G.F. *291*
REESE, P.B. *77*
REGUERO, T. *303*
REHER, G. *308*

REICHELT, J. *304*
REISCH, J. *302, 308, 309*
RETEY, J. *74, 77*
REYES, R.E. *296*
REYNOLDS, K.A. *74*
REYNOLDS, W.F. *293*
RICHARDSON, M.A. *80*
RINGELMAN, E. *79*
RIPKA, S. *79, 80*
RISLEY, J.M. *74*
RITCHIE, E. *297*
RITCHIE, G.A.F. *73, 75*
RIZZACASA, M.A. *309*
ROBERTS, G.A. *78, 81*
ROBERTS, L.M. *79*
ROBERTS, M.F. *307*
ROBERTS, R.J. *306*
ROBIEN, W. *296*
ROBINSON, H. *288, 289, 299, 315*
ROBINSON, J.A. *1, 73, 74–77*
ROCK, C.O. *78*
RODIGHIERO, P. *309*
RODRÍGUEZ, B. *295, 307*
RODRIGUEZ, F. *290*
RODRIGUEZ-LUIS, F. *289*
RODRIGUEZ-UBIS, J.C. *289, 290, 292*
ROJAHN, W. *34, 76*
ROMÁN, L.U. *300*
ROMERO, A.G. *30, 75*
ROMERO, M.L. *293*
ROQUE, N.F. *291*
ROSENFELD, I.S. *77*
ROUESSAC, F. *285*
ROY, S. *290*
ROZHKOVA, L.I. *295, 300*
RÓZSA, Z. *309*
RUANGRUNGSI, N. *313*
RUBIO, M. *292*
RÜCKER, G. *292*
RUPP, R.H. *286*
RUSSELL, S.T. *75, 76*
RUSTAIYAN, A. *309*

SADAKANE, N. *77, 78*
SADAVONGVIVAD, C. *309*
SAGITDINOVA, G.V. *309*
SAIDA, I. *298*
SAIDKHODZHAEV, A.I. *295, 300, 306, 309*
SAIKI, Y. *294, 298, 304, 309*
SAIMOTO, H. *309*
SAITO, I. *315*
SAITOH, T. *301, 309, 311*
SAKAI, S. *309*
SAKAI, T. *309*
SAKAKIBARA, H. *74*
SAKAKIBARA, I. *285, 309, 310*

SALKELD, I.C. *285*
SALVÁ, J. *294, 303*
SAMMAKIA, T. *75*
SAMMES, P.G. *291*
SÁNCHEZ-VIESCA, F. *310*
SANDUJA, R. *301, 307*
SANDUJA, S.K. *307*
SAN FELICIANO, A. *292*
SANKAWA, U. *298*
SANTHANAM, P.S. *297*
SARADA, A. *308*
SARADHI, K.P. *307*
SARASWARTHY, A. *308*
SARG, T.M. *303*
SARGENT, M.V. *309*
SARKAR, S. *290*
SAROJA, T. *304*
SASAKI, H. *310*
SASHIDA, Y. *311*
SATAKE, T. *298, 304*
SATI, O.P. *310*
SATI, S.P. *310*
SATO, K. *299*
SATYANARAYANA, P. *310*
SATYSPEROVA, I.F. *301*
SAVINA, A.A. *286, 314*
SAXENA, V.K. *310*
SCHACHT, U. *313*
SCHEINMANN, F. *290*
SCHEUER, P.J. *298*
SCHILLTZ, E. *79*
SCHMID, H. *293, 311*
SCHNEIDERS, G.E. *310*
SCHONER, B.E. *80*
SCHREIBER, F.G. *310*
SCHREIBER, S.L. *30, 75*
SCHULMAN, M.D. *75*
SCHULTE, G. *75*
SCHUSTER, A. *299*
SCHWEIZER, E. *79*
SCHWEIZER, M. *79*
SDOBNINA, L.I. *293*
SECUNDINA, M. *292*
SEDMERA, P. *75*
SEITMURATOV, E. *287, 304*
SEKIGUCHI, F. *299*
SEKIGUCHI, Y. *76*
SELIGMANN, O. *291, 316*
SELLIAH, S.S. *286, 297*
SEMENOV, A.A. *295*
SEMPLE, J.E. *290*
SEN, M. *310*
SEN, R. *290*
SENGUPTA, G.C. *303*
SENGUPTA, P. *301, 310*
SENGUPTA, S. *291, 310*

SENIOR, R.G.  297
SENO, E.T.  80
SEPULVEDA, S.  311
SERAFIM NETO, A.  292
SERKEROV, S.V.  310
SERVETTAZ, O.  287
SESHADRI, T.R.  284, 286, 287, 300, 304, 305, 310
SEVENET, T.  298
SEYEA, H.  306
SHABANA, M.  303
SHAH, G.D.  310
SHAH, R.C.  310
SHAH, V.R.  310
SHAIKH, S.  314
SHAMMA, M.  287
SHAMSUDDIN, K.M.  285, 294, 310
SHAMSUDDIN, T.  294, 310
SHANG, T.-M.  310
SHANK, R.C.  289
SHARMA, B.R.  310
SHARMA, D.K.  310
SHARMA, M.  311
SHARMA, N.D.  284, 285, 295, 299
SHARMA, P.  310, 311
SHARMA, P.N.  308, 311
SHARMA, R.B.  311, 312
SHARMA, R.P.  286, 291, 311
SHAW, G.J.  288, 311
SHAW, S.C.  290
SHE, Q.-L.  295
SHEICHENKO, V.I.  286, 307
SHEN, F.-L.  314
SHEN, M.-S.  302
SHERIDAN, J.B.  304
SHERMAN, D.H.  79, 81
SHERMAN, M.M.  74, 77
SHERRINGHAM, J.A.  76
SHIBATA, S.  285, 300, 301, 304, 306, 309–312
SHIBUKAWA, N.  300
SHIGEMATSU, N.  311
SHIMAZAKI, Y.  76
SHIMIZU, Y.  74, 75
SHIMOMURA, H.  311
SHIOTSU, M.  302
SHIOZAWA, T.  311
SHIPCHANDLER, M.  311
SHIRATAKI, Y.  301
SHOEB, A.  285, 311
SHOOLERY, J.N.  286, 290
SHU, Y.-Z.  297
SHUKLA, V.S.  311
SIALER DE ZAPATA, D.  292
SIBANDA, S.  311
SIDDIQUI, I.R.  284

SIDDIQUI, S.  290
SIDOROVA, I.P.  283
SIEGNER, A.  79
SIGGAARD-ANDERSEN, M.  78
SIH, C.J.  76
SIL, P.  295
SILVA, M.  288, 291
SIMONI, D.  286
SIMONITSCH, E.  311
SINGH, H.  311
SINGH, J.  284, 299
SINGH, K.R.R.P.  290
SINGH, M.M.  311
SINGH, R.  287
SINGH, R.P.  284, 311
SINGH, S.  307
SINNWELL, V.  308
SKELTON, B.W.  295
SKLYAR, Y.E.  286, 287, 293, 301, 307, 311, 314
SLACK, D.A.  287, 291
SMITH, D.M.  292
SMITH, J.  307
SMITH, P.W.  2, 72
SMITH, S.  79
SMITH III, A.B.  299
SNAPE, E.W.  73
SOINE, T.O.  311
SOKOLOVA, A.I.  307, 311
SOLOMON, J.  316
SOMVICHIEN, N.  292
SONDENGAM, B.L.  306
SOOD, G.R.  77
SOTHEESWARAN, S.  292
SOTO, R.  312
SPAVOLD, Z.  74
SPECTOR, D.M.  79
SPENCER, J.B.  79
SPENCER, R.R.  288, 312
SPINO, C.  76
SPREAFICO, F.  73
SPRING, M.S.  312
SRIMANNARAYANA, G.  308
STAHMANN, M.A.  312
STAINTON, P.  300
STANISLAS, E.  314
STANLEY, W.L.  312
STANZAK, R.  80
STAUNTON, J.  23, 73, 75
STAVER, M.J.  81
STECK, W.  312
STEFANOVIĆ, M.  305, 312
STEINMEYER, A.  288
STEVENSON, R.  293, 310
STICHER, O.  293, 294, 306, 314
STILL, W.C.  2, 30, 72, 75

STOECKLI-EVANS, H. *297*
STOKER, J.R. *312*
STOLOW, D.T. *79*
STONE, M.J. *72*
STONESIFER, J. *80*
STREICHER, S.L. *80*
STROBEL, H. *309*
STROSHANE, R. *73*
STUART, A.D. *295*
STUTZMAN-ENGWALL, K.J. *80*
SU, Y. *315*
SUBRAHAMANYAM, P. *310*
SUBBA RAO, N. *308*
SUBBA RAO, N.V. *294*
SUDGEN, F.S. *80*
SUGA, H. *313*
SUGDEN, D.A. *80*
SUHADOLNIK, J.C. *75*
SULTANBAWA, M.U.S. *286, 297*
SUN, H. *312*
SUORTTI, T. *312*
SUPAVILAI, P. *309*
SUTER, P.J. *73*
SUTHERLAND, I.O. *306*
SWADA, T. *315*
SWAGER, T.M. *312*
SWAROOP, D. *311, 312*
SWEENY, J.G. *302*
SYRCHINA, A.I. *295*
SZABO, G. *298, 312*
SZENDREI, K. *308, 309*

TABATA, M. *285*
TABATA, Y. *285, 286*
TAGLIAPIETRA, S. *285*
TAGUCHI, H. *310*
TAHARA, S. *299*
TAKABAYASHI, K. *79*
TAKAHASHI, K. *306*
TAKAHISHI, N. *300*
TAKANO, S. *76*
TAKATA, M. *306, 312*
TAKEDA, A. *309*
TAKENAKA, H. *306*
TAKENAKA, S. *76*
TAKIZAWA, T. *306*
TALAPATRA, B. *313*
TALAPATRA, S.K. *313*
TAM, Y.Y. *301*
TAMEZ R., S. *292*
TANAHASHI, S. *299*
TANAKA, H. *313*
TANAKA, N. *304*
TANAKA, O. *310*
TANAKA, T. *298*

TANAKA, Y. *74, 78*
TANDON, S. *287*
TANDON, S.P. *310*
TANI, S. *294*
TANIGUTI, M. *286*
TATUM, J.H. *313*
TAWIL, B.A.H. *294*
TAYLOR, D.A.H. *292*
TAYLOR, H.L. *304*
TAYLOR, P.B. *77*
TAYLOR, W.C. *297*
TEREM, B. *313*
THAKUR, R.S. *301*
THAKUR, S. *295*
THASTRUP, O. *303, 313*
THEBTARANONTH, C. *313*
THOMPSON, C.R. *288*
THOMPSON, E.B. *313*
THOMPSON, H.J. *313*
THOMPSON, J.C. *293*
THOMSON, R.H. *297*
THYAGARAJAN, G. *286, 311*
TILVE, S.G. *303*
TIWARI, K.P. *310*
TOMIMATSU, T. *315*
TOMLINSON, J.A. *298*
TOMODA, H. *77*
TOUBIANA, R. *308*
TOWERS, G.H.N. *293*
TOYONAGA, T. *306*
TRAMA, L.A. *294*
TRIPATHY, R.N. *287*
TSCHESCHE, R. *313*
TSOU, H.-R. *75*
TSUDA, R. *300*
TSUGE, O. *313*
TSUKAMOTO, Y. *78, 79*
TSUZUKI, K. *74*
TUAN, J.S. *81*
TURNER, D.L. *73, 74, 75*
TURNER, J.R. *80*
TUZLACI, E. *313*
TYRELL, P.M. *78*

UCHIYAMA, S. *78*
UDAGAWA, S. *306*
UEDA, E. *299*
UENO, A. *302, 309*
ULBRICH, B. *294*
ULUBELEN, A. *292, 296, 313*
UPADHYE, B.K. *314*
URATA, S. *311*
URBATSCH, L.E. *314*
URZUA, A. *294*
USGAONKAR, R.N. *314*

VAGELOS, P.R.  77
VAISIRIROJ, V.  313
VALLE, M.G.  314
VANAJA, K.  308
VANCE, D.  77
VAN DER M. BRINK, C.  289
VANDYSHEV, V.V.  307
VAN ETTEN, R.L.  20, 74
VANEK, Z.  75
VANMIDDLESWORTH, F.  76
VANNIER, S.H.  312
VAN WAGENEN, B.C.  314
VAQUETTE, J.  298
VARA, J.  80
VAZQUEZ-BUENO, P.  289
VDOVIN, A.D.  314, 315
VEDERAS, J.C.  20, 74, 77
VEERAVALLI, J.  307
VENTURELLA, G.  287
VENTURELLA, P.  287, 314
VERDE S., J.  293
VERENČEVIĆ, M.  312
VESCHAMBRE, H.  75
VESELOVSKAYA, N.V.  286, 307, 314
VEZHKHOVSKA-RENKE, K.  307
VIEIRA, D.M.  295
VINCIERI, F.F.  291
VINCIGUERRA, V.  292
VISHWAPAUL  310
VISWANATHAN, N.  314
VOGELOS, P.R.  78
VÖGTLE, F.  73
VOKOUN, J.  75
VOLPE, J.J.  78
VON WARTBURG, A.  73
VON WETTSTEIN-KNOWLES, P.  78
VON WRIGHT, A.  312
VOYER, N.  291
VUORELA, H.  314

WADA, Y.  297
WADHWA, V.  311
WADIA, M.S.  307
WAGNER, H.  291, 316
WAIGHT, E.S.  292, 297
WAKHARKAR, R.D.  314
WAKIL, S.J.  77–79
WALBA, D.M.  34, 72, 76
WALKER, C.B.  76
WALL, M.E.  304
WALTER, H.  79
WALTER, J.A.  74, 77
WAND, M.D.  76
WANDJI, J.  294
WANG, C.  314
WANG, L.  310

WANG, M.-S.  314
WANG, Q.-H.  294
WANG, S.-X.  297
WANG, Y.-G.  80
WANI, M.C.  304
WANNIGAMA, G.P.  292
WANZLICK, H.-W.  314
WARREN, M.M.  80
WARRINGTON, P.J.  290
WATANABE, K.  311
WATERMAN, P.G.  287, 299, 301
WAYKOLE, P.  314
WEBER, H.P.  73
WEBER, J.M.  80, 81
WEEKS, R.E.  73
WEILER, L.  76
WEINREB, S.M.  289
WEISSENSTEINER, W.  298
WEITZ, J.C.  289
WENKERT, E.  310
WESTLEY, J.W.  15, 20, 23, 43, 44, 46, 72, 73, 77
WHALLEY, W.B.  293
WHEELER, R.E.  293, 294
WHITE, A.H.  295
WHITING, D.A.  287, 291
WICKRAMARATNE, D.B.M.  302, 314
WIDHALM, M.  298
WIENHOLD, C.  288
WIESNER, P.  80
WILD, G.M.  76
WILKES, M.C.  76
WILLENBROCK, F.  74
WILLIAMS, B.  79
WILLIAMS, D.H.  72
WILLIAMS, D.J.  75
WILZER, K.A.  314
WIRASUTISNA, K.R.  314
WITKOWSKI, A.  79
WITT, S.C.  288, 303, 312
WOGAN, G.N.  289
WOLFF, I.  312
WOLFRUM, C.  314
WONG, E.  314
WONG, H.  79
WONG, R.  73
WONG, S.  291
WOODWARD, R.B.  7, 73
WORTH, G.K.  295
WU, T.-S.  298, 315
WULFF, W.D.  315

YADAEV, V.J.  303
YAGUDAEV, M.R.  287, 306, 314, 315
YAHIRO, K.  300
YAMADA, F.  310

YAMAHARA, J. *315*
YAMAMOTO, Y. *76*
YAMAWAKI, A. *309*
YANAKA, Y. *77*
YANEZ O., R. *312*
YANG, C.-C. *76, 77*
YANG, J. *315*
YANG, J.-S. *315*
YANG, P. *315*
YASUHARA, T. *297*
YATES, M.K. *311*
YE, J. *315*
YODO, M. *297*
YOKOE, I. *301*
YOKOI, T. *294*
YOSHI, E. *73*
YOSHIOKA, H. *304*
YOSHIZAWA, Y. *77*
YOSIOKA, I. *310*
YOUNG, C.-Y. *78*
YUAN, C. *315*
YUAN, Z. *79*
YUE, S. *74, 76*
YULDASHEV, M.P. *315*

ZAGORSKI, M.G. *74*
ZAKARIA, M.B. *315*
ZAMAN, A. *285*
ZAMUDIO, A. *293*
ZAWARE, R.N. *303*
ZDERO, C. *288, 289, 315, 316*
ZEGHDI, S. *305*
ZEMELMAN, R. *291*
ZENG, G.F. *295*
ZHANG, G. *290*
ZHANG, H. *315*
ZHANG, L.G. *316*
ZHANG, Y. *297*
ZHELEVA, A. *302*
ZHENG, Q.-T. *314*
ZHONG, S. *299*
ZHU, T.-R. *297*
ZILG, H. *316*
ZILLI, A.B. *292*
ZMEIKOV, V.P. *283*
ZMIJEWSKI, M.J. *73*
ZOGHBI, B. *291*

# Subject Index

A23187  8, 9, 43, 44
Acanthifolicin  12, 13
Acetate  7–12, 15, 20, 22, 23, 26, 38, 43, 44, 50–52
Acetic acid  30
Acetoacetyl-ACP synthase  58
7-Acetoxycoumarin  85
(E)-ω-Acetoxyferprenin  197, 275
(Z)-ω-Acetoxyferprenin  197, 275
(E)-ω-Acetoxyferulenol  196, 275
(Z)-ω-Acetoxyferulenol  196, 275
7-Acetoxy-4-methylcoumarin  85
7-Acetoxypectanone  169, 275
Acetyl-CoA  7, 51, 55–58, 61, 62
Acetyldaphnoretin  234, 275
Acetyldeparnol  167, 275
Acetyldihydromelin A  108, 275
Acetyldrimartol A  170, 275
– B  170, 275
Acetylisodrimartol A  170, 275
Acetylpectachol  168, 275
– B  168, 275
Acetyl transacylase  56
– transferase  59, 61, 62
Acetylumbelliferone  101, 275
Achillea depressa  166, 167
– ochroleuca  166, 167, 169–171
– pseudopectinata  167, 168
– schischkinii  181
Acrimarine A  109, 275
– B  109, 275
– C  109, 275
Actinorhodin  54, 55, 64–66
Acyl carrier protein  57, 58, 60, 62–64, 66, 69, 70
Acyl transferase  51, 52, 69
S-Adenosylmethionine  8
Adicardin  100, 275
Adina cordifolia  100
Aegelinol  112, 275
– benzoate  112, 275
Aegle marmelos  112
Albartin  168, 275
Albartol  167, 275

Aleurites fordii  158
– moluccana  150
Aleuritin  158, 275
Alyxia lucida  195, 215
Amino acids  15
Ammi majus  124, 135
Ammirin  249, 275
Amyris balsamifera  103, 175
– elemifera  110, 175, 176
– madrensis  107
Angelica apaensis  163
– archangelica  111, 155, 161–164
– dahurica  161–163
– edulis  125
– keiskei  124
– officinalis  135
– pachycarpa  162
– pubescens  105–107, 140, 141, 163
Angelicin  252, 275
Angelin  140, 275
Angelol  86, 248, 275
– A  86, 248, 275
– B  106, 275
– C  105, 275
– D  107, 275
– E  105, 275
– F  105, 275
– G  107, 275
– H  106, 275
– I  106, 275
Angeloside A  106, 275
Angustifolin  174, 275
Anhydronotoptol  136, 275
Anhydrorutaretin  155, 275
Anisocoumarin A  174, 275
– B  133, 275
– C  175, 275
– D  175, 275
– E  117, 275
– F  116, 275
– G  118, 275
Anisolactone  136, 275
(+)-Anomalin  253, 276
Anpubesol  107, 276

*Anthemis cretica* 169, 171
Antibiotic biosynthesis 15, 52, 54
Apaensin 163, 276
Apetalolide 190, 276
*Aphyllocladus denticulatus* 205, 211, 212
Aphyllodenticulide 211, 276
Apigravin 153, 276
*Apium graveolens* 134, 153, 154, 161, 249
– *leptophyllum* 154, 155
– *petroselinum* 102
Apiumetin 154, 276
Apiumoside 153, 276
Aplasmomycin 10, 11
Apterin 251, 276
*Aquilaria agallocha* 173
Aquillochin 173, 276
Arnocoumarin 111, 276
Arnottiacoumarin 156, 276
Arnottinin 250, 276
*Arracacia nelsonii* 131
Artanin 159, 276
Artelin 261, 276
*Artemisia alba* 168
– *apiacea* 151, 152
– *armeniaca* 151, 159, 160
– *caruifolia* 151
– *dracunculoides* 153
– *lacianata* 174
– *laciniata* 152, 159, 160
– *persica* 148, 149
– *pontica* 169
– *tanacetifolia* 159
– *vestita* 166
– *vulgaris* 262
Asacoumarin A 90, 276
– B 92, 276
*Aspergillus glaucus* 232
– *niger* 232
– *parasiticus* 20
Assafoetidin 91, 276
*Aster prealtus* 88, 89
*Atalantia ceylanica* 164
– *racemosa* 164
Aurantiumal 254, 276
Auraptenal 237, 276
Auraptenol 250, 276
Aureol 227, 276
Avermectin 7, 23, 24, 64
Averufin 20, 21

*Baccharis darwinii* 88, 89
Badrakemin 241, 276
– acetate 241, 276
Badrakemone 241, 276
Balsamiferone 86, 175, 276
Benahorin 259, 276

Bethancorin 149, 276
Bethancorol 149, 276
Bhubaneswin 230, 276
Bicoumol 230, 276
Bocconin 128, 276
*Boenninghausenia albiflora* 103, 230
– *sessilicarpara* 176
*Bothriocline laxa* 210, 213, 217, 218
– *longipes* 213, 214
– *ripensis* 213
Bothrioclinin 210, 276
*Brachyclades megalanthus* 203, 205, 206
Brachycoumarin 203, 276
Brevetoxin A 11–13, 25, 29
– B 11–13, 25, 29
*Brocchia cinerea* 150, 168
Bruceol 254, 276
Bungediol 148, 276
Buntansin 102, 276
Butyrate 7, 8, 10, 15, 22, 38, 43, 44, 50–52
[1-$^{13}$C]n-Butyrate 12
[1-$^{13}$C, $^{18}$O$_2$]Butyrate 20
n-Butyryl-CoA 12
( – )-Byakangelicin 161, 276
( ± )-Byakangelicol 161, 276

*Cachrys libanotis* 112
– *sicula* 110
Calaustralin 189, 276
Calcimycin 8
Calophyllolide 189, 276
*Calophyllum apetalum* 190
– *cordatooblongum* 181
– *inophyllum* 189, 191, 192
– *soulattri* 191
– *tomentosum* 191
Campestrinol 130, 276
Campestrinoside 127, 276
Campestrol 128, 276
Candicanin 236, 276
Candicidin 64, 65
Capensin 165, 276
Capnolactone 238, 276
Carbomycin 64, 65, 67
Carboxylic acid 52
Carriomycin 45, 47
*Casearia graveolens* 121
Casegravol 121, 276
– isovalerate 121, 276
Cauferidin 94, 276
Cauferin 93, 276
Cauferinin 98, 276
Cauferoside 93, 276
Cauloside 93, 276
Celereoin 134, 276
Celereoside 134, 276

Celerin 153, 276
*Cephalosporium caerulens* 54
Cerulenin 54, 55, 58
(2*R*,3*S*)-Cerulenin 55
Ceylantin 164, 276
Chalepin 263, 276
Chinchircumine 203, 276
*Chloroxylon swietenia* 103, 132, 143, 156, 175
Chloticol 119, 276
Citric acid 12
Citrubuntin 105, 276
*Citrus flavedo* 118, 123
– *funadoko* 102–104, 109
– *grandis* 102, 105, 142, 144
– *junos* 128
– *medica* 102, 114, 141, 142, 144, 164
– *paradisi* 135
– *sinensis* 104
– *tamurana* 114
– *unshiu* 114, 121, 142
Citrusal 250, 276
Clausarin 177, 276
*Clausena anisata* 116–118, 133, 136, 137, 155, 174, 175
– *indica* 155
– *lansium* 155
– *pentaphylla* 176, 177
– *wampi* 155
Clausmarin A 176, 276
– B 176, 276
*Cleome icosandra* 173
– *viscosa* 172, 173
Cleomiscosin A 173, 276
– B 172, 276
– C 173, 276
– D 172, 276
Cleosandrin 173, 276
CM-c$_2$ 117, 276
*Cneorum tricoccum* 149
*Cnidium monnieri* 117, 124
Cniforin A 251, 276
– B 124, 276
Coccidia 6
Coccidiostats 6
*Coelopleurum gmelinii* 106
*Coleonema album* 88, 151
– *calycinum* 88, 151
Colladin 241, 276
Colladocin 245, 276
Colladonin 240, 276
– isovalerate 93, 276
Compositae 85
Conferdione 244, 276
Conferin 243, 276
Conferol 242, 276

– acetate 242, 276
Conferone 242, 276
Conferoside 97, 276
*Conocliniopsis prasiifolia* 149
*Conyza obscura* 147, 148
Cordatolide A 181, 276
– B 181, 276
Corylidin 227, 276
Coumarsabin 217, 276
Coumestrin 226, 276
Coumestrol 223, 276
Coumurrayin 255, 276
Coumurrin 118, 276
*Coutarea hexandra* 182–186, 194
– *latiflora* 186
CP-44661 46, 48, 49
Cyclase 66, 67
Cyclobisuberodiene 235, 276
Cyclobrachycoumarin 206, 276
Cycloethuliacoumarin 208, 277
Cycloisobrachycoumarin 205, 277
Cyclolongipesin 213, 277
Cyclolycoserone 212, 277

*Dalbergia baroni* 193
– *latifolia* 187
– *melanoxylon* 193
– *oliveri* 178
– *sissoo* 192, 193
– *stevensonii* 193
– *volubilis* 180, 185, 187
Dalbergin 192, 277
*Daphne gnidiodes* 101, 233
– *mezereum* 233
– *tangutica* 153
Daphneticin 153, 277
Daphnoretin 233, 277
Daphnorin 233, 277
*Dasycladus vermicularis* 220
Daunorubicin 64, 65
Dauroside A 101, 277
– B 101, 277
– C 147, 277
– D 134, 277
Deacetylkellerin 99, 277
Deacetyltadshikorin 90, 277
Decuroside I 109, 277
– II 109, 277
– III 110, 277
– IV 109, 277
– V 111, 277
( – )-3′R-Decursinol 112, 277
Dehydrase 51, 59, 66, 70
Dehydratase 56, 61
*exo*-Dehydrochalepin 86, 175, 277
10′,11′-Dehydrocyclolycoserone 212, 277

Dehydrogeijerin  248, 277
Dehydroindicolactone  155, 277
cis-Dehydroosthol  119, 277
trans-Dehydroosthol  251, 277
Dehydropectanone  169, 277
Demethyldaphnoretin  233, 277
Demethylluvangetin  156, 277
3-O-Demethylmonensin A  22, 24
Demethylsuberosin  247, 277
Dentatin  256, 277
Deoxybruceol  254, 277
20-Deoxy-17-epi-salinomycin  16–18
6-Deoxyerythronolide A  38
 – B  38, 67–71
 – biosynthesis  70
26-Deoxymonensin A  22, 24
20-Deoxysalinomycin  16, 17
Deparnol  166, 277
Derris glabrescens  221
 – robusta  221
 – scandens  222, 223
 – spruceana  222
Derrusnin  221, 277
Desertorin A  231, 277
 – B  231, 277
 – C  231, 277
Desmethylkotanin  232, 277
Desoxylacarol  151, 277
Dianemycin  23
1,3-Dibromopropane  60
Dicoumarol  234, 277
Dihydromelin A  108, 277
 – B  108, 277
Dihydrosuberenol  103, 277
Dihydroxanthyletin  112, 277
Dinoflagellates  11
Diospyros ismailii  236
 – sapota  146
Diospyroside  146, 277
Dioxyanomalin  253, 277
Diversin  237, 277
trans-Diversin  237, 277
Dolichlasium lagascae  219
Dracunculin  262, 277
Drimachone  171, 277
Drimanthone  171, 277
Drimartol A  169, 277
 – B  170, 277

Echinops niveus  182
Eclipta alba  227, 228
Edgeworin  232, 277
Edgeworoside A  236, 277
Edgeworthia chrysantha  232, 236
 – gardneri  215, 233, 234
Edgeworthin  233, 277

Edulisin I  125, 277
 – II  125, 277
Ekersenin  270, 277
Elemiferone  175, 277
Emericella desertorum  231
Enoyl reductase  51, 56, 59, 61, 70
2'-Epicycloisobrachycoumarin  206, 277
1'-Epi-6',7'-dehydrocyclolycoserone  212, 277
1'-Epilycoserone  204, 277
6',7'-Epoxyaurapten  237, 277
3',6'-Epoxycycloaurapten  237, 277
Epoxyfarnochrol  166, 277
Eriobrucinol  85, 254, 277
Eriocephaloside  234, 277
Erioside  165, 277
Erlangea fusca  205, 207
 – rogersii  218
Erlangeafusciol  207, 277
Erosnin  227, 277
Eryngium campestre  112
 – ilicifolium  110
Erythrina stricta  86
Erythromycin A  23, 38, 52, 53, 64, 67, 68
 – B  38–40
Escherichia coli  55–61, 63, 64, 67
Ethulia conyzoides  197, 207–210
Ethuliacoumarin  208, 277
Ethylmalonyl-CoA  10, 51
Ethylsuberenol  104, 277
Eupatorium lancifolium  149
Euphorbetin  231, 277
Euphorbia lathyris  231
 – lunulata  149
 – paralias  216
 – terracina  216
Exostema caribaeum  182–184, 186, 193
Exostemin  193, 277

Farnesiferol C  238, 277
Farnochrol  166, 277
Fatagarin  233, 277
Fatty acid synthase  50, 54–61, 63, 64, 67, 69
Fatty acids  7, 15, 38, 43, 46, 50–52, 54, 56, 60
Fecarpin  100, 277
Fekolin  91, 277
Fekolone  91, 277
Fekrol  91, 277
Fekrynol  92, 277
 – acetate  92, 277
Fepaldin  98, 277
Ferilin  94, 277
Fernolin  163, 277
Ferocaulicin  96, 277

Ferocaulidin 96, 277
Ferocaulin 96, 277
Ferocaulinin 96, 277
*Feronia limonia* 163
Feropolidin 242, 277
Feropolin 239
Feropolol 239, 277
Feropolone 239, 278
Ferprenin 196, 278
Ferruol A 265, 278
Ferucrinone 99, 278
Ferujol 151, 278
Ferukrin 98, 278
– acetate 99, 278
– isobutyrate 99, 278
*Ferula assafoetida* 90–92, 100, 195
– *communis* 195–197
– *conocaula* 93, 94, 96–98
– *diversivittata* 95
– *foetidissima* 99
– *foliosa* 92
– *galbaniflua* 95, 97
– *iliensis* 94
– *jaeschkeana* 151
– *karatavica* 90
– *kokanica* 97, 99
– *kopetdaghensis* 91, 92, 99
– *krylovii* 91, 92, 98
– *lehmanni* 91, 94
– *loscosii* 93
– *malacophylla* 89
– *marmarica* 93
– *microcarpa* 100
– *microloba* 99, 100
– *nevskii* 98
– *pallida* 98
– *schtschurowskiana* 97
– *sinaica* 94
– *tadshikorum* 90
– *teterrima* 94
*Ferulago granatensis* 110
Feselol 241, 278
– angelate 95, 278
Feshurin 97, 278
Feterin 94, 278
FK506 7
Flemichapparin C 226, 278
*Flemingia chappar* 226
Floribin 86
Foetidin 195, 278
Foliferin 92, 278
*Fraxinus floribunda* 86, 150, 216
– *sp.* 86
– *stylosa* 165
Fries rearrangement 86

Funadonin 102, 278
Furopinnarin 255, 278

Galbanic acid 240, 278
*Galipea trifoliata* 119
Galipein 119, 278
Gancaonin F 228, 278
Geijerin 86, 248, 278
Geiparvarin 85, 238, 278
Geldanamycin 9
Geranyl acetate 34
*Gerbera anandria* 197, 198
– *jamesonii* 197, 216
– *lanuginosa* 235
Gerberinol 235, 278
Glabrescin 221, 278
*Glehnia littoralis* 115
Gleinadiene 139, 278
Gleinene 137, 278
Glucose 9
Glycerate 9
Glycerol 11
$(1R, 2R)$-$[1$-$^2H_1,^3H]$Glycerol 11
*Glycine max* 226
Glycollate 9
Glycycoumarin 179, 278
Glycyrin 179, 278
Glycyrol 228, 278
*Glycyrrhiza sp.* 228
– *spp.* 179, 228
– *uralensis* 179, 228
*Gmelina arborea* 100
*Gnidia lamprantha* 234
Gnidiacoumarin 234, 278
*Gomortega keule* 218, 219
Granaticin 64–66
Grandivittin 112, 278
Grandivittinol 105, 278
*cis*-Grandmarin 141, 278
*trans*-Grandmarin isovalerate 142, 278
Gravelliferone 263, 278
*Gynura crepioides* 113
Gynuron 113, 278
Gypothamniol 211, 278
*Gypothamnium pinifolium* 204, 205, 211

Hainanmurpanin 122, 278
*Haplopappus multifolius* 147
Haploperoside A 146, 278
– B 147, 278
– C 146, 278
– D 146, 278
– E 147, 278
*Haplophyllum bungei* 133, 148

– *dauricum*   101, 133, 134, 147
– *obtusifolium*   145, 147, 148, 165, 166, 173, 177, 262
– *pedicellatum*   148
– *perforatum*   146, 147
– *ramosissimum*   177
– *schelkovnikovii*   220
– *tenue*   102, 152
– *versicolor*   123
– *villosum*   152
Haplopinol   147, 278
Haptusinol   165, 278
Hassanon   114, 278
*Helianthus heterophyllus*   166
*Helichrysum diosmifolium*   157
– *serpyllifolium*   157
– *stirlingii*   149
Heliclactone   218, 278
*Helicteres angustifolia*   218
*Helietta parvifolia*   176
Heraclenol acetonide   154, 278
– 2'-*O*-isovalerate   155, 278
– 2'-*O*-senecioate   155, 278
Heraclesol   158, 278
*Heracleum candicans*   236
– *granatense*   154, 162
– *lanatum*   141
– *leskovii*   158
– *mantegazzianum*   125
– *thomsonii*   141, 143, 144
Heratomin   258, 278
Heratomol   258, 278
Hermandiol   125, 278
Heterocyclic ring formation   15, 31
*Hintonia latiflora*   183
*Hippomarathrum cristatum*   104
Honydusin   144, 278
Hopeyhopin   107, 278
*Hortia arborea*   141
Hortinone   141, 278
Hortiolone   256, 278
7-Hydroxycoumarin   85, 86
10'-Hydroxycyclolycoserone   212, 278
Hydroxyeriobrucinol   254, 278
(*E*)-ω-Hydroxyferprenin   196, 278
(*Z*)-ω-Hydroxyferprenin   196, 278
12'-Hydroxyferulenol   195, 278
ω-Hydroxyferulenol   195, 278
(*E*)-ω-Hydroxyferulenol   195, 278
(*Z*)-ω-Hydroxyferulenol   195, 278
5-Hydroxy-6-methoxycoumarin   86
8-Hydroxypereflorin   217, 278
[2,3,-$^{13}$C$_2$]β-Hydroxythioester   40
*Hymenodictyon excelsum*   146
Hymexelsin   146, 278

ICI139603   6, 8, 9, 23, 28, 43
Imperatorin   259, 278
Indicolactone   155, 278
Indicolactonediol   155, 278
Inophyllolide   192, 278
*cis*-(+)-Inophyllolide   192, 278
*trans*-(+)-Inophyllolide   192, 278
Inophyllum A   191, 278
– B   191, 278
– C   192, 278
– D   191, 278
– E   192, 278
Ipomopsin   230, 278
*Ipomopsis aggregata*   230
Ismailin   236, 278
Isoangelol   107, 278
Isobergaptol   141, 278
[1-$^{13}$C]Iso-butyrate   12
Iso-butyryl-CoA   12, 15
Isobyakangelicol   161, 278
Isodalbergin   193, 278
Isodrimachone   171, 278
Isodrimartol A   169, 278
Isoerlangeafusciol   205, 278
Isoethuliacoumarin A   208, 278
– B   208, 278
– C   210, 278
Isoeuphorbetin   231, 278
Isofraxetin   260, 278
Isofraxoside   262, 279
Isogerberacoumarin   270, 279
Isoglycyrol   228, 279
Isolasalocid   15, 16, 44
Isolycoserone   205, 279
Isomammeigin   189, 279
Isomexoticin   140, 279
Isomurralonginol acetate   115, 279
– nicotinate   115, 279
Isomurranganon senecioate   122, 279
Isopeucenidin   251, 279
Isophlojodicarpin   122, 279
Isorutarin   259
Isosabandin   174, 279
Isosamarcandin   244, 279
– angelate   244, 279
Isosibiricin   255, 279
Isosojagol   224, 279
Isothamnosin A   235, 279
Isotriptiliocoumarin   206, 279
Isotriptospinocoumarin   210, 279

*Jatropha curcas*   157, 172
– *glandulifera*   172
Jatrophin   172, 279
Jayantinin   230, 279

Jumutinol 134, 279
*Jungia herzogiana* 203, 213
*Juniperus sabina* 217, 220
Junosmarin 128, 279

Kamolonol 100, 279
Karatavic acid 240, 279
Karatavicin 90, 279
Karatavicinol 238, 279
Kellerin 246, 279
β-Ketoacyl reductase 51, 56, 59, 61–63
– synthase 51, 56–62, 66, 69, 70
Ketoreductase 66, 67, 69
*Kielmeyera pumila* 189
Kinocoumarin 144, 279
Kiyomal 114, 279
Kokanidin 97, 279
Kopeolin 239, 279
Kopeolone 92, 279
Kopeoside 239, 279
Kotanin 232, 279
Kuhlmannin 194, 279

Lacarol 159, 279
Lacinartin 152, 279
– epoxide 152, 279
Lacinartindiol 152, 279
*Lactarias necator* 85, 215
Lanatin 141, 279
Lariside 145, 279
Lasalocid A 2, 4, 5, 8, 12, 15, 16, 20–22, 43,
  44, 46, 48, 49, 52, 53
Laserpitin 253, 279
Lasiocephalin 233, 279
Lasioerin 234, 279
*Lasiosiphon eriocephalus* 165, 233, 234
Latilobinol 90, 279
Lehmferidin 94, 279
Lehmferin 91, 279
Lenoremycin 23, 24, 27
*Leptodactylon californicum* 159
Leptodactylone 159, 279
Leptophyllidin 154, 279
Leptophyllin 259, 279
Leptophylloside 259, 279
Leucomycin 50
– A₃ 9
Leuseramycin 23
Libanotin A 86, 252, 279
*Libanotis buchtormensis* 86
Licopyranocoumarin 179, 279
*Limonia acidissima* 103
– *crenulata* 101
Linalool 31
(3R)-Linalool 34
(3S)-Linalool 34

LL-F28249α 23, 24
Lomatin 86, 252, 279
*Lomatium dissectum* 123, 124
Lonchocarpenin 223, 279
Lonchocarpic acid 222, 279
*Lonchocarpus sp.* 222
Lonomycin A 45
Lucernol 229, 279
*Lycoseris latifolia* 204, 211, 212
Lycoserone 204, 279
Lysocellin 46, 49

MAB 1 188, 279
*Machaerium kuhlmannii* 194
Macrolide antibiotics 37, 52
Macrolides 7
*Macrothelypteris torresiana* 180, 216
Maduramicin 23, 24
*Magydaris panacifolia* 161
Malonyl transacylase 56
– transferase 59, 61, 62
Malonyl-CoA 7, 11, 51, 55–57, 61
Mammea A/AA 188, 279
– A/AB 188, 279
– A/AD 187, 279
– A/BA 187, 279
– A/BB 187, 279
– A/BB cyclo D 189, 279
– B/AA 266, 279
– B/AA cyclo F 267, 279
– B/AB 266, 279
– B/AB cyclo D 268
– B/AB cyclo F 267, 279
– B/AC 266, 279
– B/AC cyclo F 267, 279
– B/BA 265, 279
– B/BA cyclo E 268, 279
– B/BA cyclo F 267, 279
– B/BB 86, 264, 279
– B/BB cyclo E 268, 279
– B/BB cyclo F 267, 279
– B/BC 264, 279
– B/BC cyclo E 268, 279
– B/BC cyclo F 266, 279
– B/BD 264, 279
– C/AB cyclo D 269, 279
– C/BB 265, 279
– D/BB 265, 279
– E/BA 86, 269, 279
– E/BB 86, 269, 279
*Mammea africana* 188
– *americana* 187, 188, 190
– *longifolia* 180
Mammeigin 190, 279
Mammeisin 188, 279
Manganese 35

Maoyancaosu 149, 279
Marmaricin 93, 279
Marmelide 259, 279
Marmesin acetate 110, 280
– isovalerate 110, 280
Marmin 237, 280
Matsukazelactone 230, 280
*Medicago sativa* 223–226, 229
Medicagol 225, 280
*Melampodium divaricatum* 151, 152
Melanettin 193, 280
Melannein 193, 280
*Melilotus alba* 234
Meranzin hydrate 250, 280
Merillin 117, 280
*Merrillia caloxylon* 117
*Mesua ferrea* 187, 189, 190
– *thwaitesii* 188
Mesuagin 189, 280
Mesuarin 190, 280
Mesuol 187, 280
Methionine 8–12
[$^{13}$C-Me]Methionine 12
8-Methoxycoumarsabin 220, 280
7-Methoxy-8-(15-hydroxypentadecyl)coumarin 86
( – )-8-Methoxyobliquin 166, 280
*O*-Methylcedrelopsin 150, 280
Methyldalbergin 193, 280
( – )-Methyldecursidinol 112, 280
Methylene chloride 17, 19
7,8-Methylenedioxycoumarin 86
5-Methylethuliacoumarin 213, 280
Methyl galbanate 240, 280
*trans-O*-Methylgrandmarin 142, 280
Methyllacarol 159, 280
9-Methyllongipesin 214, 280
Methylmalonyl-CoA 7, 10, 12, 15, 51–54, 67, 68
(*S*)-Methylmalonyl-CoA 15
Methyl robustate 221, 280
6-Methylsalicylic acid 61
– synthase 61, 62, 67
(*E*)-Methylsuberenol 103, 280
(*Z*)-Methylsuberenol 104, 280
O-Methyltransferase 66, 67
Mevalonolactone 12
Mexolide 235, 280
( – )-Mexoticin 140, 280
*syn*-Michael addition 23
Microlobiden 99, 280
Microlobin 100, 280
Micromelin 248, 280
*Micromelum minutum* 108, 115, 120
Microminutin 115, 280
*Micromonospora griseorubida* 41

*Microsorium fortunei* 177
*Milletia thonningii* 223
Minumicrolin 120, 280
Mirificoumestan 225, 280
– glycol 225, 280
– hydrate 225, 280
Mogoltacin 243, 280
Mogoltavicin 246, 280
Mogoltavidin 246, 280
Mogoltavin 243, 280
Mogoltavinin 243, 280
Mogoltin 243, 280
Moluccanin 150, 280
Monensin A 1–4, 7, 8, 12, 14, 20–23, 33–35, 37, 38, 45, 47, 52–54
– B 2, 4, 8, 14, 22, 30, 43
– biosynthesis 31, 42–44
*Morus lhou* 134
Moyukamycin 23
Mulberroside B 134, 280
Murpanicin 120, 280
Murpanicol 122, 280
Murpanidin 120, 280
( – )-Murracarpin 120, 280
Murraculatin 139, 280
Murragleinin 158, 280
Murralongin 251, 280
Murranganon 122, 280
Murrangatin 251, 280
– acetate 121, 280
– palmitate 121, 280
Murraol 117, 280
Murraxocin 120, 280
*Murraya exotica* 114, 115, 117–122, 235
– *gleinei* 119, 137–140, 158
– *omphalocarpa* 139
– *paniculata* 114, 115, 118, 120–123, 137, 139, 140, 158
Murrayacarpin A 114, 280
– B 137, 280
Murrayanone 158, 280
Murrayatin 118, 280
*Musineon divaricatum* 126–129
Mutalomycin 45, 47
Mutase activity 12
*Mutisia acumeata* 203
– *acuminata* 207
– *orbignyana* 198–202, 206, 207
– *spinosa* 204, 206
Mutisicoumarin 206, 280
Mutisifurocoumarin 207, 280
Mycinamicin 50
– IV 41
Myrsellin 115, 280
Myrsellinol 116, 280
*Myrtopsis sellingii* 115, 116

Nachsmyrin 111, 280
Nanaomycin 54
Naphthoherniarin 85, 108, 280
Narasin 17, 23, 25, 44, 46, 48, 49
Nargenicin 40
*Nassauvia argentea* 202, 203
– *lagascae* 202, 203
– *magellanica* 210
Nassauvirevolutin A 202, 280
– B 202, 280
– C 203, 280
2-*epi*-Nassauvirevolutin C 203, 280
Necatorin 85, 215, 280
Neoartanin 160, 280
Neoartanindiol 160, 280
Neoartaninepoxide 160, 280
Neofolin 179, 280
*Neorautenia ficifolia* 179
*E*-Nerolidol 33
(3*RS*)-Nerolidol 35
*Z*-Nerolidol 33
Neryl acetate 34
Nevskin 244, 280
Nevskone 98, 280
Nieshoutin 257, 280
Nigericin 31, 34, 45, 47
Nivegin 182, 280
Noboritomycin A 46, 48, 49
*Nocardia argentinensis* 40
Nonactic acid 9
( + )-(2*S*,3*S*,6*R*,8*R*) Nonactic acid 10
( – ) Nonactic acid 10
Nonactin 9, 10, 23, 26
9′,13′-Norcyclolycoserone 211, 280
Nordalbergin 192, 280
Nordentatin 256, 280
Norhalicondrin A 12, 13
Norwedelolactone 227, 280
Notopterol 136, 280
*Notopterygium incisum* 136
Notoptol 136, 280

Obliquetin 257, 280
Obliquetol 257, 280
Obliquin hydrate 149, 280
Oblongulide 181, 280
Obtusicin 165, 280
Obtusidin 177, 280
Obtusifol 262, 280
Obtusifolin 177, 280
Obtusin 166, 280
Obtusinin 147, 281
Obtusinol 147, 281
Obtusiprenin 173, 281
Obtusiprenol 173, 281
Obtusoside 148, 281

*Ochrocarpus siamensis* 188, 190
Octacosanyl 3′-hydroxy-4′-methoxycinna-
mate 86
Officinalin 247, 281
– acetate 247, 281
– isobutyrate 247, 281
Okadaic acid 12, 13
Omphalocarpin 140, 281
Omphamurin 139, 281
Onognaphalin 209, 281
*Onoseris gnaphaloides* 209
– *hyssipifolia* 203
Oreojasmin 234, 281
Orlandin 232
Oroselol 252, 281
Oroselone 252, 281
Osthenon 114, 281
*cis*-Osthenon 114, 281
Ostruthin 249, 281
Oxofarnochrol 166, 281
(*E*)-ω-Oxoferprenin 197, 281
(*E*)-ω-Oxoferulenol 196, 281
Oxyanomalin 253, 281
Oxycapnolactone 238, 281
Oxytetracycline 64, 65

Pachyrrhizin 178, 281
*Pachyrrhizus aureus* 227
– *erosus* 178
Palmitate 55, 61
Palmitic acid 57
Palmitoyl-CoA 56, 61
Panial 123, 281
Paniculal 114, 281
Paniculin 115, 281
Paniculonol isovalerate 122, 281
*Passiflora serratodigitata* 182
Pd–Ib 131, 281
Pd–III 129, 281
Pd–C–I 112, 281
Pd–C–II 113, 281
Pd–C–III 113, 281
Pd–C–IV 113, 281
Pd–C–V 113, 281
Pectachol 167, 281
– B 168, 281
Pectanone 169, 281
Pectenotoxin-1 12, 13
*Pelea barbigera* 137
*Penicillium jensenii* 195
– *patulum* 61, 63, 67
Pereflorin 270, 281
– B 220, 281
*Perezia alamani* 217
– *coerulescens* 220
– *multiflora* 216, 217, 219, 220

Peroxyauraptenol  119, 281
Peroxymurraol  120, 281
Peucedanol  248, 281
*Peucedanum arenarium*  112
– *austriaca*  130
– *decursivum*  109–113
– *hispanicum*  114
– *japonicum*  130
– *mogoltavicum*  97
– *oreoselinum*  125
– *palustre*  164
– *praeruptorum*  126, 127, 129, 131, 153
– *ruthenicum*  156
– *turcomanicum*  135
– *turgeniifolium*  128, 130, 131
Peuruthenicin  247, 281
Phaseol  227, 281
*Phaseolus aureus*  224, 227
Phebalin  235, 281
*Phebalium alstonii*  123
– *nudum*  235
– *squaleum*  89
( – )-Phebalosin  119, 281
7-Phenylacetoxycoumarin  85
Phlojodicarpin  121, 281
*Phlojodicarpus sibiricus*  117, 121, 122, 126
– *turczaninovii*  104
Phosphatidic acid  56
Phosphoenolpyruvate  11
Phosphoglyceric acid  11
*Phyllosma capensis*  165
Pig liver esterase  31, 32
Piloselloidan  203, 281
Piloselloidol isovalerate  213, 281
Polyepoxide cyclisation cascades  31
Polyether antibiotic biosynthesis  36
– antibiotics  1, 2, 5–7, 11, 24, 46
– biosynthesis  14, 15, 30, 31, 43
*Polygala paniculata*  121, 127
Polyketide biogenesis  6
– chain assembly  36
– metabolites  7
– synthase  15, 40–42, 51, 52, 54, 55, 61, 63, 64, 67, 69, 70
*Poncirus trifoliata*  144
Ponfolin  86, 144, 281
Ponnalide  189, 281
Praeroside I  153, 281
– II  127, 281
– III  127, 281
– IV  126, 281
– V  126, 281
( + )-Praeruptorin A  129, 281
– B  253, 281
Praeruptorin E  129, 281
Prandiol  249, 281

– senecioate  281
*Prangos latiloba*  90
– *pabularia*  135
Prangosin  249, 281
Prealtin A  88, 281
– B  89, 281
– C  89, 281
– D  89, 281
Preethuliacoumarin  207, 281
Prenyllacarol  160, 281
8-Prenylnodakenetin  132, 281
Proline  8
Propacin  172, 281
Propionate  7–10, 12, 15, 22, 23, 26, 38, 43, 44, 50–52
[1-$^{13}$C, $^{18}$O$_2$]Propionate  20
($R$)-[2-$^2$H$_1$]-Propionate  52, 54
($S$)-[2-$^2$H$_1$]-Propionate  52, 54
[2-$^2$H$_2$]-Propionate  52, 54
[2-$^2$H$_2$, 2-$^{13}$C]Propionate  52
Propionyl-CoA  53, 54, 67, 68
*Protium opacum*  172
*Pseudomonas oleovorans*  15
*Psoralea corylifolia*  226, 227
Psoralen  249, 281
Psoralidin  226, 281
– oxide  226, 281
Pterybinthinone  111, 281
*Pteryxia terebinthina*  111
Pubesinol  248, 281
*Pueraria mirifica*  225
– *tuberosa*  226, 229
Puerarol  227, 281
Puerarostan  229, 281
*Pueronia labata*  227
Pygmaeoherin  215, 281
*Pygmaeopremna herbacea*  215

Racemol  259, 281
Racemosin  164, 281
Ramosin  250, 281
Ramosinin  177, 281
Reoselin  238, 281
Repensol  225, 281
Robustic acid  221, 281
Robustin  221, 281
*Rumex conglomeratus*  133
Ruminants  6
*Ruta angustifolia*  174
– *chalepensis*  86, 181, 233
– *graveolens*  108, 161, 175, 233
– *oreojasme*  233, 234
– *pinnata*  103, 104, 135
– *sp. Tene 29662*  174, 235
Rutalpinin  86, 181, 281
Rutamarin  263, 281

Rutarensin   233, 281
Rutaretin   259, 281
– methyl ether   259, 281
Rutarin   259, 281

*Saccharopolyspora erythraea*   67
Salinomycin   16–20, 23, 46
– acetate   19
17-*epi*-Salinomycin   17, 19
– acetate   19
*Salsola laricifolia*   145
Samarcandin   86, 245, 282
– acetate   245, 282
Samarcandone   245, 282
( + )-Samidin   130, 282
*Sapium sebiferum*   261
*Sargentia greggii*   116
Sativol   229, 282
Saxicolon   125, 282
Scandenin   222, 282
Scopodrimol A   149, 282
Scopofarnol   148, 282
Scopoletin   86
Secodrial   167, 282
Secodriol   167, 282
Secrolin   124, 282
Seravschanin   127, 282
Serratin   182, 282
Sesebrin   138, 282
Sesebrinol   140, 282
*Seseli bocconi*   128
– *campestre*   127, 128, 130
– *grandivittatum*   105, 112
– *jomuticum*   134
– *libanotis*   85, 126, 130, 154
– *montanum*   104
– *mucronatum*   124
– *peucedanoides*   156
– *saxicolum*   125
– *seravschanicum*   127
– *sibiricum*   138–140
– *tortuosum*   110–112, 116–119, 127, 129, 136, 138
Seselinal   139, 282
Seseloside   156, 282
Seshadrin   185, 282
Sesibiricol   139, 282
Shijiaocaolactone A   176, 282
Shikimate   9
Shikimic acid   8
( – )-Sibiricin   138, 282
Sibiricol   138, 282
Sibirinol   139, 282
Sisafolin   187, 282
Smirnioridin   249, 282
Smirniorin   249, 282

*Smyrniopsis aucheri*   111
Sodium [1-$^{13}$C]propionate   12
*Soja hispida*   224
Sojagol   224, 282
Sophoracoumestan A   224, 282
– B   229, 282
*Sophora franchetiana*   224, 229
*Soulamea soulameoides*   173
Soulattrolide   191, 282
Sphondin   258, 282
Spiramycin   50, 64, 65, 67
Sponges   11
Sporozoa   6
( – )-Sprengelianin   110, 282
Stevenin   193, 282
*Streptomyces albus*   16, 31, 34
– *cinnamonensis*   4, 25, 33, 35, 43
– *erythraea*   40, 54, 55
– *fradiae*   42
– *griseus*   10
– *lasaliensis*   15
– *sp.*   7, 11, 12, 15, 31
Streptomycete screening programmes   7
Streptomycetes   7, 14, 15, 64, 65, 70
Stylosin   165, 282
(Z)-Suberenol   103, 282
Suberosin   247, 282
Succinate   10, 12, 23, 26
Surangin A   265, 282
– B   86, 269, 282
– C   180, 282
Swietenocoumarin A   132, 282
– B   143, 282
– C   132, 282
– D   143, 282
– E   132, 282
– F   143, 282
– G   143, 282
– H   132, 282
– I   175, 282
Swietenol   103, 282

Tadshiferin   90, 282
Tadshikorin   90, 282
Tamarin   103, 282
*Tamarix troupii*   181
*Tanacetum heterotumum*   171
Tavimolidin   97, 282
Tenudiol   152, 282
Tenuidin   102, 282
*Tephrosia hamiltonii*   229
– *villosa*   229
Tephrosol   229, 282
Tetracenomycin   64–66
[$^3$H]Tetrahydrocerulenin   55
Tetronomycin   6, 43

*Thamnosma montana* 235
– *texana* 131
Thioesterase 52, 56, 59, 61, 69
Thiolactomycin 58
Thonningine A 223, 282
– B 223, 282
Toddalenone 137, 282
*Toddalia aculeata* 133, 134, 159
– *asiatica* 133, 134, 137, 235
Toddanol 133, 282
Toddanone 134, 282
Toddasin 235, 282
Tomenin 260, 282
Tomentin 157, 260, 282
Tomentolide A 191, 282
Tortuoside 118, 282
Tortuosidin 116, 282
Tortuosin 136, 282
Tortuosinin 110, 282
Tortuosinol 282
Tortuosinol Prandiol senecioate 111
Trachyphyllin 256, 282
*Trachyspermum roxburghianum* 101, 102
Trichoclin 154, 282
*Trichocline incana* 154
Trifluoracetic acid 19
Trifoliol 225, 282
*Trifolium repens* 225, 230
Trigocoumarin 219, 282
Trigoforin 85, 195, 282
*Trigonella foenumgraecum* 180, 195, 219
3,4,7-Trimethylcoumarin 85
Tripartol 171, 282
*Triphasia trifoliata* 117
Triphasiol 117, 282
Triptiliocoumarin 209, 282
*Triptilion benaventei* 206, 207, 209, 210
– *spinosum* 210, 218
Triptispinocoumarin 210, 282
*Triticum sativum* 216
Troupin 181, 282
Tuberostan 226, 282

Turgeniifolin A 131, 282
– B 128, 282
– C 130, 282
Tylactone 37–41
– polyketide synthase 37
Tylosin 12, 23, 37, 38, 42, 50, 64, 67
– biosynthesis 55
– polyketide synthase 38

Ulismoncadin 85, 131, 282
– A 176, 282
Ulopterol 248, 282
Umckalin 260, 282

*Ventilago calyculata* 215, 217
Ventilatone A 215, 282
– B 217, 283
*Vernonia cinarescens* 207
Versicolin 123, 283
*Viburnum awabuki* 145
– *suspensum* 145
Villosin 152, 283
Voludal 187, 283

Wairol 226, 283
Wampetin 155, 283
*Wedelia calendulacea* 228
Wedelolactone 228, 283
WhiE spore pigment 64, 66

X14547A 43, 44
Xanthoxyletin 255, 283
*Xanthoxylum arnottianum* 111, 156
Xanthyletin 250, 283
Xeroboside 146, 283
*Xeromphis obovata* 146
– *spinosa* 146
*Xi-bai licorice* 179

*Zanthoxylum suberosum* 235
– *usambarense* 150

*Fortschritte der Chemie organischer Naturstoffe*

# Progress in the Chemistry of Organic Natural Products

## Volume 57:

1991. 26 figures and 2 plates. X, 212 pages. DM 210,–, öS 1470,–.
ISBN 3-211-82245-3

*Contents:* P. Metzger, C. Largeau, E. Casadevall: Lipids and Macromolecular Lipids of the Hydrocarbon-rich Microalga *Botryococcus braunii.* Chemical Structure and Biosynthesis. Geochemical and Biotechnological Importance. – D. P. Chakraborty and S. Roy: Carbazole Alkaloids III. – G. R. Pettit: The Bryostatins.

## Volume 56:

1991. 8 figures. X, 188 pages. Cloth DM 220,–, öS 1540,–.
ISBN 3-211-82188-0

*Contents:* J. Asselineau: Bacterial Lipids Containing Amino Acids or Peptides Linked by Amine Bonds. – J. Kagan: Naturally Occurring Di- and Trithiophenes.

## Volume 55:

1989. 41 figures. X, 208 pages. Cloth DM 190,–, öS 1330,–.
ISBN 3-211-82087-6

*Contents:* M.T. Davies-Coleman and D. E. A. Rivett: Naturally Occurring 6-substituted 5,6-dihydro-α-pyrones – K. Krohn: Building Blocks for the Total Synthesis of Anthracyclinones – M. Lounasmaa and J. Galambos: Indole Alkaloid Production in Catharanthus Roseus Cell Suspension Cultures – C. E. James, L. Hough, and R. Khan: Sucrose and Its Derivatives.

## Volume 54:

1988. VII, 353 pages. Cloth DM 320,–, öS 2240,–.
ISBN 3-211-82086-8

*Contents:* T. Murakami and N. Tanaka: Occurrence, Structure and Taxonomic Implications of Fern Constituents.

## Volume 53:

1988. 72 figures. VIII, 311 pages. Cloth DM 275,–, öS 1930,–.
ISBN 3-211-82074-4

*Contents:* L. F. Alves: Chemical Ecology and the Social Behavior of Animals –
T. Nomura: Phenolic Compounds of the Mulberry Tree and Related Plants –
A. Chimiak and M. J. Milewska: N-Hydroxyamino Acids and Their Derivatives.

## Volume 52:

1987. 65 figures. VIII, 224 pages. Cloth DM 210,–, öS 1470,–.
ISBN 3-211-81989-4

*Contents:* U. Weiss, L. Merlini, and G. Nasini: Naturally Occurring Perylene-
quinones – H. Achenbach: The Pigments of the Flexirubin-Type. A Novel
Class of Natural Products – T. Goto: Structure, Stability and Color Variation of
Natural Anthocyanins – P. Bhattacharyya and D. P. Chakraborty: Carbazole
Alkaloids.

## Volume 51:

1987. VII, 317 pages. Cloth DM 280,–, öS 1960,–.
ISBN 3-211-81972-X

*Contents:* M. Gill and W. Steglich: Pigments of Fungi (Macromycetes).

All Volumes and Cumulative Index 1–20 available

*Price reduction for subscribers: 10%*

**Special reduced price (20% reduction) for the complete Series Vols. 1–58
incl. the Cumulative Index to Vols. 1–20**

*Springer-Verlag* **Wien New York**

Sachsenplatz 4–6, A-1201 Wien
175 Fifth Avenue, New York, NY 10010, U.S.A.
Heidelberger Platz 3, D-1000 Berlin 33
37-3, Hongo 3-chome, Bunkyo-ku, Tokyo 113, Japan